3rd International Conference on Rheology and Modeling of Materials 2017 (IC-RMM3)

Journal of Physics: Conference Series
Volume 1045

Miskolc-Lillafured, Hungary
2 – 6 October 2017

ISBN: 978-1-5108-6787-1
ISSN: 1742-6588

Printed from e-media with permission by:

Curran Associates, Inc.
57 Morehouse Lane
Red Hook, NY 12571

Some format issues inherent in the e-media version may also appear in this print version.

Printed by Curran Associates, Inc. (2018)

For permission requests, please contact the Institute of Physics
at the address below.

Institute of Physics
Dirac House, Temple Back
Bristol BS1 6BE UK

Phone: 44 1 17 929 7481
Fax: 44 1 17 920 0979

techtracking@iop.org

Additional copies of this publication are available from:

Curran Associates, Inc.
57 Morehouse Lane
Red Hook, NY 12571 USA
Phone: 845-758-0400
Fax: 845-758-2633
Email: curran@proceedings.com
Web: www.proceedings.com

Table of contents

Volume 1045

3rd Internatiomal Conference on Rheology and Modeling of Materials (ic-rmm3)

2–6 October 2017, Miskolc-Lillafüred, Hungary

Accepted papers received: 29 May 2018
Published online: 17 July 2018

Preface

Papers

IOP Conf. Series: Journal of Physics: Conf. Series **1045** (2018) 011001 doi:10.1088/1742-6596/1045/1/011001

ic-rmm3

3RD INTERNATIONAL CONFERENCE ON RHEOLOGY AND MODELING OF MATERIALS

October 2-6, 2017/ Miskolc-Lillafüred, Hungary

Organized by the International Scientific Advisory Board (**ISAB**),

International Organizing Committee (**IOC**) and

Session Chairs

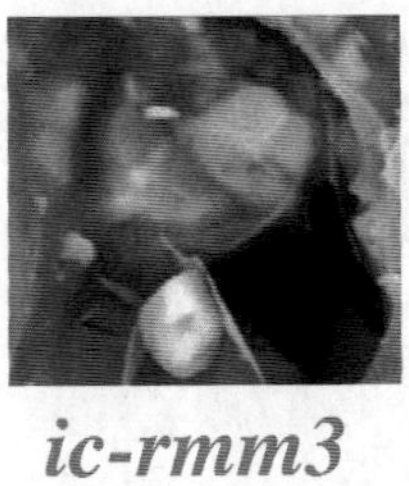

ic-rmm3

IOP Conf. Series: Journal of Physics: Conf. Series **1045** (2018) 011001 doi:10.1088/1742-6596/1045/1/011001

ic-rmm3

Abstract

Understanding the rheological properties of materials and their rheological behaviors during their manufacturing processes and in their applications in many cases can help to increase the efficiency and competitiveness not only of the finished goods and products but the organizations and societies also. The more scientific supported and prepared organizations develop more competitive products with better thermal, mechanical, physical, chemical and biological properties and the leading companies apply more competitive knowledge, materials, equipment and technology processes. The goals of **ic-rmm3** the **3rd International Conference on Rheology and Modeling of Materials** are the followings:

- Promote new methods and results of scientific research in the fields of modeling and measurements of rheological properties and behavior of materials under processing and applications.
- Change information between the theoretical and applied sciences as well as technical and technological implementations and applications.
- Promote the communication between the scientist of different disciplines, nations, countries and continents.

The international conference **ic-rmm3** provides a platform among the leading international scientists, researchers, PhD students and engineers for discussing recent achievements in measurement, modeling and application of rheology in materials technology and materials science of liquids, melts, solids. crystals and amorphous structures. Among the major fields of interest are the influences of material structures, mechanical stresses, temperatures and deformation speeds on rheological and physical properties, phase transformation of foams, foods, polymers, plastics and other competitive materials like ceramics, nanomaterials, medical- and biomaterials, cosmetics, coatings, light metals, alloys, glasses, films, composites, hetero-modulus, hetero-viscous, hetero-plastic complex materials, petrochemicals and hybrid materials, … etc.

Multidisciplinary applications of rheology and rheological modeling in material science and technology encountered in sectors like alloys, ceramics, glasses, thin films, polymers, clays, construction materials, energy, aerospace, automotive and marine industry. Rheology in food, chemistry, medicine, biosciences and environmental sciences are of particular interests.

In accordance to the program of the rheology conference **ic-rmm3** **134** registrations were received from all the **6** continents. Finaly **115** papers were accepted for presentation in the 3rd International Conference on Rheology and Modeling of Materials. Together with New Zealand the scientists and researchers have arrived to Miskolc-Lillafüred (Hungary) from **34** different countries of Europe, Asia, Africa, North- and South America and Australia.

If the co-authors are also considered then more then **350** scientists and researchers have presented their research works of in the **ic-rmm3** conference on rheology and in this book.

Prof. Dr. László A. Gömze
chair, ic-rmm3

IOP Conf. Series: Journal of Physics: Conf. Series **1045** (2018) 011001 doi:10.1088/1742-6596/1045/1/011001

ic-rmm3

THE INTERNATIONAL SCIENTIFIC ADVISORY BOARD (ISAB)

Prof. Dr. Igor EMRI - *Chair (2012-16) of the International Committe of Rheology,* ***(Slovenia)***
Prof. Gabriel Ascanio GASCA - *Universidad Nacional Autonóma de Mexico,* ***(Mexico)***
Prof. Dr. László A. GÖMZE - *University of Miskolc and IGREX Ltd.* ***(Hungary)***
Academician Prof. Valerii KULICHIKHIN - *Presid. of Vinogradov Society of Rheology,* ***(Russia)***
Prof. Dr. Sergei N. KULKOV - *Tomsk State University,* ***(Russia)***
Prof. Dr. Blaise NSOM - *Université de Bretagne Occidentale,* ***(France)***
Prof. Valentina PAVLOVA - *University St. Kliment Ohridski,* ***(Macedonia)***
Prof. Shabu THOMAS - *Mahatma Gandhi University,* ***(India)***
Prof. Youjiang WANG- *Georgia Institute of Techmology,* ***(USA)***

THE INTERNATIONAL ORGANIZING COMMITTEE (IOC)

Dr. Fares BEAINY - Volvo Construction Equipment ***(USA)***

Dr. Fatiha BOUDJEMAA - Université de Khemis Miliana, (***Algeria***)

Dr. Ivica DURETEK - Montanuniversity Leoben, (***Austria***)

Dr. Nicky ESHTIAGHI - RMIT University Melbourne (***Australia***)

Dr.Saeed FARROKPAY - The University of Queensland, (***Australia***)

Prof. Gabriel Ascanio GASCA - Universidad Nacional Autonóma de Mexico, (***Mexico***)

Prof. Dr. Maxim KHRAMCHENKOV - Kazan Federal University, (***Russia***)

Prof. Anna KNYAZEVA - Tomsk Polytechnic University, (***Russia***)

Dr. Tünde KOVÁCS - University Óbuda, (***Hungary***)

Prof. Milan KRACALIK - Johannes Kepler University, (***Austria***)

Academician Prof. Valerii KULICHIKHIN - President of Vinogradov Society of Rheology, (***Russia***)

Prof. Dr. Sergei N. KULKOV - Tomsk State University, (***Russia***)

Dr. Martin LEDERER - Vienna University of Technology, (***Austria***)

Prof. Dr. Blaise NSOM - Université de Bretagne Occidentale, (***France***)

Prof. Dr. Grigory PYSHNOGRAI - Altai State Technical University, (***Russia***)

Prof. Dr. Alexander SLOBODOV - ITMO University, (***Russia***)

Dr. Nina STOPPE - Christian Albrechts University, (***Germany***)

As. Prof. Ursula WINDBERGER - Medical University of Vienna, (***Austria***)

Prof. Dr. Ichiro YONENAGA - *Tohoku University Sendai* (***Japan***)

CHAIR OF THE INTERNATIONAL ORGANIZING COMMITTEE

Prof. Dr. László A. GÖMZE, *University of Miskolc,* ***(Hungary)***

The financial organizer of the conference: **IGREX Engineering Service Ltd** *(Hungary)*

ic-rmm3

SESSIONS AND SYMPOSIUMS

1. Earth Rheology and Applied Rheology of Clays, Minerals, Rocks and Soils (ERARCMRS)
2. Rheology and Modeling of Polymers (RMP)
3. Food Rheology and Texture of Foods (FRTF)
4. Avalanche Dynamics in Rheology and Phase Transitions (ADRPT)
5. Experimental Study and Modeling of Complex Flows (ESMCF)
6. Rheology and Modeling of Asphalt Mixtures and Asphalt Pavements (RMAMAP)
7. Multiscale Mechanics of Materials (MMM)
8. Rheological Fluid Mechanics Including Micro and Nano Scale Phenomena (RFM)
9. Rheology of Biomaterials and Biological Systems (RBBS)
10. Rheology and Modeling of Concretes, Limes and Mortars (RMCLM)
11. Rheology of Emulsions, Suspensions, Colloids and Foams (RESCF)
12. Rheology of Ferrofluids and Magneto - Rheological Materials (RFMRM)
13. Rheology of Fluid Interfaces and Modeling (RFIM)
14. Rheology of Melts and Molten Materials (RMMM)
15. Rheology of Nanofluids (RNF)
16. Rheology of Cosmetics and Pharmaceutical Materials (RCPM)
17. Rheological Measurement – Methods, Equipments and Errors (RMMEE)
18. Rheology of Carbon Structures (RCS)
19. Rheology and Modeling of Complex Materials and Composites
20. Rheology and Molecular Dynamics Simulation of Ionic Liquids (RMDSIL)
21. Rheology of Nano-Powder Suspensions and Gels (RNPSG)
22. Ratio of the Rheological and Physicochemical Properties of Materials

is-emm1: The 1st International Symposium on Experimental Mechanics of Materials
is-fms1: The 1st International Symposium on Fractures and Mechanics of Solids
is-pim2: The 2nd International Symposium on Powder Injection Molding

Some of the presented works of the ic-rmm3 conference were selected by the Editorial Board of scientific journal **Építőanyag-JSBCM** and were published in it as the followings:

V. MAKSAROV – A. KHALIMONENKO – J. OLT 2017 *Építőanyag-JSBCM* **69** (2) 66 http://dx.doi.org/10.14382/epitoanyag-jsbcm.2017.12

G. M. AYININOULA - O. A. ADEKITAN 2017 *Építőanyag-JSBCM* **69** (2) 34 http://dx.doi.org/10.14382/epitoanyag-jsbcm.2017.7

Z. SOÓS – R. GÉBER – Cs. TÓTH – Zs. IGAZVÖLGYI – B. UDVARDI 2017 *Építőanyag-JSBCM* **69** **(3)** 89 http://dx.doi.org/10.14382/epitoanyag-jsbcm.2017.15

M. G. KHRAMCHENKOV – R. M. USMANOV 2017 *Építőnyag-JSBCM* **69** (4) 110 http://dx.doi.org/10.14382/epitoanyag-jsbcm.2017.19

A *MELLAK* 2017 *Építőanyag-JSBCM* **69** (4) 114 http://dx.doi.org/10.14382/epitoanyag-jsbcm.2017.20

M Kracalik 2015 *Építőanyag-JSBCM* **69** (4) 116 http://dx.doi.org/10.14382/epitoanyag-jsbcm.2017.21

M. BENDOUKHA – M. MOSBAH 2017 *Építőanyag-JSBCM* **69** (4) 122 http://dx.doi.org/10.14382/epitoanyag-jsbcm.2017.22

M. LEDERER – G. KHATIBI 2017 *Építőanyag-JSBCM* **69** (4) 127 http://dx.doi.org/10.14382/epitoanyag-jsbcm.2017.23

B. I. KANEV – T. M. BEZNOSOVA – V. A. MATVEEV – L. A. GÖMZE 2017 *Építőanyag-JSBCM* **69** (4) 132 http://dx.doi.org/10.14382/epitoanyag-jsbcm.2017.24

ic-rmm3

ACKNOWLEDGEMENT

In the name of **Conference Board** I would like acknowledge and say many thanks to our following sponsors for their support in press-campaign and contributions the news and information about the **3rd International Conference on Rheology and Modeling of Materials** between their members and in their media:

AIPEA – Internatonal Association for the Study of Clays
AIR – Associazione Italiana di REOLOGIA, (Italy)
biomat.net – The Biomaterials Network, (Portugal)
GFC – Groupe Francais de la Céramique, (France)
ICG – International Commission of Glass
ICR - International Committee of Rheology and its member organizations
IGREX Engineering Service Ltd, (Hungary)
Journals: Academic Journals
APPLIED RHEOLOGY
cfi - Ceramic Forum International
ÉPÍTŐANYAG, Journal of Silicate Based and Composite Materials
refractiries WORLDFORUM
Rheology Bulletin

ME – University of Miskolc (Hungary)
NRC – Nordic Rheology Society (Europe)
RAS – Institute of Strength Physics and Materials Science (Russia)
RGCCS – Rheology Group of the Czech Chemical Society, (Czech Republic)
SMR – Sociedad Mexicana de Reologia, (Mexico)
SSR – Slovenia Society of Rheology, (Slovenia)
SOR – The Society of Rheology (USA)
TSU - Tomsk State University (Russia)
VSR – Vinogradov Society of Rheology, (The Russian Federation)

Many thanks to colleagues of **ISAB, IOB** and to **chairs** of the SESSIONS for their support and help in organization work and in successful transaction of conference **ic-rmm3** the **3rd International Conference on Rheology and Modeling of Materials**. I would like to say many thank personally to *Prof. Dr. Igor* ***Emri*** (the 2012 - 2016 president of the International Committee of Rheology), *Dr. Nicky* ***Eshtiaghi*** (RMIT University Melbourne *Prof. Dr. Sergey N.* ***Kulkov*** (Tomsk State University), *Dr. Giséle* ***Lecomte-Nana*** (ENSCI Limognes University), *Dr. Martin* ***Lederer*** (University of Viena) and *Prof. Dr. Blaise* **Nsom** (Université de Bretagne Oc.) for their excellent organization works in their countries and in continents of Africa, America, Asia, Australia and Europe. Especially many thanks to editors of journal **Applied Rheology** (Switzerland) **Építőanyag-JSBCM** (Hungary) and **Rheology Bulletin** (USA) for their strong support in scientific press and media.

Prof. Dr. László A. Gömze
chair of **ic-rmm3** conference board

Peer review statement

All papers published in this volume of *Journal of Physics: Conference Series* have been peer reviewed through processes administered by the proceedings Editors. Reviews were conducted by expert referees to the professional and scientific standards expected of a proceedings journal published by IOP Publishing.

IOP Conf. Series: Journal of Physics: Conf. Series **1045** (2018) 012001 doi:10.1088/1742-6596/1045/1/012001

Changing the rheo-mechanical models of light metal Ti and Ti-alloy powders under uniaxial compaction

László A Gömze[1, 2, 3], Emese Kurovics[1] and Ludmila N Gömze[2]

[1]Intsitute of Ceramics and Polymer Engineering, University of Miskolc, Hungary
[2] IGREX Engineering Service Ltd, Igrici, Hungary
[3]Tomsk State University, Tomsk, Russia

E-mail: femgomze@uni-miskolc.hu, mse205@gmail.com and igrex2009@yandex.ru

Abstract. On the basis of uniaxial compacting process the authors have examined the rheological and mechanical properties of light metal Ti and Ti-alloy powders as function of forming pressures and times. For the examination of the compacting properties and create the rheo-mechanical model a special uniaxial instrument was developed and used by the authors. To determine the time dependence behavior of these light metal alloy powders a combine rheo-tribometer was used which was patented, developed and produced by IGREX Ltd. in Hungary. Using these two instruments the authors could successfully illustrate how the rheological models of light metal Ti and Ti-alloys are changed depending on forming pressures and times of loading forces.

In this work the rheological behaviors and models of different Ti alloys are shown as function of compaction times and pressures. The values of modulus of elasticity of Voight-Kelvin and Hooke components as well as values of viscosity of the viscous-plastic components are also determined and given by the authors.

1. Introduction

In our days rheology is playing very essential role in different segments both of biology and human life [1-8], industrial technology [9-20] and materials science [21-30]. The physical, mechanical and rheological properties of raw, semi finished and finished materials are very important in development of new materials for different applications [31-34]. The rheological properties are playing especially important role during forming and compacting materials under pressures [35-42].

The powder compacting technology is one of the most efficient forming technologies not only in ceramic industry [43-47] but in metallurgy [48-53] and machine industry [54-58] also. In ceramic industry to prepare units and parts from powders with increased physical and mechanical properties very often high temperature pressing [59-60], isostatic pressing [61-63], hot isostatic pressing or compaction under high speed mechanical forces and shock-waves [64-66] are used. The powder pressing technology is especially popular in production of electrical components [67] and machine parts not only with axis-symmetrical shapes but with complicated geometrical forms. At the same time to make metallic machine elements and parts with the required final geometrical parameters are impossible without understanding the changes of rheological and mechanical properties of the used metal powders during the pressing processes.

2. Materials and experiments

The knowledge of rheological properties of titanium alloy powders and flakes is necessary to interpret the processes occurring in materials during their compacting and understand the changes of geometrical sizes of the manufactured pre-compacted specimens after their compaction.

For rheological tests the titanium alloy powders were obtained from IPAX Ltd. and the tests were realized on a 100 kN mechanical pull-press. Thus, it was possible to compact cylindrical specimens with diameter of D = 15 mm, providing static compaction of titanium alloy powders according to the principle shown in Fig. 1. During the compacting the compression pressures were linearly increased and kept by 10 seconds at pressure values of 10MPa, 100MPa, 150MPa, 200MPa, 250MPa, 300MPa, 350MPa and 400MPa. The results of load-displacement (pressure-compaction) values were recorded automatically in diagrams on computer (11).

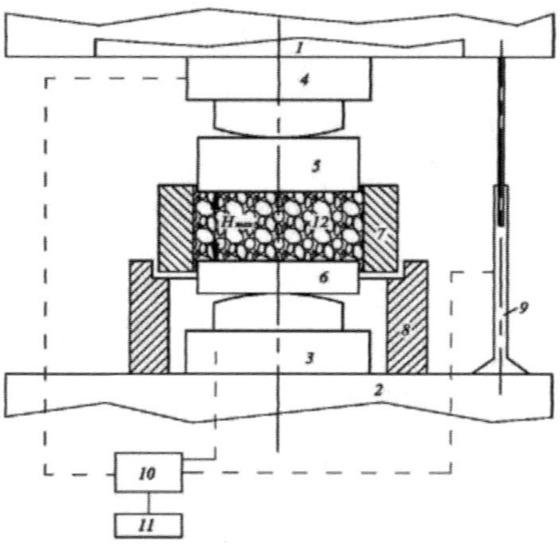

Figure 1. The compacting based rheology measuring device
1- top table (movable); 2- lower table (movable); 3, 4- pressure gauge CTA – 276; 5- upper punch; 6- lower punch; 7- compacting die cavity with Ø15 mm; 8- tool holder frame; 9- way (displacement meter) HBM WA-300; 10- measuring data collector (Spider 8); 11- computer; 12- tested powder or flake

It is well known from the literature [37, 68-71] that the rheological properties and parameters of materials can be determined from the deformation-time curves taken at different shear stresses and rates. To determine rheological parameters of pre-compacted specimens made from different titan alloy powders the combined rheo-tribometer of IGREX Ltd was used. The schematic outline of this apparatus is shown in Figure 2.

IOP Conf. Series: Journal of Physics: Conf. Series **1045** (2018) 012001 doi:10.1088/1742-6596/1045/1/012001

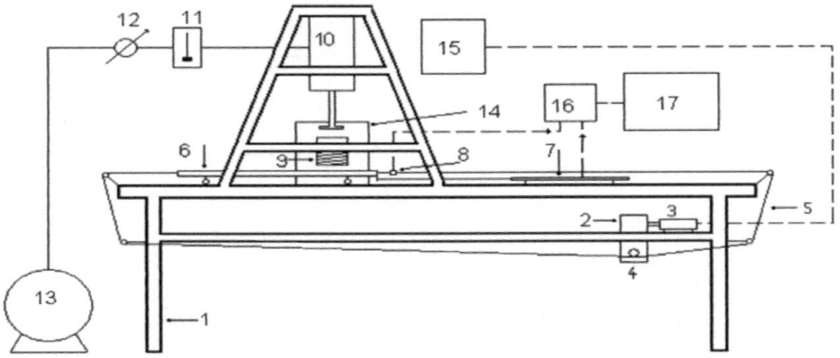

Figure 2. Scheme of combined rheo-tribometer
1-instrument table, 2-small drive, 3-electric motor, 4-cable drum, 5-cableway, 6-batching car (with the shearing plate), 7-inductive displacement detector, 8-force-meter (spider), 9-heatable specimen holder, 10-pneumatic power cylinder, 11-magnetic valve, 12-pressure gauge, 13-compressor, 14-thermostat, 15-control unit, 16-data recorder (spider 8), 17-computer (capturing and processing data)

3. Results and discussions

The compaction displacement (deformation) as function of load pressures were taken on the rheology measuring device (Fig. 1) and were recorded automatically in diagrams (Fig. 3) on the computer. The analyze of these diagrams shows that the compacting process of the titan alloy powders and flakes is starting with the filling of the compacting die cavity thanking to the arrangement of the powder grains after closing the die cavity. This intensive compaction is continuing until the compression pressures are reaching the values of 8-10MPa. Such behavior of alloy powders can be described with Maxwell's rheological material model. Because of the intensive compacting process at low compression pressures of the titanium alloy powders in the cavity of the dies, approx. 25% of bulk density growths are taken place under the above low compacting pressure. Despite of the significant compaction, however, the examined titanium alloy powders have flown away from the compacting die after withdrawn the lower punch.

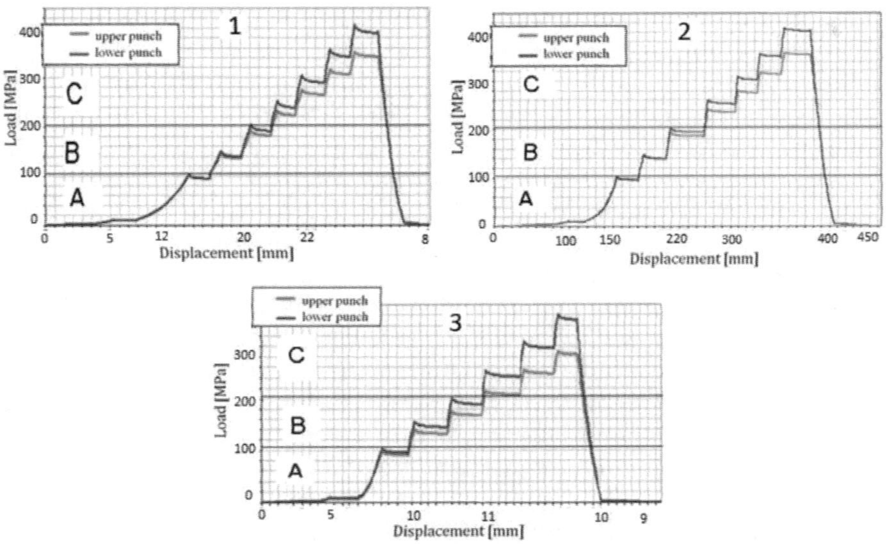

Figure 3. Compaction displacement (deformation) as function of load pressures
1 - powders of titanium sponge, 2- Ti50Ni alloy powders, 3- Ti6Al2,5Fe

From the diagrams obtained, it can be seen well that with the maintenance by 10 seconds the different pressing pressures the compression processes continuing thanking to the decreases of gaps between the powder particles. This kind of compaction of the granules is demonstrating a non-elastic deformation as function of time.

From the achieved compaction displacement (deformation) curves (Fig. 3) it is easy to recognize that in case when the values of loaded pressures are above 8-10MPa, the compacting process consists of the following three relatively distinct sections.

A. At this stage of the compaction the grains of titanium alloy powders are arranged in the cavity of the compacting die, the majority of the volumes of gaps (pores) between the particles are disappearing step by step and the degree of compactness or specific deformation reaches 70-75% of the total final compaction. After removing the lower and upper punches, the pre-compacted specimens remains in the compacting die cavity thanking to the accumulated elastic energy of the metallic particles and pressured airs in the gaps (pores) between the grains. At this stage of compression the degree of the residual deformation is so great that in the rheological model viscous and "quasi-plastic" parts must be present together with the elastic components to characterize the mechanical properties of the material. At this stage of compaction however the compacted pre-compacted specimens cannot be removed without disruption. When the compacting pressure less than 100MPa the titan alloy powders of pre-compacted specimens will be dissipated during removing from the compacting die cavity due to the shear stress generated by the friction force between the surfaces of die cavity and the granules. Due to this phenomenon and to the significant compaction observed at this stage the deformation properties of the titanium alloy powders can be characterized by the rheological model shown in Figure 4 (A).

B. At this stage of the compaction the granules of the titanium alloy powders are further arranged in the compacting die cavity the volumes of gaps (pores) between the particles continue to decrease or a substantial part of the particles are disappeared. The degree of compactness or specific deformation reaches 95-97% of the total final compaction. In this stage the pre-compacted specimens remains in the compacting die cavity not only after withdrawn the lower and upper punchers, but about 50-60% of the weight of the pressed rods will stay together after remove it from the die cavity. At this stage of compacting the deformation properties of titanium alloy powders can be characterized with rheological model given in Figure 4 (B). In this case the titanium alloy powders behave as a "parallel" plastic viscous body in line with a parallel viscous-elastic Voight-Kelvin's "body", giving a Burgers type of rheological model.

C. In the last stage of compaction, the granules of the titanium alloy powders in the compacting die cavity are settled only by destruction of the grain particles; the "permeable" gaps between the particles disappear as they become "pores" or "capillaries". The compression and specific deformation of particles are already not intense and very small and they reach nearly 100% of the available compression. The pre-compacted specimens can be taken out from the compacting die cavity without injury and surface defects, meanwhile leaving the die cavity their geometric dimensions are immediately increased measurably thanking to the elastic deformation of the alloy particles. At the same time the geometrical sizes of the pre-compacted specimens are growing continuously during in a few hours after the pressing. Because of this deformation properties of specimens in the final compression stage the rheological properties of titanium alloy powders can be characterized by Figure 4 (C). In this stage of compacting the titanium alloy powders behave themselves like an "elastic-plastic-viscous body" thanking to the destructed particles under pressure and a viscous-elastic Voight-Kelvin's body of the non-destructed particles.

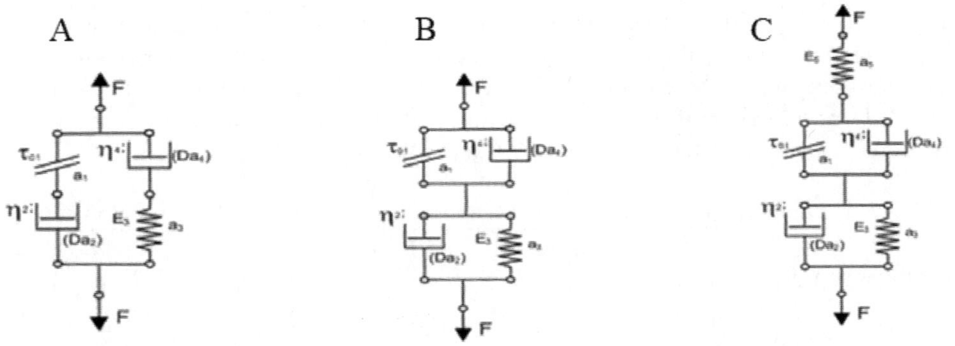

Figure 4. The rheological models of titanium alloy powders in the different stages of compaction

The rheological equation of materials having rheological model as C in Figure 4 is well known in the literature [68, 69, 71 and 72]. Its general form is:

$$\tau(t) = \tau_0 + \eta_1\dot{\varepsilon} + \eta_1 t_r\ddot{\varepsilon} - \dot{\tau}\left(t_{fr} + t_r - t_{fr}\cdot t_r\frac{\eta_1}{\eta_2}\right) - t_{fr}t_r\ddot{\tau} \quad \text{[MPa]} \tag{1}$$

Where:

- $\tau(t)$ – time dependent shear stress developed in the material during compaction [MPa];
- τ_0 – the static yield point [MPa];
- $\dot{\tau}$ – the first derivative of the shear stress [$MPa\cdot s^{-1}$];
- $\ddot{\tau}$ – the second derivative of the shear stress [$MPa.s^{-2}$];
- η_1 – viscosity of the plastic-viscous body [MPa·s];
- η_2 – the elasticity modulus of the Voight-Kelvin test [MPa·s];
- $\dot{\varepsilon}$ – the first derivative of the deformation developing in material or shear rate[s^{-1}];
- $\ddot{\varepsilon}$ – the second derivative of the deformation developing in the material or first derivative of shear rate [s^{-2}];
- t_r – the delay time of the elastic deformation in the material [s];
- t_{fr} – the time of relaxation of mechanical stresses in material [s].

The combined rheo-tribometer of IGREX Ltd (Fig. 2) was used to determine the rheological parameters of pre-compacted specimens made from different titan alloy powders. The deformation-time curve of Ti50Ni alloy powders at 3.5MPa normal pressure and 72N car moving force is shown in Figure 5.

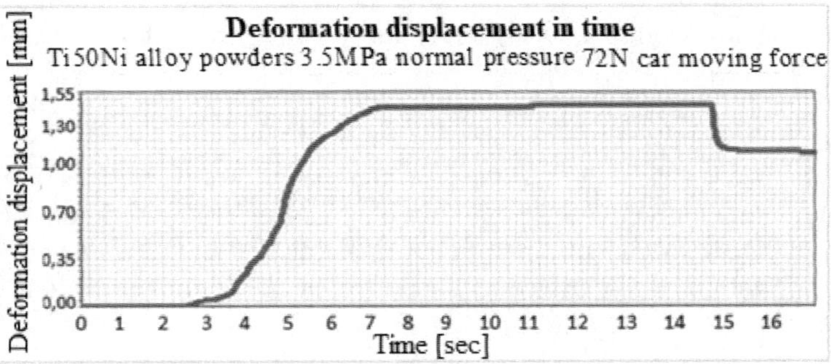

Figure 5. The deformation-time curve of Ti50Ni alloy powders pre-compacted at 400MPa

The rheological equation (Eq.1) is capable to characterize complexly the mechanical process in the titanium alloy powders during their uniaxial compacting. The parameters of equation (Eq.1) were determined experimentally on the combined rheo-tribometer and it provides a corresponding answer to the mechanical state and behavior of powders during their uniaxial compacting in the die cavities and the pre-compacted specimens after they leave the die (Table 1).

Table 1. The parameters of equation (1) measured by combined rheo-tribometer

The measured parameters		**Ti sponge powder**	**Ti50Ni alloy powders**	**Ti6Al2,5Fe alloy powders**
Specific Hooke deformation; Y_R	[-]	0,075	0,225	0,68
Specific elastic deformation; Y_E	[-]	0,035	0,070	0,090
Specific plastic deformation, Y_{pl}	[-]	0,615	1,150	0,360
Specific yield deformation Y_0	[-]	0,110	0,295	0,158
Static yield stress τ_0	[MPa]	0	0	0
The instantaneous shear stress determining the elasticity modulus of the Hooke body; τ_{Inst}	[MPa]	0,188	1,907	0,179
The elastic shear stress determining the elasticity modulus of the Voight-Kelvin body; τ_{Elast}	[MPa]	0,122	0,170	0,167
The elastic module of Hooke body; E_1	[MPa]	2,510	8,474	2,630
The elastic module of Voight-Kelvin body; E_2	[MPa]	3,495	2,423	1,859
The viscosity of the plastic-viscous body; η_1	[MPas]	1,399	4,860	1,592
The viscosity of the Voight-Kelvin body; η_2	[MPas]	0,020	0,371	0,019
The slope of the plastic deformation; α	[°]	5°	2°	6°
The slope of elastic deformation; β	[°]	84°	79°	84°
Shear rate in the broken material structure; v_1	[1/s]	0,087	0,035	0,105
Shear rate in the non-destructed material structure; v_2	[1/s]	9,514	5,145	9,514
The delay time of the elastic deformation in the material; t_r	[s]	0,557	0,573	0,605
Relaxation time of mechanical stresses in material; t_{fr}	[s]	0,014	0,076	0,012

4. Conclusions

The time and pressure dependences of rheological behaviors and models of different Ti alloys are shown in this work. A rheological equation (Eq.1) and the values of elasticity modulus of Voight-Kelvin and Hooke components as well as values of viscosity of the viscous-plastic components are also determined and given by the authors in the presented work. With this work the authors contributed considerable to develop and establish the powder technology of light metal alloys in Hungary.

Acknowledgement

The described article was carried out as part of the EFOP-3.6.1-16-2016-00011 "Younger and Renewing University – Innovative Knowledge City – institutional development of the University of Miskolc aiming at intelligent specialization" project implemented in the framework of the Szechenyi 2020 program. The realization of this project is supported by the European Union, co-financed by the European Social Fund.

References

[1] Bansil R, Hardcastle J, Constantino M 2015 *Építőanyag-JSBCM* **67** (4) 150 http://dx.doi.org/10.14382/epitoanyag-jsbcm.2015.25

[2] Osorio F, Roman A and Ortiz J 2015 *J. Phys.: Conf. Ser.* **602** 012016 https://doi.org/10.1088/1742-6596/602/1/012016

[3] Beliaev A, Svistkov A, Iziumov R, Osorgina I, Kondyurin A, Bilek M and McKenzie D 2016 *IOP Conf. Ser.: Mater. Sci. Eng.* **123** 012001 https://doi.org/10.1088/1757-899X/123/1/012001

[4] Reshetnikov A, Urakov A, Kasatkin A, Soiher M G and Kopylov M 2016 *IOP Conf. Ser.: Mater. Sci. Eng.* **123** 012007 https://doi.org/10.1088/1757-899X/123/1/012007

[5] Bendoukha M, Mosbah M 2017 *Építőanyag-JSBCM* **69** (4) 122 https://dx.doi.org/10.14382/epitoanyag-jsbcm.2017.22

[6] Torres-Cadenas S, Bravo-Patino A, Zarate-Medina J and Contreras-García M E 2017 *Építőanyag-JSBCM* **69** (1) 2 https://dx.doi.org/10.14382/epitoanyag-jsbcm.2017.1

[7] Kumbár V, Nedomová Š, Votava J and Buchar J 2017 *J. Phys.: Conf. Ser.* **790** 012016 https://doi.org/10.1088/1742-6596/790/1/012016

[8] Reyes-Ocampo I, Aguayo-Vallejo J P, Ascanio G and Córdova-Aguilar M S 2017 *J. Phys.: Conf. Ser.* **790** 012028 https://doi.org/10.1088/1742-6596/790/1/012028

[9] Szemán J 2006 *Építőanyag-JSBCM* **58** (1) 15 http://dx.doi.org/10.14382/epitoanyag-jsbcm.2006.4 (in Hungarian)

[10] Kulkov S, Savchenko N 2008 *Építőanyag-JSBCM* **60** (4) 62 http://dx.doi.org/10.14382/epitoanyag-jsbcm.2008.10

[11] Gömze A L, Gömze N L 2009 *Építőanyag-JSBCM* **61** (2) 38 http://dx.doi.org/10.14382/epitoanyag-jsbcm.2009.7

[12] K Katahira, H Ohmori 2009 *Építőanyag-JSBCM* **61** (4) 114 http://dx.doi.org/10.14382/epitoanyag-jsbcm.2009.21

[13] R Nagy, A Elekes, L Bartha, Á Vágó 2016 *Építőanyag-JSBCM* **68** (4) 98 http://dx.doi.org/10.14382/epitoanyag-jsbcm.2016.17

[14] Gömze A L and Gömze N L 2017 *J. Phys.: Conf. Ser.* **790** 012001 https://doi.org/10.1088/1742-6596/790/1/012001

[15] Burik P, Pesek L, Kejzlar P, Andrsova Z and Zubko P 2017 *J. Phys.: Conf. Ser.* **790** 012003 https://doi.org/10.1088/1742-6596/790/1/012003

[16] Emira Binti Azaman N, Rafi Raza M, Muhamad N, Niaz Akhtar M and Bakar Sulong A 2017 *J. Phys.: Conf. Ser.* **790** 012008 https://doi.org/10.1088/1742-6596/790/1/012008

[17] Izak P, Mastalska-Poplawska J, Lis J and Stempkowska A 2017 *J. Phys.: Conf. Ser.* **790** 012011 https://doi.org/10.1088/1742-6596/790/1/012011

[18] Koshelev K, Kuznetcov A, Merzlikina D, Pyshnograi G, Pyshnograi I and Tolstykh M Y 2017 *J. Phys.: Conf. Ser.* **790** 012014 https://doi.org/10.1088/1742-6596/790/1/012014

[19] Mastalska-Popiawska J and Izak P 2017 *J. Phys.: Conf. Ser.* **790** 012021 https://doi.org/10.1088/1742-6596/790/1/012021

[20] Khramchenkov M G, Usmanov R M 2017 *Építőanyag-JSBCM* **69** (4) 110 http://dx.doi.org/10.14382/epitoanyag-jsbcm.2017.19

[21] Ershova N I, Kelina I Yu 2009 *Építőanyag-JSBCM* **61** (2) 34 http://dx.doi.org/10.14382/epitoanyag-jsbcm.2009.6

[22] Borosnyói A, Kara P, Mlinárik L, Kaše K 2013 *Építőanyag-JSBCM* **65** (3) 90 http://dx.doi.org/10.14382/epitoanyag-jsbcm.2013.17

[23] Slobodov A, Uspenskiy A, Yavshits S, Mischenko A 2015 *Építőanyag-JSBCM* **67** (4) 164 http://dx.doi.org/10.14382/epitoanyag-jsbcm.2015.28

[24] Wang Y, Lakho D A, Yao D 2015 *Építőanyag-JSBCM* **67** (4) 132 http://dx.doi.org/10.14382/epitoanyag-jsbcm.2015.21

[25] Remiš T 2017 *J. Phys.: Conf. Ser.* **790** 012027 https://doi.org/10.1088/1742-6596/790/1/012027

[26] Sedlacik M and Pavlinek V 2017 *J. Phys.: Conf. Ser.* **790** 012031 https://doi.org/10.1088/1742-6596/790/1/012031

[27] Stoppe N and Horn R 2017 *J. Phys.: Conf. Ser.* **790** 012032 https://doi.org/10.1088/1742-6596/790/1/012032

[28] Zaragoza J, Chang A and Asuri P 2017 *J. Phys.: Conf. Ser.* **790** 012037 https://doi.org/10.1088/1742-6596/790/1/012037
[29] Gömze A L 2016 Applied Materials Science I. Published by IGREX Ltd. ISBN 978-963-12-6600-9
[30] Kracalik M 2017 *Építőanyag-JSBCM 69* (4) 116 http://dx.doi.org/10.14382/epitoanyag-jsbcm.2017.21
[31] Silva F D, Oliveira F M C, Souza J C 2010 *Építőanyag-JSBCM* **62** (1) 27 http://dx.doi.org/10.14382/epitoanyag-jsbcm.2010.06
[32] Miranda-Hernández J G, Rocha-Rangel E, Díaz De La Torre S 2010 *Építőanyag-JSBCM* **62** (1) 2 http://dx.doi.org/10.14382/epitoanyag-jsbcm.2010.01
[33] Khare S, Sharma M, Venkateswarlu K 2010 *Építőanyag-JSBCM* **62** (2) 39 http://dx.doi.org/10.14382/epitoanyag-jsbcm.2010.08
[34] Rocha-Rangel E, Hernández-Silva D, Terréz-Rojas E, Martínez-Franco E, Díaz De La Torre S 2010 *Építőanyag-JSBCM* **62** (3) 75 http://dx.doi.org/10.14382/epitoanyag-jsbcm.2010.15
[35] Csányi J 2009 *Építőanyag-JSBCM* **61** (1) 6 http://dx.doi.org/10.14382/epitoanyag-jsbcm.2009.2
[36] Kalatur E S, Kozlova A V, Buyakova S P and Kulkov S N 2013 *IOP Conf. Ser.: Mater. Sci. Eng.* **47** 012004 https://doi.org/10.1088/1757-899X/47/1/012004
[37] Gömze A L, Gömze N L, Kulkov S N, Shabalin I, Gotman I, Pedraza F, Lecomte-Nana G, Mayorova T, Kurovics E and Hamza A 2015 *Építőanyag-JSBCM* **67** (4) 143 http://dx.doi.org/10.14382/epitoanyag-jsbcm.2015.24
[38] Rodríguez Torres G M, Zarate Medina J, Contreras García M E 2016 *Építőanyag-JSBCM* **68** (4) 120 http://dx.doi.org/10.14382/epitoanyag-jsbcm.2016.21
[39] Abad-Javier M E, Cajero-Juárez M and Contreras García M E 2016 *Építőanyag-JSBCM* **68** (4) 125 http://dx.doi.org/10.14382/epitoanyag-jsbcm.2016.22
[40] Duretek I and Holzer C 2017 J. Phys.: Conf. Ser. **790** 012007 https://doi.org/10.1088/1742-6596/790/1/012007
[41] Mir A El and Nehme S G 2017 *Építőanyag-JSBCM* **69** (3) 83 http://dx.doi.org/10.14382/epitoanyag-jsbcm.2017.14
[42] Binti Azaman N E, Rafi Raza M, Muhamad N, Niaz Akhtar M and Bakar Sulong A 2017 *J. Phys.: Conf. Ser.* **790** 012008 https://doi.org/10.1088/1742-6596/790/1/012008
[43] Egész Á, Gömze A L 2013 *Építőanyag-JSBCM* **65** (4) 107 http://dx.doi.org/10.14382/epitoanyag-jsbcm.2013.20
[44] Egész Á, Gömze A L 2014 *Építőanyag-JSBCM* **66** (4) 125 http://dx.doi.org/10.14382/epitoanyag-jsbcm.2014.23
[45] Rácz Á 2014 *Építőanyag-JSBCM* **66** (1) 13 http://dx.doi.org/10.14382/epitoanyag-jsbcm.2014.3
[46] Egész Á, Gömze A L 2015 J. Phys.: Conf. Ser. **602** 012023 https://doi.org/10.1088/1742-6596/602/1/012023
[47] Ayininoula G M, Adekitan O A 2017 *Építőanyag-JSBCM* **69** (2) 34 https://dx.doi.org/10.14382/epitoanyag-jsbcm.2017.7
[48] Svéda M, Benke M, Roósz A 2013 *Építőanyag-JSBCM* **65** (2) 39 http://dx.doi.org/10.14382/epitoanyag-jsbcm.2013.8
[49] Trusova E A 2013 *IOP Conf. Ser.: Mater. Sci. Eng.* **47** 012043 https://doi.org/10.1088/1757-899X/47/1/012043
[50] Papp S, Kelemen A, Jakab-Farkas L, Vida-Simiti I and Biró D 2013 *IOP Conf. Ser.: Mater. Sci. Eng.* **47** 012016 https://doi.org/10.1088/1757-899X/47/1/012016
[51] Lutsyk V and Vorob'eva V 2013 *IOP Conf. Ser.: Mater. Sci. Eng.* **47** 012049 https://doi.org/10.1088/1757-899X/47/1/012049
[52] Kis D, Béres G, Dugár Zs and Hansághy P 2015 *J. Phys.: Conf. Ser.* **602** 012024 https://doi.org/10.1088/1742-6596/602/1/012024

[53] Machado López M M, Espitia Cabrera M I, Faure J and Contreras García M E 2016 *IOP Conf. Ser.: Mater. Sci. Eng.* **123** 012006 https://doi.org/10.1088/1757-899X/123/1/012006
[54] Tisza M 2013 *IOP Conf. Ser.: Mater. Sci. Eng.* **47** 012010 https://doi.org/10.1088/1757-899X/47/1/012010
[55] Kovacs-Coskun T, Volgyi B and Sikari-Nagl I 2015 *J. Phys.: Conf. Ser.* **602** 012026 https://doi.org/10.1088/1742-6596/602/1/012026
[56] Adamczyk-Cieslak B, Zdunek J and Mizera J 2016 *IOP Conf. Ser.: Mater. Sci. Eng.* **123** 012019 https://doi.org/10.1088/1757-899X/123/1/012019
[57] Isaev A, Grechishnikov V, Kozochkin M, Pivkin P, Petuhov Y and Romanov V 2016 *IOP Conf. Ser.: Mater. Sci. Eng.* **123** 012045 https://doi.org/10.1088/1757-899X/123/1/012045
[58] Maksarov V, Khalimonenko A and Olt J 2017 *Építőanyag-JSBCM* **69** (2) 66 http://dx.doi.org/10.14382/epitoanyag-jsbcm.2017.12
[59] Kovacs-Coskun T 2016 *IOP Conf. Ser.: Mater. Sci. Eng.* **123** 012010 https://doi.org/10.1088/1757-899X/123/1/012010
[60] Baideldonova A, Ksandopulo G and Mukhina L 2016 *IOP Conf. Ser.: Mater. Sci. Eng.* **123** 012032 https://doi.org/10.1088/1757-899X/123/1/012032
[61] Hampshire S, Park H K, Thompson D P and Jack K H 1978 *Nature* **274** 880
[62] Zharikov E V, Zaramenskikh K S, Popova N A, PFaikov P, Iskhakova L D, Gerke M N, Kutrovskaya S V and Nogtev D S 2011 *Glass and Ceramics* **68** 80
[63] Zharikov E V, Duong T T T, Faikov P P, Popova N A and Sovyk D M 2016 *Inorganic Materials: Applied Research* **7** 163
[64] Abe K, Hamada J, Miyazawa M, Kuwahara H and Itatani K 2013 *IOP Conf. Ser.: Mater. Sci. Eng.* **47** 012024 https://doi.org/10.1088/1757-899X/47/1/012024
[65] Hakeem A S, Khan R M A, Al-Malki M M, Patel F, Bakare A I, Ali S, Hampshire S and Laoui T 2014 *Advancesin Science and Technology* **89** 63
[66] Kinelovskii S A and Maevskii K K 2017 *Építőanyag-JSBCM* **69** (1) 29 http://dx.doi.org/10.14382/epitoanyag-jsbcm.2017.6
[67] Kaygorodov A, Khrustov V 2010 *Építőanyag-JSBCM* **62** (4) 116 http://dx.doi.org/10.14382/epitoanyag-jsbcm.2010.21
[68] Gömze A L 1983 *Építőanyag* **35** (5) 173
[69] Gömze A L, Kovács Á 2005 *Építőanyag-JSBCM* **57** (2) 34 http://dx.doi.org/10.14382/epitoanyag-jsbcm.2005.7
[70] Devecseri G 2010 *Építőanyag-JSBCM* **62** (1) 23 http://dx.doi.org/10.14382/epitoanyag-jsbcm.2010.5
[71] Gömze A L 2015 Rheology Compilation of Scientific Papers I. Published by IGREX Ltd. ISBN 978-963-12-3088-8
[72] Gömze A L, Gömze N L 2010 *Építőanyag-JSBCM* **62** (4) 98 http://dx.doi.org/10.14382/epitoanyag-jsbcm.2010.18

IOP Conf. Series: Journal of Physics: Conf. Series **1045** (2018) 012002 doi:10.1088/1742-6596/1045/1/012002

Effects of Sliding Velocity and Thermal Conduction of the Tool on X20Cr4 Steel Friction Coefficient and Structure in Nanostructuring Burnishing

V P Kuznetsov, A S Skorobogatov, M L Lobanov, A S Yurovskih, M S Khadyev, M S Karabanalov

Institute of New Materials and Technologies, Ural Federal University, 28 Mira St., 620078 Ekaterinburg, Russia

E-mail: v.p.kuznetcov@urfu.ru

Abstract. The developmental study has succeeded in finding how the sliding velocity of an indenter affects the friction coefficient and changes the structure and phase state in the surface layer of a X20Cr13 stainless steel when nanostructuring burnishing is done with a tool with no heat removal and, alternatively, when the tool is equipped with a cooling system. It has been shown that structural dispersion of the treated material results in obtaining 20…80 nm nanocrystallites if the friction coefficient of the spherical synthetic diamond indenter is within 0.15…0.18 nm. Application of a compact cooling system, based on Peltier's thermoelectric module, made it possible to stabilize the friction coefficient at 0.17 and to increase the sliding velocity from 13 m/min, this being the case of no heat removal, to 45 m/min in the case when the cooling system having a cooling performance of 120 W was used. TEM and SEM analyses of the surface layer structure confirmed that there is a correlation between the friction coefficient and the size of nanocrystallites and the thickness of the dispersed layer. EBSD analysis of the structure showed that a maximum permissible sliding velocity can be established as referenced to the nucleation and growth of γ-phase grains in the nanostructured layer caused by heating of the material under deformation and reaching the temperature beyond the point $\alpha' \rightarrow \gamma$ phase transition as well as by behavior of dynamic recrystallization. It was established that the heat removal ensures suppression of dynamic recrystallization when the sliding velocity is increased up to 50 m/min.

1. Introduction

Fragmenting the grain structure in structural materials until nanocrystallites are obtained significantly improves their physical, mechanical and functional properties. Nanocrystalline structures are predominantly achieved by severe plastic deformation techniques (SPD) such as multi axial forging, rolling, ECAP, HPT, etc. Consistent results of forming a nanostructural state in the surface layer are yielded by nanostructural burnishing when finishing is done on machines.

Kuznetsov et al. showed that the nanostructural state in a thin surface layer of heat-strengthened structural steels is achieved by applying a sliding spherical indenter with a friction force of 20…70 N in the contact; it ensures obtaining a degree of accumulated deformation of 2…9 [1]. It was established that consistent forming of nanocrystallites measuring 20…50 nm in the surface layer with a thickness up to 8 μm is possible if the sliding velocity is 4…12 m/min. Kuznetsov et al. determined that increasing the velocity beyond 12 m/min results in a considerably lower microhardness, worse roughness and destruction of the surface layer caused by a loss of the shear stability of the material. Limiting the linear velocity of the sliding indenter can be attributed to changing mechanisms of the contact friction due to friction heat; it brings about higher costs of nanostructural burnishing [2].

Korzynski et al. studied effects of burnishing on the fatigue strength of 42CrMo4 and 41Cr4 chrome-coated steels [3]. Burnishing was done with a spherical indenter. It was proved that sliding burnishing improves smoothing of the surface roughness. What is more, it contributes to reaching a high level of compressing residual stresses which in the case of cyclic symmetrical deflecting gives a better fatigue strength up to 40%. To minimize the friction coefficient and heat emission an indenter with a natural diamond was used. Therefore, no attempts were made to establish the effects of friction loading and thermal state of the material under deformation.

Kuznetsov et al. determined values of the friction coefficient for indenters fitted with PCD synthetic diamonds and DBN cubic boron nitride in tribological tests performed on X20Cr13 and X20Cr4 steels applying a reciprocating pattern at velocities of 0.07 and 0.035 m/s in dry conditions and with lubrication [4]. It was defined that the value of the friction coefficient in the cases of sliding the PCD on X20Cr13 steel and the DBN on X20Cr4 steel was 0.14…0.30 and 0.13…0.17 in dry conditions and ~0.12 and 0.07…0.09 in lubrication, respectively. However, no research was done into determining the friction coefficient and discovering structural transformations in the material at sliding velocities exceeding 0.07 m/s. All the more so, the reciprocating pattern of tribological tests offers no opportunity to simulate real conditions for sliding the indenter simultaneously in the direction of the velocity and the feed of burnishing.

Rabinowicz and Imai researched into the friction of a stainless steel pin on a stainless steel disk coated with low-melting-point materials under gradually increasing temperatures [5]. It was ascertained that there occurs a 1.5…7-fold increase of the friction coefficient when the temperature approaches the melting point of the coating material. The authors associated these abrupt increases of the friction coefficient with a growing effect of adhesion though no consideration was given to alterations of either the phase state, or the structure of the low-melting-point material. What is more, throughout the tests the velocity remained constant standing at 0.1 m/s. Bonnet et al. looked into the effect of the sliding velocity on the friction coefficient of a +TiN coated spherical pin on AISI 316L steel [6]. The research showed that while the sliding velocity is below 60 m/min, the value of the friction coefficient is constant and is equal to 0.4. As the velocity rises beyond this value, the friction coefficient begins to go down until it stabilizes at 0.27 with the velocity being 120 m/min. Similar research was undertaken by Ben Abdelali et al. [7], which demonstrated that the steady values of the friction coefficient were 0.54 and 0.23 at velocities of 60 m/min and 120 m/min respectively. Somehow, the researchers did not take into consideration the heating of the material in the tribological contact as a factor influencing the friction coefficient.

Bonnet et al. [6] and Ben Abdelali et al. [7] also studied branching of frictional heat for a carbide+TiN coated pin sliding on AISI 316L and AISI 1045 steels. They proved that the fraction of heat going into the pin decreases from 22 to 15% and from 60 to 10%, respectively, when the sliding velocity goes over 60 m/min. Meanwhile, they did no research into the correlation between the structural state of the material and the friction coefficient in respect to alterations of the sliding velocity and the pin's thermal state.

From our point of view, dynamic recrystallization of the nanostructured layer is likely to originate resulting from a combined impact of SPD and the temperature, which bring about a critical density of dislocations and the diffusion activity of the material. McQueen and Jonas determined that when active loading is applied, dynamic recrystallization can be observed as soon as the temperatures are 0.3…$0.35T_m$ [8]. Pu et al. showed an evolving mechanism of dynamic recrystallization in the case of cryogenic burnishing of an AZ31B magnesium alloy [9]. At the same time, no research was revealed that would have looked into recrystallization of nanostructuring friction finishing of structural steel.

Kuznetsov et al. discovered a possibility to increase the rate of nanostructuring burnishing for X20Cr4 steel utilizing liquid cooling of the tool indenter with a capacity of heat removal ~10 W [10]. Application of a dual-cycle cooling system succeeded in obtaining consistent formation of the surface layer, its roughness being Ra<170 μm and its microhardness almost reaching ~1200 $HV_{0.5}$ if the sliding velocity of the indenter stayed below 25 m/min. No correlation was investigated between the sliding velocity, on the one hand, and the friction coefficient and the surface layer structure, on the other hand, in the case of applying liquid cooling.

The goal of the present research is to establish patterns representing the changes of the friction coefficient and the structure and phase state in the surface layer of X20Cr13 stainless steel occurring when the sliding velocity of the spherical indenter equipped with a tool cooling system is increased.

2. Experimental setup

The developmental research on nanostructuring burnishing made use of disk-type samples, manufactured of X20Cr13 thermo-strengthened steel (C <0.20, Cr 12.7, Si <1.0, Mn <1.0 Mo 0.69, V 0.22, Fe rest (wt.%)) and having a diameter of 60 mm and a thickness of 12 mm. After hardening at 1050 °C and soaking for 2 hours, followed by low tempering at 150 °C, the material had a lath martensite structure with laths measuring 150…200 nm, its hardness being 45…46 HRC.

The flat surfaces of the samples were machined on a KNUTH V-Turn 410 lathe by cutting and consequent nanostructuring burnishing on two sides. Cutting was done with a Sandvik WNMG 080408-SM carbide insert at a cutting speed of 80 m/min, while the feed was 0.06 mm/rev. After the cutting had been done, the microhardness was 690…750 $HV_{0.5}$. Nanostructuring burnishing of the surfaces followed the scheme presented at (Figure 1, a); it was dry burnishing applying a normal force of 340 N on the indenter F_b, and the tool feed rate was f_b=0.025 mm/rev. As the tool was being fed, the sliding velocity increased linearly from 3 to 75 m/min.

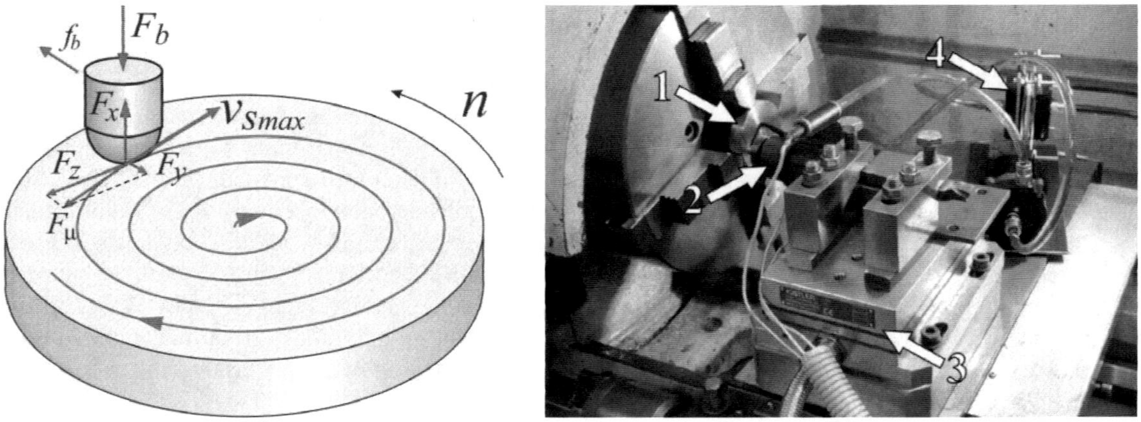

(a) (b)

Figure 1. Machining of the experimental sample: a – kinematic scheme of nanostructuring burnishing; b – photo showing the experimental unit mounted on a KNUTH V-Turn 410 lathe

A special tool was developed that enabled to research into the processing procedure without heat removal and with an indenter cooling system (Figure 2, a). The principle of operation of the cooling system is based on applying the Peltier thermoelectric module (Peltier TEC) with a cooler [10]. Heat is removed from the indenter by a circulating coolant with a pump (pump1) in a closed circuit through special channels in the tool and in the cooler. To maintain a consistent cooling performance of the TEC another loop was integrated into the system for water cooling. Maximum cooling performance primarily depends on the overall dimensions of the tool. The designed tool with a compact cooler enabled to provide a cooling performance of 120 W.

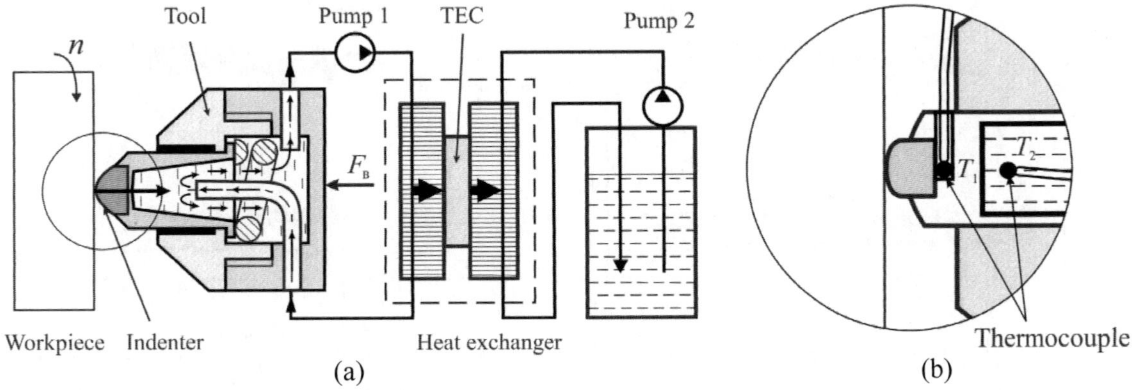

Figure 2. Illustration of the tool dual loop cooling system (a) and the sensing unit (b) for measuring the temperature of the tip and the cooling agent

In the process of the research one face of the sample was burnished by the tool with no heat removal from the indenter, while the other face was done with the tool with the cooling system in operation. The indenters, used for burnishing, were fitted with 2mm radius spherical synthetic diamond tips, the roughness of their working surface being Ra≤0.05 μm. Sample 1 (Figure 1, b) was burnished on the KNUTH V-Turn 410 lathe. Burnishing tool 2 was fitted into a three-axis Kistler 4257BA dynamometer 3. The Peltier module with the cooler 4 was mounted on the slide assembly of the machine. Type K thermocouples were used to measure the temperature T_1 of the indenter tip and the temperature T_2 of the cooling agent circulating in the closed loop (Figure 2, b).

TEM, SEM and EBSD techniques were applied and JEOL JEM 2100 and Zeiss AURIGA CrossBeam microscopes were used for studying the surface layer after finish machining and nanostructuring burnishing had been completed. Microhardness was measured with the Vickers technique on a Leica VMHT microhardness tester applying 0.25 N on the indenter.

3. Results and Discussion

3.1. Friction coefficient and indenter temperature

Dynamometry of nanostructuring burnishing (Figure 3), done with the tool which had no heat removal, showed that a normal force in contact F_x fluctuates with an amplitude of 340±70 N. Application of the indenter cooling system made it possible to reduce the amplitude of these fluctuations to 50 N. Initially at a sliding velocity of 11 m/min the contact force F_z initially goes down from 120 N to 50 N in the case of burnishing with no heat removal with peaks of short duration within 200 seconds of the experiment. As the velocity increases, the force F_z rises in an abrupt to ~100 N, and after 300 seconds it starts to fluctuate again in the range from 50 to 100 N. The fluctuations lead to wearing out the indenter when the total time of the experiment reaches the 370-th second; it corresponds to the sliding velocity v_s≈20 m/min. In the case of applying the indenter cooling system the force F_z is consistent throughout the length of the experiment until the velocity is 75 m/min.

The friction coefficient was calculated making use of the obtained dynamometric data:

$$\mu = \frac{\sqrt{F_y^2 + F_z^2}}{F_x} = \frac{F_\mu}{F_x}, \tag{1}$$

where F_x is normal reaction force; F_y is reaction force in the direction of sliding velocity v_s and F_z is normal reaction force in the direction of feed f_b.

IOP Conf. Series: Journal of Physics: Conf. Series **1045** (2018) 012002 doi:10.1088/1742-6596/1045/1/012002

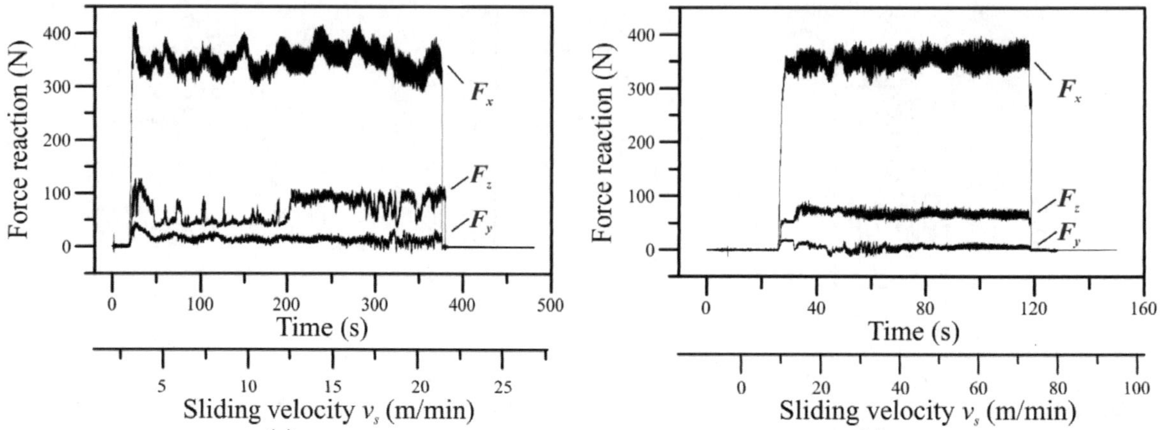

(a) (b)

Figure 3. Dynamometry of contact forces in nanostructuring burnishing: a – no heat removal applied; b – cooling system applied.

The pattern of the manner, in which the coefficient of friction changes depending on the value of the indenter sliding velocity, is identical in both cases, for the application with no heat removal and for the other case when the cooling system is in operation (Figure 4). Initially the coefficient of friction goes down reaching the steady-state value μ=0.13 when burnishing with no heat removal and μ=0.17 when burnishing with the cooling system. The steady-state value of the friction coefficient remains until the critical velocity v_{crit} is reached; it is ~13 m/min without heat removal and ~45 m/min in the case of cooling (Table 1). As soon as the critical velocity is exceeded, a sharp increase of the friction coefficient occurs. It starts to go down again when the velocity continues to grow.

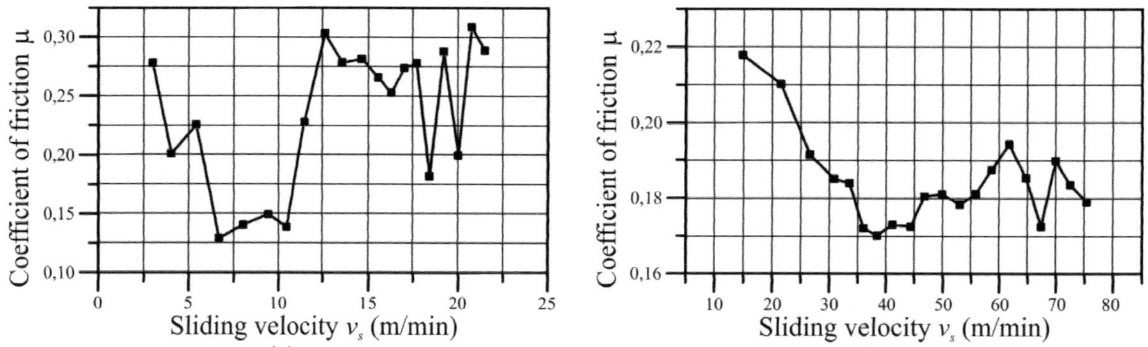

(a) (b)

Figure 4. Variations of the friction coefficient in relation to the tool sliding velocity: a – no heat removal applied; b – cooling system applied.

Table 1. Analytical values of friction coefficient.

Tool	Friction coefficient μ			Sliding velocity v_s (m/min)	
	μ_{max}	μ_{min}	μ_{stable}	Prior to steady-state value of friction coefficient v_{stable}	Critical v_{crit}
No heat removal	0.31	0.125	0.13…0.14	7…11	13
With cooling system	0.215	0.165	0.17…0.18	30…55	60

During the nanostructuring burnishing with no heat removal the temperature T_1 of the indenter increased continuously (Figure 5, a) finally reaching ~180 °C at a velocity of 20 m/min; it resulted in wearing out the indenter tip and putting an end to the operation. Application of the cooling system made it possible to do machining at a velocity of 60 m/min. In this case the temperature T_1 did not exceed 60 °C (Figure 5, b). During the nanostructuring burnishing by the tool with the cooling system temperature T_2 was maintained within 0 and +3 °C.

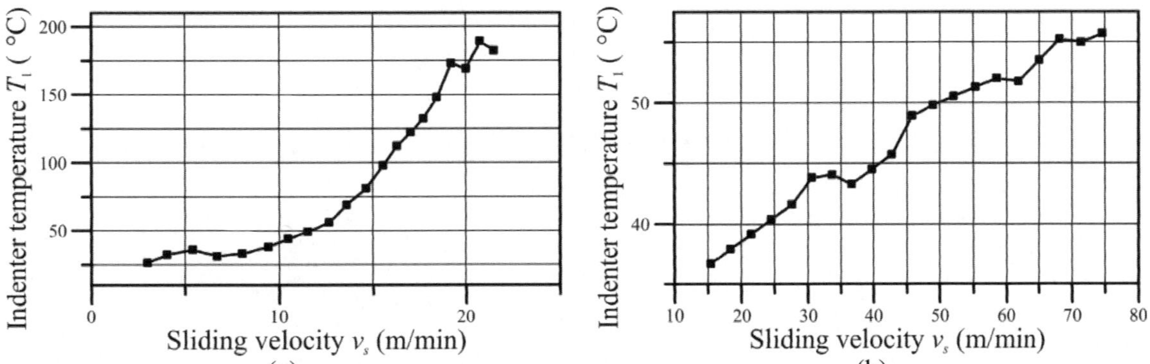

(a) (b)

Figure 5. Variations of temperature T_1 at the indenter tip in relation to the tool sliding velocity: a – no heat removal applied; b – cooling system applied.

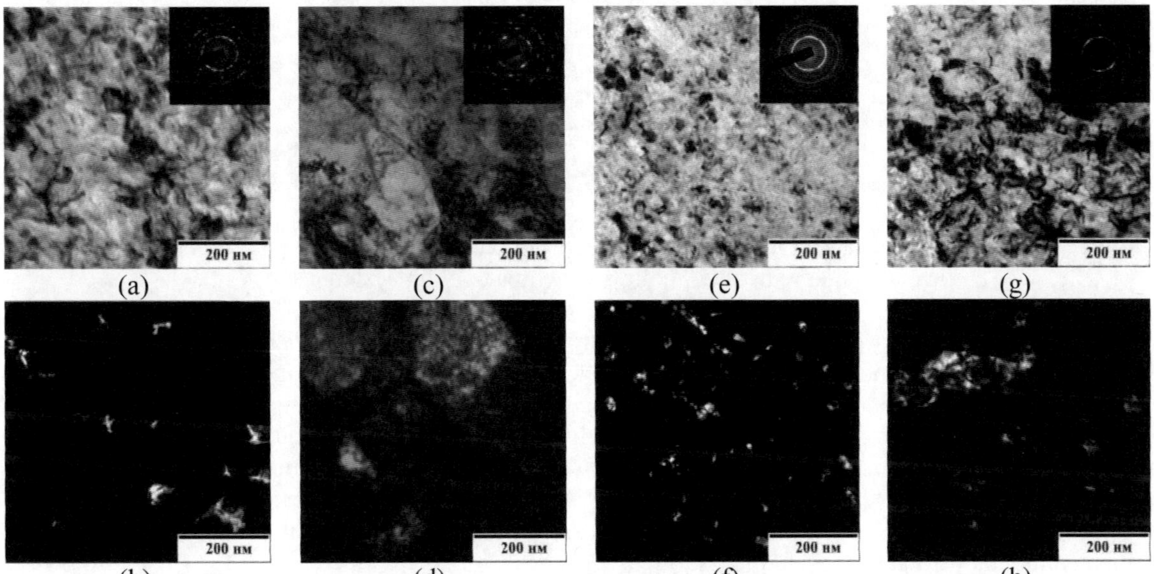

(a) (c) (e) (g)
(b) (d) (f) (h)

Figure 6. Structure of the surface layer of the sample after nanostructuring burnishing by the tool: a, b and c, d – no heat removal at 10 and 13 m/min, respectively; e, f and g, h – with a cooling system at 45 and 60 m/min.

3.2. Microstructure

TEM was done making use of foils made from the areas of the surface layer corresponding to the middle and boundaries in the range of a stable friction coefficient. It was determined that a nanocrystalline structure forms in the surface layer without losing shear stability when burnishing is done with no heat removal at a velocity below 15 m/min (Figure 6, a and b) and with the cooling system at a velocity below 65 m/min (Figure 6, c and d). High dispersion of the structure is manifested by a circular pattern of α'-phase microdiffraction with a pronounced azimuthal reflection diffusion and

IOP Conf. Series: Journal of Physics: Conf. Series **1045** (2018) 012002 doi:10.1088/1742-6596/1045/1/012002

a large number of nanocrystallites measuring 20…80 nm in the dark-field images (figure 6, a, b and e, f).

After nanostructuring burnishing had been completed, two zones were distinguished in the surface layer based on the SEM crossection (Figure 7). The material in zone 1 can be characterized as nanostructured. In zone 2 plastic deformation has a wave shaped pattern with localized shear bands, however, its magnitude is not sufficient for generating a nanostructure. After reaching the velocity v_s=15 m/min with no heat removal the material of the surface layer loses shear stability and is finally destroyed.

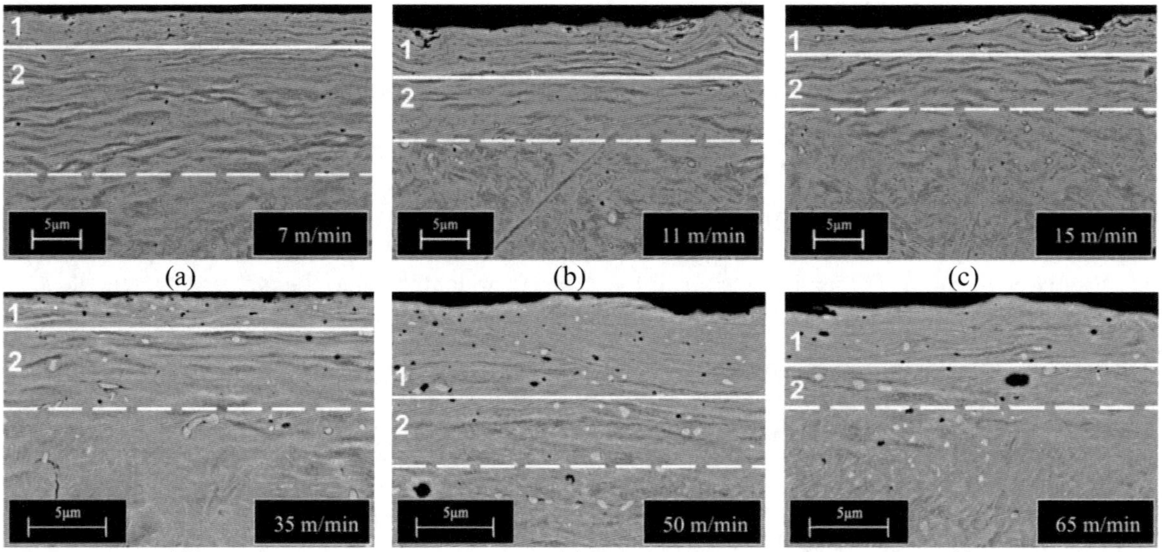

(a) (b) (c)
(d) (e) (f)

Figure 7. SEM of the surface layer after nanostructuring burnishing with a tool with no heat removal (a, b, c) and with a tool equipped with a cooling system (d, e, f).

SEM images of the surface layer after nanostructuring burnishing in the BSE mode show dark zones, which can identified as pores. The number of pores in zone 2 is lower than in zone 1; however, their size is significantly larger. The biggest pores, some reaching a size of 1 μm, appear when burnishing is done by the tool with the cooling system at a velocity of 45 m/min and above this value. It proves that zone 2 was heated up to a high temperature. In addition to the pores, the modified layer contains light precipitates of circular shape. When the sliding velocity of the indenter is below critical, these precipitates are predominantly in zone 2. At the same time, as the velocity is increased above its critical value, they are also present in zone 1.

When burnishing is done with no heat removal at a sliding velocity of 7 m/min and a coefficient of friction of 0.135, the thickness in zone 1 is 1.5…2 μm (Figure 7, a), while in zone 2 it is 10…11 μm. In zone 2 there are pores, their size being 0.2…0.9 μm. Increasing the velocity to 11 m/min results in evolution of microcracks, elongated in the direction of the indenter feed (Figure 7, b). Meanwhile, the thickness in zone 1 goes up and reaches 4…5 μm, and in zone 2 it goes down and amounts to 7…8 μm. When the sliding velocity is above 11 m/min, light precipitates measuring from 0.2 to 1.3 μm appear in zones 1 and 2; also, microcracks emerge reaching the surface (Figure 7, c). Increasing the velocity from 15 to 20 m/min led to extreme wear of the indenter and deterioration of the surface.

When burnishing is done by the tool equipped with the cooling system at a sliding velocity of 35 m/min and a friction coefficient of 0.165, the thickness in zones 1 and 2 is 1…1.5 μm and 5…6 μm, respectively (Figure 7, d). In zone 2 some light precipitates are as big as 2 μm. If the velocity is raised to 50 m/min and μ≈0.18 is obtained, the thickness in zone 1 increases to 5…6 μm, while in zone 2 it reaches 8…9 μm (Figure 7, e). In zone 2 there appear light precipitates, their size reaching 1.3 μm. When the sliding velocity is 65 m/min and the coefficient of friction is 0.17, the thickness in zone 1 and 2 turns out to be smaller, it goes down to 2…3 μm and 5…6 μm (Figure 7, f). Large light precipitates, their sizes reaching 1.4 μm, emerge in zone 2 and in the boundary of zone 1.

EBSD analysis of the structure in the surface layer (Figure 8) after nanostructuring burnishing had been completed revealed that the light precipitates are a phase with a FCC crystal lattice, i.e. austenite. It can be presumed that in the process of burnishing the surface layer material was heated up to the critical temperature A_{c3} (≥900 °C), which corresponds to the phase recrystallization $\alpha' \rightarrow \gamma$ followed by a recurrence of martensitic transformation $\gamma \rightarrow \alpha'$ as a result of cooling to the temperature A_{r3} (660 °C). Taking into consideration that the major phase in the subsurface layer is martensite, there are sufficient grounds to say that the circular austenitic precipitates evolved due to short spells when the temperature of the material was 660…900 °C. In this intercritical temperature range when ferrite and austenite can coexist, pores and particles of austenite coagulated owing to intensive diffusion of carbon vacancies and atoms in ferrite. Vigorous plastic deformation of the material could also contribute to stabilizing areas of austenite in zone 2.

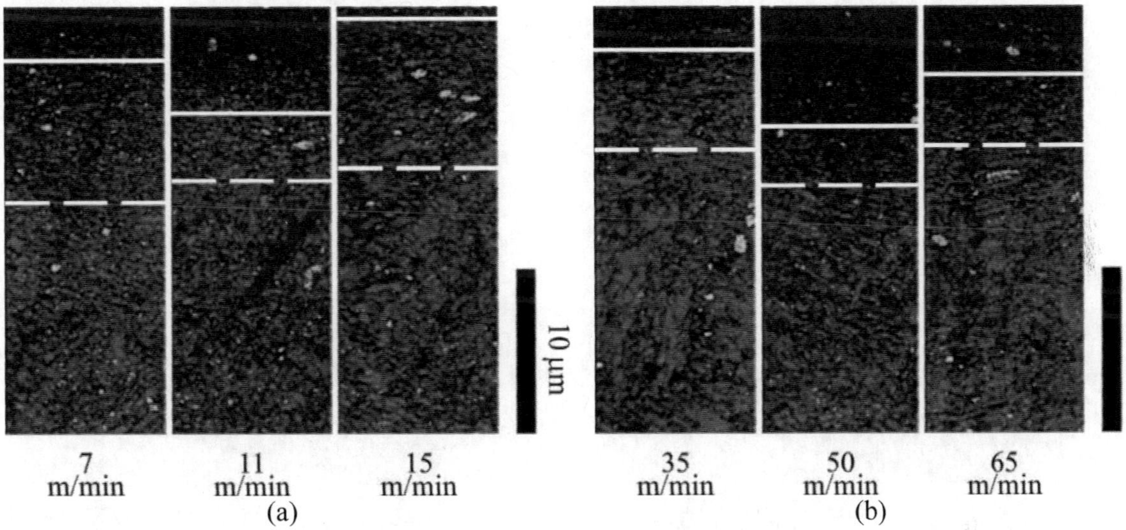

Figure 8. EBSD analysis map shows a microstructure with superimposed phase distribution (Mapping Band Contrast+Phase Contrast) in the surface layer of the sample after nanostructuring burnishing with the tool without heat removal (a) and with an indenter cooling system (b).

EBSD analysis was performed to correlate the phase composition in the material, which is being deformed, and the stability of the evolving nanostructured layer while the burnishing rate was increased. The maps of the microstructure (Figure 8) in the surface layer show selected zones that correspond to zones 1 and 2 marked in the SEM images (Figure 7).

The light precipitates, which can be observed in the marked zones and the main matrix of the deformed material, are identified as γ-phase grains. It can be presumed that emergence and growth of the austenite grains are related to the material under deformation reaching the phase transition temperature (~860 °C) and dynamic recrystallization.

Formation and growth of the γ-phase grains in zone 1 begins in the case of burnishing with the tool with no heat removal as soon as the velocity reaches 11 m/min. If the velocity grows to 15 m/min, relatively large austenite grains measuring 0.6…1.3 μm (Figure 8, a) evolve in the nanostructured layer. Hardly any γ-phase grains can be observed in the nanostructured layer when the sliding velocity of the indenter with the cooling system is increased up to 50 m/min. But when the velocity goes up from 50 to 65 m/min (Figure 8, b), in zone 1 considerable austenite grain growth commences.

Therefore, emergence of the γ-phase grains in the material going through deformation in the nanostructured layer can be considered as a practical criterion to determine the maximum permissible (critical) sliding velocity for nanostructuring burnishing. Thus, permissible sliding velocities for burnishing with the tool with no heat removal and with the indenter cooling system are the velocities equal to 11 and 50 m/min respectively.

3.3. Surface roughness and microharndess

Emergence of large γ-phase grains in the nanostructured layer also produces a detrimental effect on the roughness of the surface microprofile. Optical 3D-profilometry showed that with permissible sliding velocities of the indenter with and without heat removal a stable friction coefficient, μ= 0.17…0.18 and 0.13…0.14 respectively, is ensured and a minimum roughness of Ra≈180…220 nm is obtained (Figure 9, a and c).

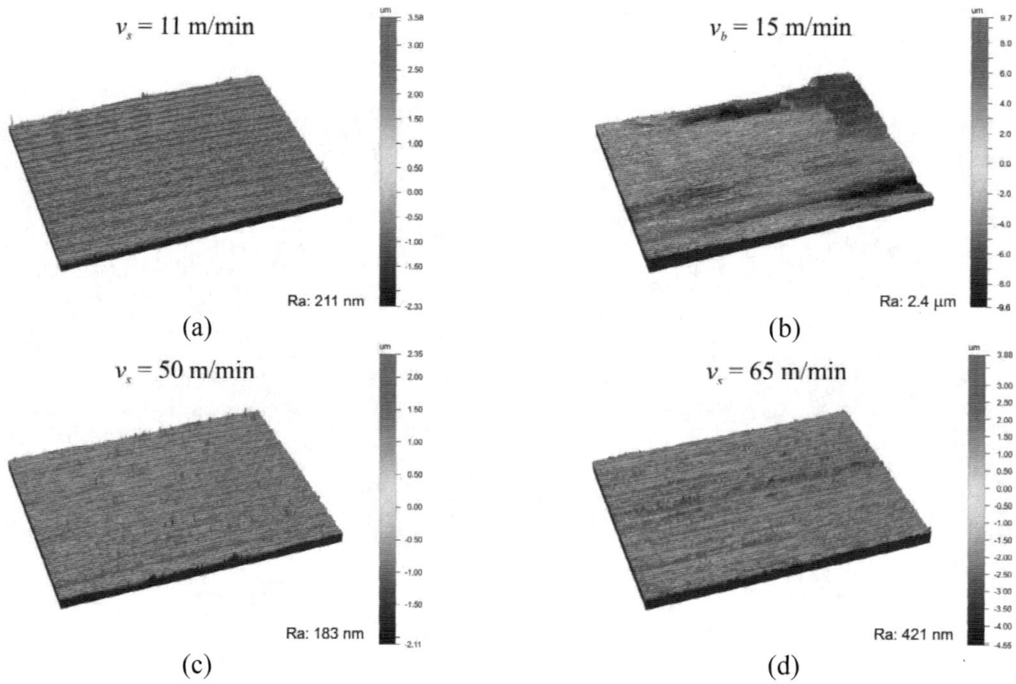

Figure 9. Microprofile of the surface after nanostructuring burnishing without heat removal (a, b) and with a cooling system (c, d).

At sliding velocities beyond permissible values large γ-phase grains form in the nanostructured layer and the roughness grows to Ra≈2.4 μm (Figure 9, b) and Ra≈0.42 μm (Figure 9, d) respectively in both cases, when burnishing with no heat removal and with the cooling system. There emerge local tears in the material of the nanostructured layer as the growth of γ-phase recrystallized grains might be accompanied by an evolving mechanism of peel-off wear. An analogous mechanism of wearing counterbodies was studied in the research (Chen et al., 2014).

Formation of relatively large γ-phase grains in the subsurface layer and nanostructured layer can also be considered to be the cause for the lower microhardness of the surface (Figure 10, a and b) because recrystallized grains have by far lower hardness compared to that of the nanocrystall martensite structure. The maximum microhardness that can be achieved in the case of burnishing without heat removal is 1100…1150 $HV_{0.5}$; in the case when the cooling system is applied, it is 1100…1210 $HV_{0.5}$.

The established character of changes to the microhardness after the critical sliding velocity is reached when burnishing without heat removal and when burnishing with a cooling system suggests that the most favorable conditions for nanostructuring exist at the temperature of the material under deformation close to the temperature of the phase transition $\alpha'\rightarrow\gamma$. However, to support this suggestion separate research is needed.

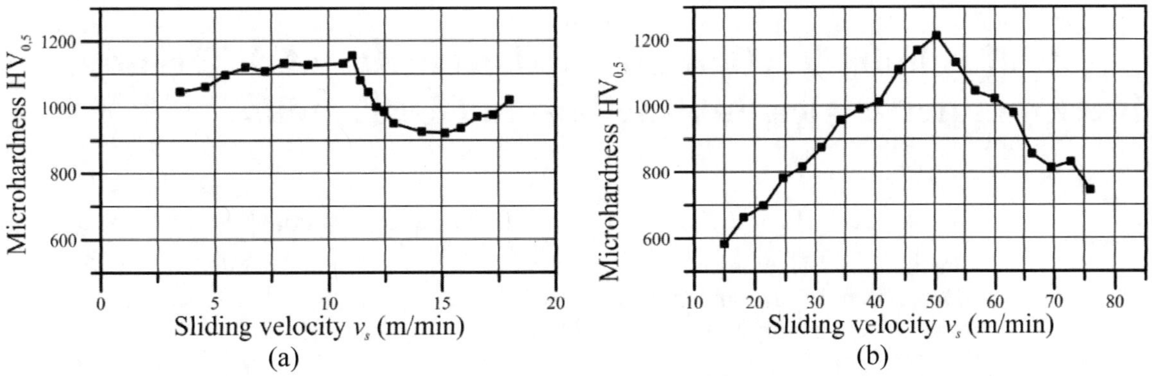

Figure 10. Correlations between the microhardness of the surface layer and sliding velocity: a – by the tool with no heat removal; b – by the tool with a cooling system.

4. Conclusions

a) The extent of dispersion of the surface layer structure to obtain nanocrystallites is to a large extent dependent on the stability and value of the friction coefficient.

b) A homogenous nanostructure containing 20…80 nm crystallites when burnishing the X20Cr13 steel with a tool with no heat removal is obtained provided the sliding velocity range from 7 to 11 m/min and the friction coefficient stays at μ=0.13…0.14. Application of a cooling system based on a Peltier TEC to maintain the temperature of the coolant within 0…+3 °C and the indenter temperature below 60 °C makes it possible to raise the sliding velocity to 65 m/min. Stabilizing the friction coefficient within 0.17…0.18 contributes to increasing the thickness of the nanostructured layer from 1.5…2 μm to 5…6 μm.

c) Emergence of the recrystallized γ-phase attributed to the intense heating of the friction contact can be applicable as a practical criterion to determine the permissible sliding velocity of the indenter.

References

[1] Kuznetsov V P, Smolin I Y, Dmitriev A I, Tarasov S Y, Gorgots V G 2016 *Surf. and Coat. Technol.* **285** 171

[2] Kuznetsov V P, Tarasov S Y, Dmitriev A I, 2015 *J. of Mater. Process. Technol.* **217** 327

[3] Korzynski M, Pacana A, Cwanek J 2009 *Surf. and Coat. Technol.* **203** 1670

[4] Kuznetsov V P, Makarov A V, Psakhie S G, Savrai R A, Malygina I Y, Davydova N A 2014 *Phys Mesomech.* **17** 250

[5] Rabinowicz E, Imai M 1963 *Wear* **6** 407

[6] Bonnet C, Valiorgue F, Rech J, Claudin C, Hamdi H, Bergheau J M, Gilles P 2008 *Int. J. of Mach. Tools and Manuf.* **48** 1211

[7] Ben Abdelali H, Claudin C, Rech J, Ben Salem W, Kapsa P, Dogui A 2012 *Wear, Tribol. in Manuf. Process.* **286–287** 108

[8] McQueen H J, Jonas J J 1985 *J. Appl. Metalwork.* **3** 410

[9] Pu Z, Song G-L, Yang S, Jr O W D, Puleo D A, Jawahir I S 2011 *Magnes. Technol.* **2011** 513

[10] Kuznetsov V P, Skorobogatov A S, Gorgots V G, Yurovskikh A S 2016 *IOP Conf. Ser.: Mater. Sci. Eng.* **124** 12127 https://doi.org/10.1088/1757-899X/124/1/012127

IOP Conf. Series: Journal of Physics: Conf. Series **1045** (2018) 012003 doi:10.1088/1742-6596/1045/1/012003

Effects of volume fraction, size and geometry of different fillers on interparticle interactions in LDPE melts

Nico Laufer[1], Harald Hansmann[1, 2], Christian Boss[1], Stefan Ofe[1]

[1] Institute for Polymer Technologies e. V., 23966 Wismar, Germany
[2] Department of Mechanical Engineering, Hochschule Wismar University of Applied Sciences, Technology, Business and Design, 23966 Wismar, Germany

Email: laufer@ipt-wismar.de

Abstract. This study focuses on the effects of volume fraction, size and geometry of different fillers on interparticle interactions in low-density polyethylene (LDPE) melts. For this purpose, the rheological properties of LDPE filled with various fillers have been investigated by means of high pressure capillary rheometry in consideration of different particle sizes. It has been shown that interparticle interactions vary depending on volume fraction of the fillers. Furthermore the aspect ratio of the particles has a great impact on tendency of particles to interact with one another. While fibrous and flake shaped particles have a great tendency to interact with one another, spherical particles, depending on the particle size, exhibit a transition from negligible to non-negligible interactions. It has been shown that on the basis of a generalized interaction function, that considers the transition from negligibly interactions to the domain of pronounced interactions, interparticle interactions can be taken into account mathematically for describing the flow behaviour of particle filled LDPE melts.

1. Introduction

Viscosity functions of particle filled polymer melts are shifted to higher values in comparison to the viscosity function of the unfilled polymer melt (Figure 1). The offset depends on the volume content of the wood particles as well as the applied shear rate.

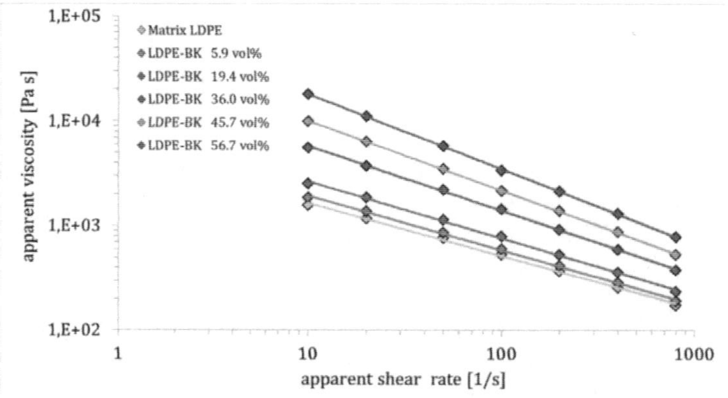

Figure 1. Viscosity functions of LDPE-based WPC with varying wood content (T = 190 °C)

W. Gleißle and M. K. Baloch developed a very vivid description for the offset of the vertical shift on the basis of the *concept of shear-stress-equivalent inner shear rate*. Within this concept, the hydraulic gap is reduced by the volume content of the solid particles. This leads to an increase in the shear rate of the continuous phase between these particles, the so-called inner shear rate [1]. Taking into account the shear thinning flow behaviour of particle filled polymer melts the relative increase of the inner shear rate to the shear rate of the suspension is expressed by the shift factor B, Eq. (1), which is influenced by a number of factors [2-5].

$$B(\tau_0)^{n_0} = \frac{K_c}{{K_0}^{\chi}} \cdot {\tau_0}^{(\chi-1)} \tag{1}$$

with:

B	shift factor	[-]
τ_0	apparent shear stress	[Pa]
K_c	consistency of suspension	[Pas^n]
K_0	consistency of matrix polymer	[Pas^n]
χ	interaction exponent = n_c/n_0	[-]

Within this study the effects of volume fraction, size and geometry of different fillers on interparticle interactions in low-density polyethylene (LDPE) melts are investigated. The aim of this study is to find a general approach describing the flow behaviour of particle filled polymer melts, which takes into account interaction effects of different types of fillers.

2. Materials and methods

The matrix polymer that has been used in this study is a low-density polyethylene (LDPE, Lupolen 2420 K), a highly branched polymer which exhibits shear thinning flow behaviour even at low shear rates. As filler particles glass beads and natural graphite have been used, in each case in two different particle size fractions. The particle size distributions were determined by laser diffraction using Mastersizer 3000 (Malvern Instruments GmbH) (Fig. 2 – 5 (a)). The geometry of the filler particles were characterized by microscopy using VHX 2000D (Keyence Microscopes) (Fig. 2 – 5 (b)).

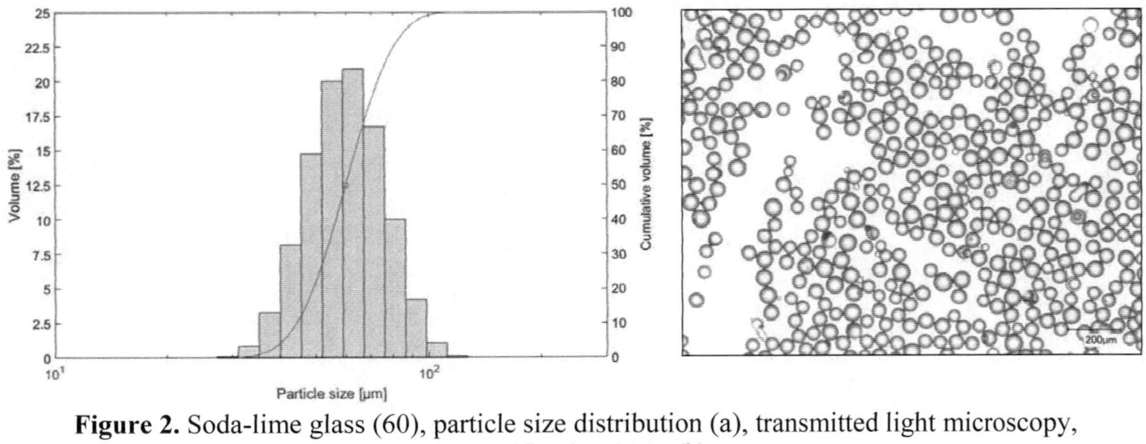

Figure 2. Soda-lime glass (60), particle size distribution (a), transmitted light microscopy, magnification 200x (b)

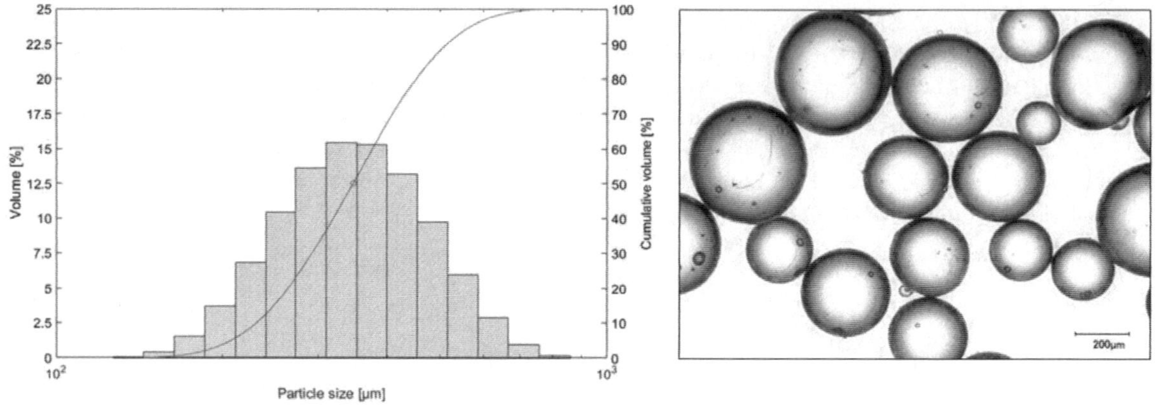

Figure 3. Soda-lime glass (346), particle size distribution (a), transmitted light microscopy, magnification 200x (b)

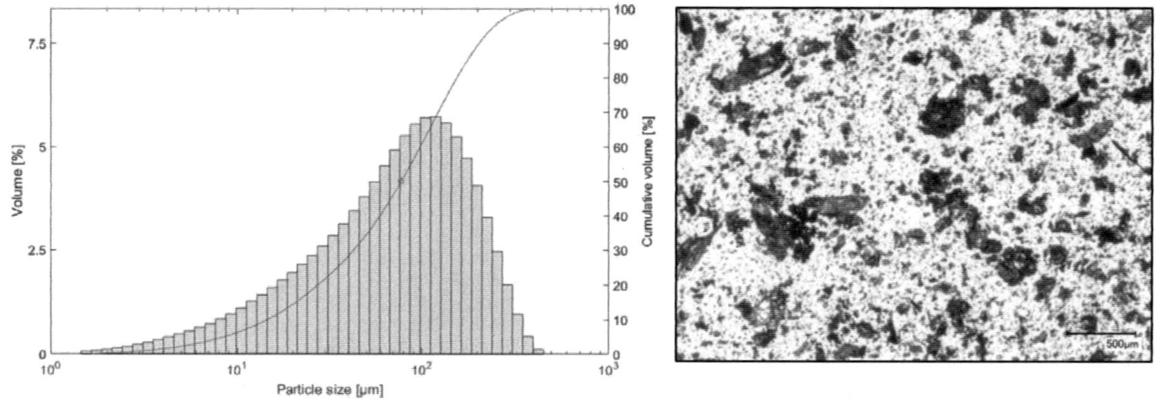

Figure 4. Natural graphite (77), particle size distribution (a), transmitted light microscopy, magnification 100x (b)

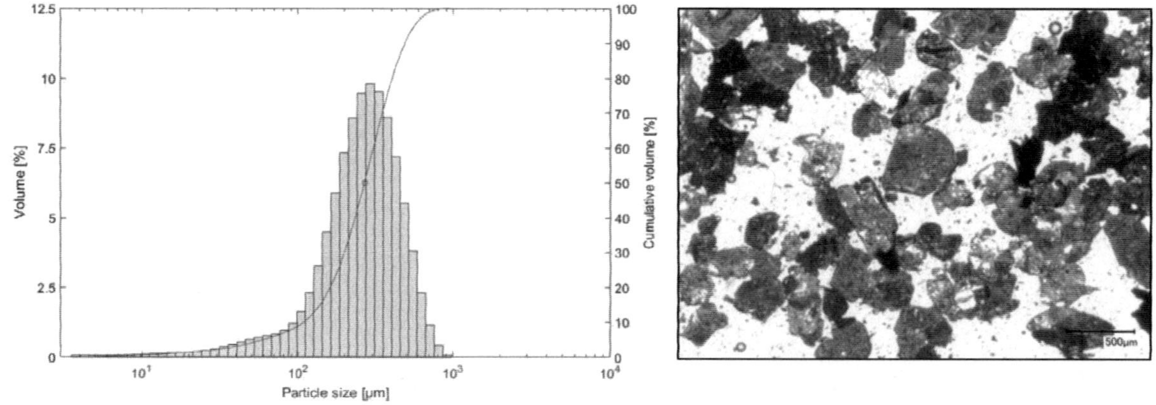

Figure 5. Natural graphite (267), particle size distribution (a), transmitted light microscopy, magnification 100x (b)

The preparations of all formulations were carried out by compounding on the co-rotating twin screw extruder, with a melt temperature of 190 °C and a mass output of 3 kg/h. All rheological analyses were performed on a high pressure capillary rheometer (HKR) Smart RHEO 5000 twin bore (CEAST,

IOP Conf. Series: Journal of Physics: Conf. Series **1045** (2018) 012003 doi:10.1088/1742-6596/1045/1/012003

Instron®) using a round die with a diameter of 2 mm and a length of 40 mm. The measured data were analysed on the basis of the apparent shear stress and shear rates.

3. Results and discussion

3.1. LDPE filled with glass beads (d50 = 346.150 µm)

Figure 6 shows the flow functions of the unfilled LDPE and LDPE which is filled with different content of glass beads. All flow functions clearly exhibit shear thinning flow behaviour. The corresponding power law model parameters are listed in Table 1.

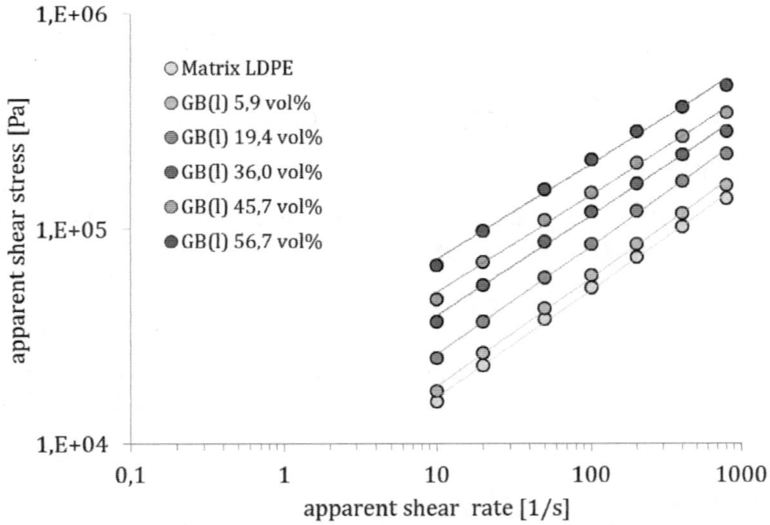

Figure 6. Flow functions of LDPE filled soda-lime glass (346) (T = 190 °C)

Table 1. Power law model parameters for LDPE filled with soda-lime glass (346)

LDPE-GB	glass beads volumetric content [%]	K [Pa s^n]	n [-]
LDPE	0	5231	0.496
LDPE-GB5.9	5.9	5806	0.502
LDPE-GB19.4	19.4	8181	0.502
LDPE-GB36.0	36.0	13490	0.465
LDPE-GB45.7	45.7	17712	0.453
LDPE-GB56.7	56.7	25936	0.443

With increasing volume content of the glass beads the flow functions are shifted to higher values. Accordingly, the consistency (K) increases with increasing volume content of the glass beads (Fig. 7 (a)).

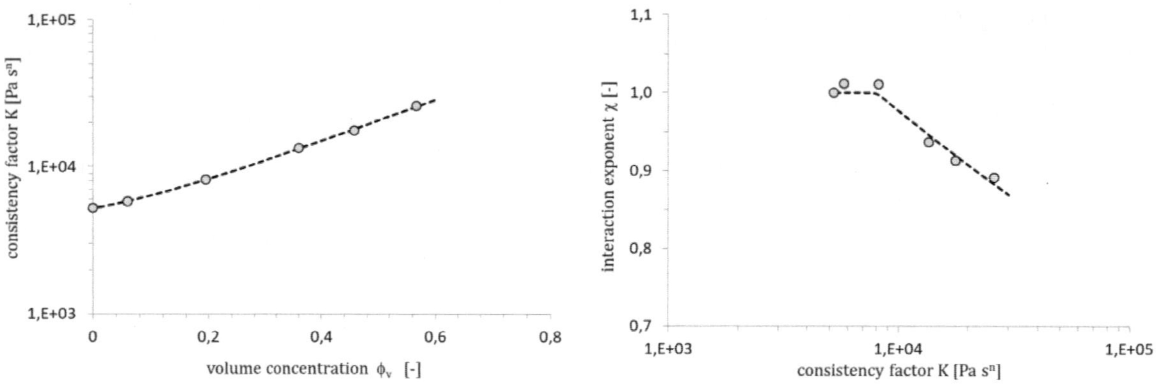

(a) (b)

Figure 7. Consistency as a function of volumetric filler content (a), interaction exponent as a function of consistency (b)

The quantitative impact of the volumetric filler concentration(ϕ_v) on the consistency can be approximated by an empirical equation, Eq. (2).

$$K_c = K_0 \cdot \left(1 + A \cdot {\phi_v}^{\alpha} \cdot e^{\beta \cdot \phi_v}\right) \tag{2}$$

The model parameters have been found by regression analysis, $A = 2.179$, $\alpha = 1.079$, $\beta = 2.114$.

Figure 7 (b) shows the correlation between interaction exponent and consistency. In case of negligible interparticle interactions only the consistency changes as a function of the volumetric filler content, while the flow index and the interaction exponent remains constant. The interaction exponent of the investigated glass bead filled LDPE melts remains almost constant up to a volume content of 19.4 vol.%. Thus, interparticle interactions are negligibly small in this region. At higher volumetric filler fractions, a transition to non-negligible interactions can be observed. In order to describe the correlation between interaction exponent and consistency with consideration of the transition from negligible interactions to the domain of non- negligible interactions a *generalized interaction function* has been used.

$$\chi = \frac{1}{\left(1 + \left(\frac{K_c}{K^*}\right)^{a}\right)^{\frac{b}{a}}} \tag{3}$$

The model parameter (K^*) describes the transition from negligible interactions to non-negligible interactions, whereas (a) expresses the width of the transition. The intensity of interaction effects is considered by parameter (b). For the glass bead filled LDPE melts the following model parameters have been found by regression analysis: $K^* = 8056$, $a = 49.238$, $b = 0.107$.

On the basis of Eq. (2) and Eq. (3) the influence of the volumetric filler concentrations and the transition of negligible to non-negligible interaction effects can be taken into account in the Equation for the shift factor B Eq. (1). Finally, the offset of the vertical shift of the flow functions of LDPE filled with different amounts of the big glass bead fraction (d50 = 346 μm) can be well described by using shift factor B in the following equation:

$$\tau_c = B^{n_0} \cdot \tau_0 = B^{n_0} \cdot K_0 \cdot \dot{\gamma}^{n_0} \tag{4}$$

Based on this approach a good match between approximated and measured flow function have been achieved (Fig. 8).

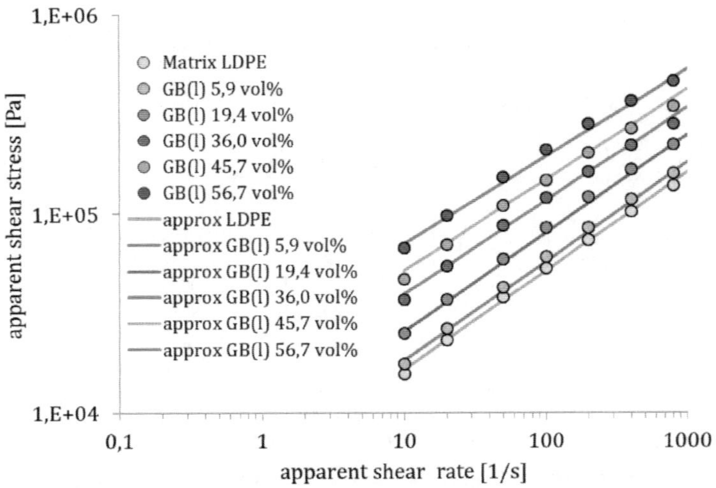

Figure 8. Comparison of approximated and measured flow functions of LDPE filled soda-lime glass (346) (T = 190 °C)

In order to validate the equations and the calculated model parameters which have been used to approximate the measured values, different compounds have been prepared in which the volume contents of the glass beads were varied. On this basis, the flow curves of the validation compounds have been estimated. The results show a good match between estimated flow curves and measured values (Fig. 9).

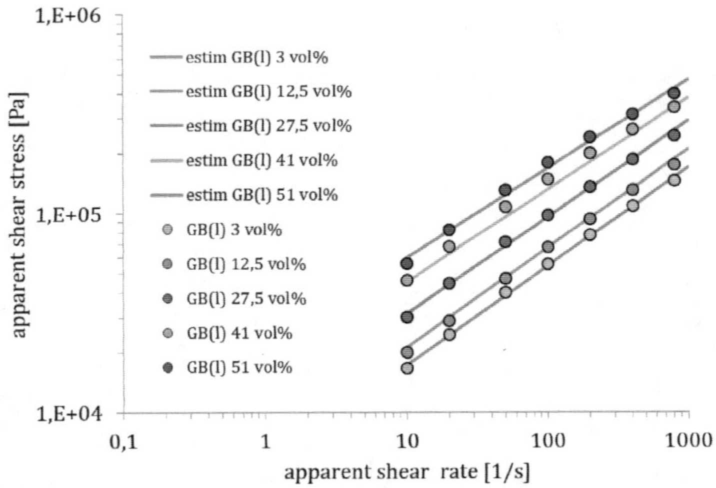

Figure 9. Comparison of estimated and measured flow functions of the validation compounds LDPE filled soda-lime glass (346) (T = 190 °C)

3.2. LDPE filled with glass beads (d50 = 60 μm)

In contrast to the large particle size fraction of the glass beads (d50 = 346 μm), the small glass beads (d50 = 60 μm) show interparticle interactions even at low filler concentrations (Fig. 10). This means that not only the particle geometry but also the particle size influence the tendency to form interparticle interactions. The generalized interaction function is also suitable for describing the correlation between interaction exponent and consistency with consideration of directly occurring interparticle interactions (Fig. 10).

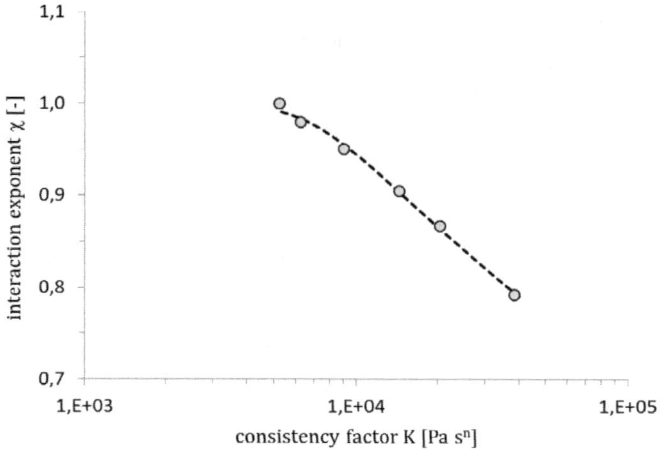

Figure 10. Interaction exponent as a function of consistency

The same model approaches Eq. (2) and Eq. (3) as for the large particle size fractions of the glass beads have been used for approximating and estimating the flow functions of LDPE filled with the small particle size fraction of the glass beads. The following model parameters have been found by regression analysis:

Equation (2):	$A = 0.276$	$\alpha = 0.180$	$\beta = 5.627$
Equation (3):	$K^* = 6744$	$a = 4.308$	$b = 0.133$

On the basis of the calculated model parameters, the flow curves have been approximated and estimated, respectively. The results show a good match between calculated flow curves and measured values (Fig. 11).

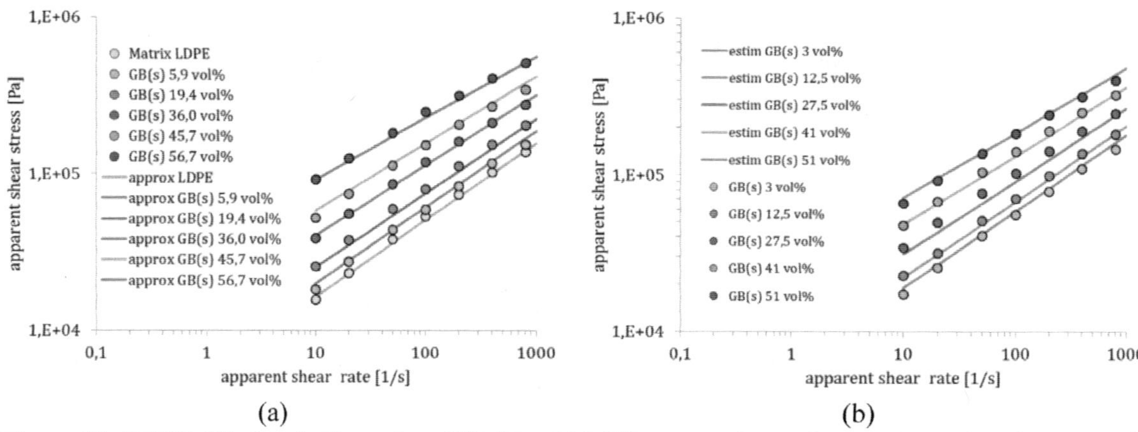

Figure 11. LDPE filled soda-lime glass (60) (T = 190 °C), comparison of approximated and measured flow functions (a), comparison of estimated and measured flow functions of the validation compounds (b)

3.3. LDPE filled with natural graphite

Lastly, the applicability of the generalized interaction function Eq. (3) on plate-shaped particles has been investigated. For this purpose, compounds with different volume contents and particle size fractions of natural graphite as filler particles in LDPE were prepared.

The following model parameters have been found by regression analysis:

Natural graphite (267) – large particle size fraction

Equation (2):	$A = 4.916$	$\alpha = 1.443$	$\beta = 3.017$
Equation (3):	$K^* = 5288$	$a = 214.295$	$b = 0.189$

Natural graphite (77) – small particle size fraction

Equation (2):	$A = 4.842$	$\alpha = 1.624$	$\beta = 3.643$
Equation (3):	$K^* = 7545$	$a = 2.841$	$b = 0.213$

Figure 12 illustrates the good match between estimated flow curves and measured values for the large fraction of natural graphite as well as for the small fraction.

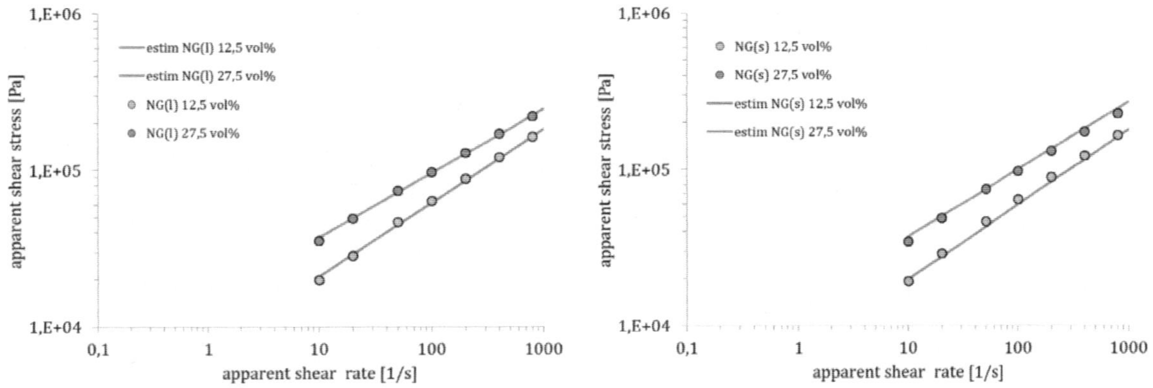

(a) (b)

Figure 12. Comparison of estimated and measured flow functions for LDPE filled with natural graphite, large particle size fraction (a) small particle size fraction (b)

4. Conclusions

In this study, the flow behaviour of LDPE which was filled with different volume and size fractions of different fillers have been investigated. On the basis of a generalized interaction function that considers the transition from negligible to non-negligible interparticle interactions, the flow behavior of the investigated compounds have been described mathematicallywith good consistency to the measured values. With this approach, flow functions for arbitrary volumetric filler concentrations can be estimated with good accuracy, independently from geometry and size of the particles. The calculated model parameters for describing the flow behavior of the particle filled LDPE melts have been successfully validated.

References

[1] Hochstein B 1997 Rheologie von Kugel- und Fasersuspensionen mit viskoelastischen Matrixflüssigkeiten, Dissertation Universität Fridericiana Karlsruhe (Technische Hochschule)

[2] Pahl M, Geißle W und Laun H-M 1995 Praktische Rheologie der Kunststoffe und Elastomere, Düsseldorf: VDI-Verlag GmbH

[3] Hansmann H, Laufer N und Kühn S 2012 Investigation on the Flow Behavior of WPC Melts, Bledzki A K and Sperber V E, Hrsg., München: Hanser, A6-1 - A6-11

[4] Laufer N, Hansmann H, Koch M 2014 High Pressure Capillary Rheometer on Wood Plastic Composites with Variation of Wood Content and Matrix Polymer, 58th Internationale Wissenschaftliche Kolloquium, Ilmenau

[5] N Laufer Hansmann H, Koch M 2017 *J. Phys.: Conf. Ser.* **790** 012017 https://doi.org/10.1088/1742-6596/790/1/012017

IOP Conf. Series: Journal of Physics: Conf. Series **1045** (2018) 012004 doi:10.1088/1742-6596/1045/1/012004

Balance of angular moments in structural mechanics of finite strains: A condition for second spatial derivatives of stress

Martin Lederer and Golta Khatibi

Institute of Chemical Technologies and Analytics, Vienna University of Technology, Christian Doppler Laboratory for Lifetime and Reliability of Interfaces in Complex Multi-Material Electronics, Getreidemarkt 9, 1060 Wien, Austria

Corresponding author: martin.lederer@tuwien.ac.at

Abstract. The conservation laws of linear and angular momentum are among the most fundamental principles of physics. In contemporary science, it is generally postulated that both principles apply to continuum mechanics of deformable solids. Nevertheless, there occasionally occurs the question whether it is necessary to formulate both conservation laws as independent principles. For the case of a statistical theory considering a large ensemble of point masses, the Nöther theorem tells that conservation of linear and angular momentum follow automatically from the existence of a Lagrange function for the energy, which is invariant with respect to translation and rotation of the reference system. If one, however, formulates equilibrium of stresses within continuum theory of solids, then it is necessary to treat the balance of angular moments as independent principle. Usually, the symmetry of the Cauchy stress tensor is derived from this postulate. In the present study, it is shown that another condition regarding the second spatial derivatives of stress may be obtained from this principle. It is suggested to use this condition in context with the theory of finite strains.

1. Introduction

Nowadays, the conservation law for angular momentum is generally postulated in mechanical sciences. However, the question whether this principle may be deduced from Newton's axioms of mechanics is by no means trivial [1]. In this context, a fundamental aspect has been emphasized by Poisson [2], who recognized that a mathematical proof for conservation of angular momentum may be given, if the forces in the system consist of pairwise equilibrated, central forces. In fact, the condition of pairwise central forces is satisfied by electrostatic interaction, and therefore this conservation law is naturally applied to mechanics of solids. Nevertheless, there still remained the question whether it is necessary to consider conservation of angular momentum as independent principle of continuum mechanics. Historically, it has for long been believed that it is sufficient to postulate conservation of linear momentum, until Euler [3] discovered that for rigid bodies conservation of angular momentum must be postulated independently. Thereafter, this conservation law has been extended to all kinds of deformable bodies through the principle of rigidification [4]. If one assumes by hypothesis that the motions of material points in a deformable body are frozen, then the body should behave like a rigid solid.

Next, we come to the question how conservation of linear and angular momentum is considered in the description of stresses. In this context, Euler proposed the principle of a free body diagram in order to visualize the applied forces. A volume element is cut free from the rest of the body, and the forces along the boundary represent the interaction between the volume element and the remaining part, schematically. Thus, one may integrate over the forces along the boundary line. Consequently, the forces leading to translation of the object must in sum be zero for equilibrium states. An equivalent statement holds for rotational forces. In sum the torques about the centre point of the volume element must vanish. On the basis of this argument, one usually proves the symmetry of the Cauchy stress tensor. However, it has been mentioned by Atanackovic and Guran [5] that one could also derive conditions where stress components depend on higher order gradients of the displacement vector. Unfortunately, they did not work out further details regarding their assessment. Instead, it has just been said that classical elasticity theory is not concerned with such cases. Nevertheless, their statement seems to be an interesting starting point for the investigations of our present study.

Here, it should also be mentioned that similar considerations regarding conservation of angular momentum have been made in the frame of Cosserat theory [6]. Within this approach orientational degrees of freedom are considered in addition to the displacements of material points. Furthermore, related theories of generalized continuum mechanics were established, including micropolar elasticity [7, 8] and couple stress theory [9, 10]. These theories have in common that the elastic response of the described materials deviates significantly from Hooke's law. But in the present study, we will restrict ourselves to deformable materials, where the elastic response is characterized by the Cauchy stress.

2. Equilibrium states of stress

In this section, the equilibrium conditions of stresses are characterized. At first equations are reemphasized, which represent the modern State of the Art. Thereafter, it is investigated whether further relations may be derived from the postulated balance of angular moments.

2.1. State of the Art

The simplest stress distributions are obtained for homogeneously strained samples consisting of one linear elastic material. In this case, the stresses are homogeneously distributed, as depicted in Fig. 1, schematically. For the moment, the investigation is restricted to 2-dimensional geometries.

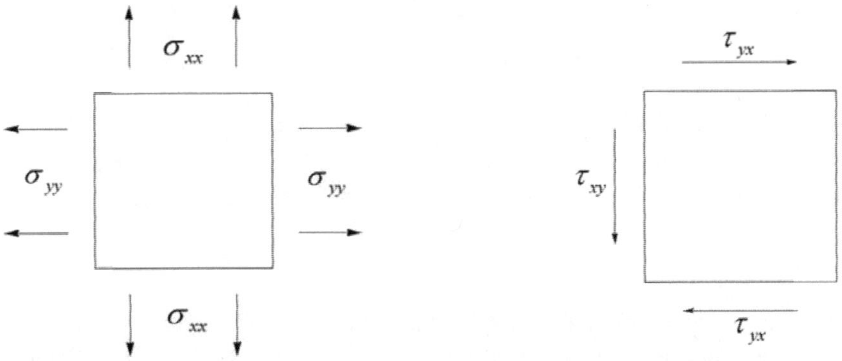

Figure 1. Free body diagram of a homogeneously strained plane volume element. The stresses are decomposed into normal stresses (left) and shear stresses (right).

Usually, stresses are decomposed into normal stresses and shear stresses as shown in Figure 1, schematically. For better clarity, the volume element is depicted twice in order to separate the contributions of different components of the stress tensor. It is obvious that the forces acting along the boundaries of the volume element do not introduce translations, because they are in sum compensating each other. However, one could imagine a resulting torque, if the shears τ_{xy} and τ_{yx} were not equal. Consequently,

$$\tau_{xy} = \tau_{yx} \tag{1}$$

IOP Conf. Series: Journal of Physics: Conf. Series **1045** (2018) 012004 doi:10.1088/1742-6596/1045/1/012004

In other words, the Cauchy stress tensor is symmetric.

The situation gets more complex, when samples are inhomogeneously strained. Here, we consider the general case where also body forces $\vec{f}$ might occur. Let us further assume that the first derivatives of stress components are constant across the volume element.The free body diagram of this case is depicted in Figure 2, where again contributions of normal stresses and shear stresses were separated.

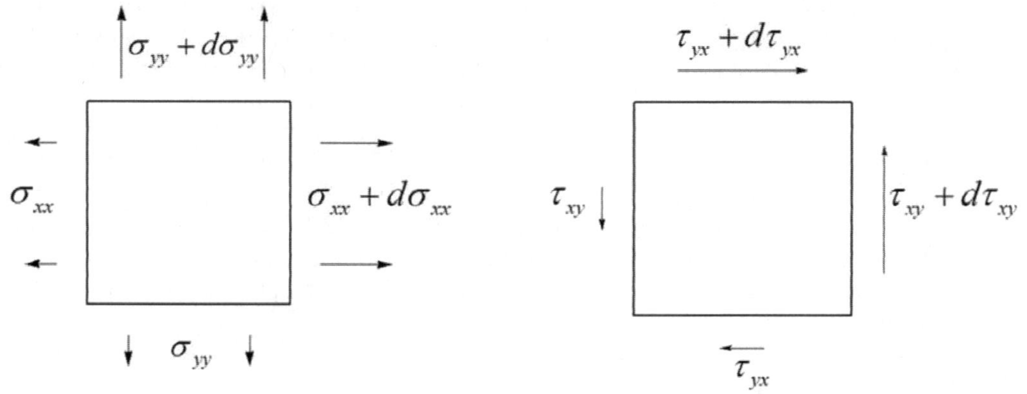

Figure 2. Free body diagram of a volume element showing the first derivatives of stresses. Normal stresses are depicted on the left while shear stresses are depicted on the right side, respectively.

Here, it is necessary to investigate forces which might introduce a translation of the volume element. For every direction of the coordinate system, one equilibrium condition is derived. Thereby, the loads carried by normal and shear stresses must be added. Thus, the equilibrium conditions read as

$$\frac{\partial \sigma_{xx}}{\partial x} + \frac{\partial \tau_{xy}}{\partial y} + f_x = 0 \tag{2 a}$$

$$\frac{\partial \sigma_{yy}}{\partial y} + \frac{\partial \tau_{xy}}{\partial x} + f_y = 0 \tag{2 b}$$

This equilibrium condition, in its 3-dimensional representation, was first suggested by Cauchy [11]. While condition (1) was motivated by conservation of angular momentum, condition (2) is due to conservation of linear momentum. In literature, the conditions (1) and (2 a, b) have been reformulated for various representations of continuum mechanics [12]. They are commonly used for elastic and elasto-plastic materials.

2.2. Further examination related to the balance of angular moment

Figure 2 demonstrates the equilibrium of linear moments. But what can be said about the angular moment related to cases, where the first derivatives of stress are considered? Obviously, the first derivatives of normal stresses do not contribute to the torque around the center point. Let us therefore look at the contribution of shear stresses. At first we treat the case where $\partial \tau_{xy} / \partial x = const > 0$, and $\partial \tau_{xy} / \partial y = 0$. Further, the shear stress at the point (x_0, y_0) is called τ_0. This condition is illustrated in the incomplete free body diagram on the left side of Figure 3. There, we have started to draw a distribution of surface tractions in vertical directions, which are in agreement with our assumptions. Since the Cauchy stress tensor is symmetric, there is only one possibility how to continue the diagram at the corners. Finally, the unique solution for the complete free body diagram is shown in Figure 3 on the right side.

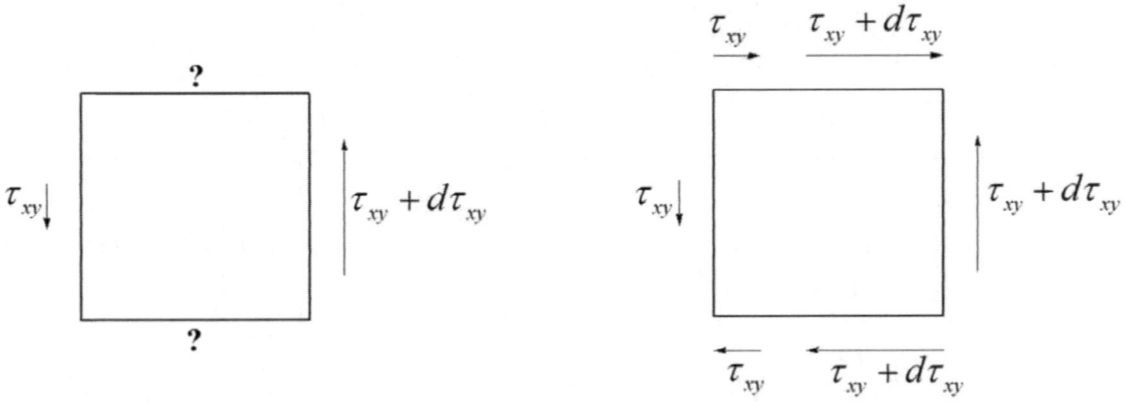

Figure 3. Free body diagram related to $\partial\tau_{xy}/\partial x = const. > 0$. The first step of the construction is depicted on the left side. The complete diagram is shown on the right side. The sum of torques is zero.

Integration of contributions to the angular moment along the boundary line yields

$$\oint_C \vec{F}\times\vec{r}ds = 0, \tag{3}$$

where $\vec{F}ds$ is the force increment related to the shear τ_{xy}, and $\vec{r}$ is the vector from a point at the boundary of the volume element to its centre. The sum of torques would not change, if $\partial\tau_{xy}/\partial y$ was also unequal to zero. Hence, the first derivatives of stress do not contribute to the angular moment.

2.3. Conditions derived for higher derivatives of stress

The balance equations including the first derivatives of stress are, however, not sufficient to guarantee balance of angular moments for arbitrary stress distributions. Historically, the application of conservation laws to higher derivatives of stress was not included in the principle of rigidification, because such methods were not needed for the description of rigid bodies either. On the other hand, conditions for the second derivatives of stress were later derived from the compatibility condition of neighbouring volume elements within the linearized theory of elasticity. But this approach uses simplifications, which are not essentially exact. Unfortunately, the linearized strain tensor does not fulfil the requirements of an objective tensor. Therefore, the present study is devoted to the purpose of evaluating conditions, which are in accordance with the principle of material frame indifference. The results obtained from our new considerations are to be used in combination with the theory of finite strains. A comparison with the simplified conditions of linear elasticity will be given in section 3.

Let us therefore continue with the case, where second derivatives of stress have constant values within the given volume element. First derivatives or constant stress values don't need to be considered here, since their contributions were already covered in previous steps of our investigation. It is easily found that components of the form $\partial^2\sigma_{xx}/\partial x^2$ or $\partial^2\sigma_{yy}/\partial y^2$ do not introduce a torque. However, the derivatives $\partial^2\sigma_{xx}/\partial x\partial y$ and $\partial^2\sigma_{yy}/\partial x\partial y$ produce a torque, as illustrated in Figure 4, schematically:

IOP Conf. Series: Journal of Physics: Conf. Series **1045** (2018) 012004 doi:10.1088/1742-6596/1045/1/012004

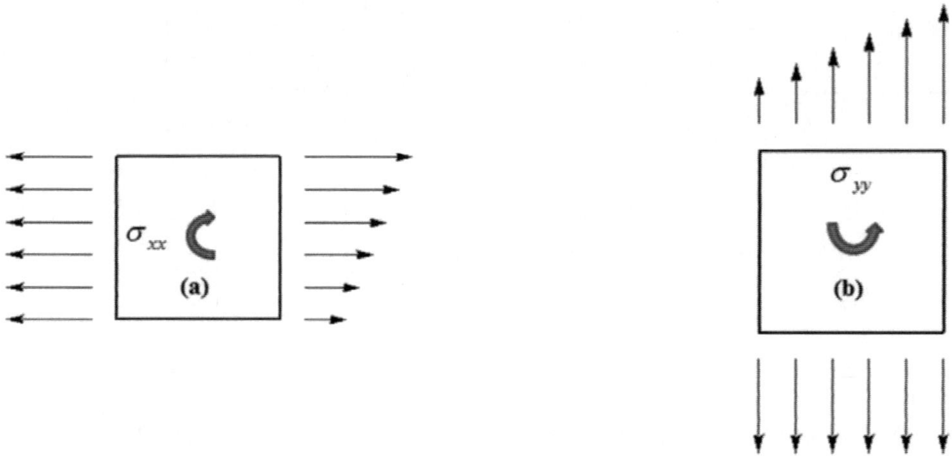

Figure 4. The second derivatives of normal stress components $\partial^2\sigma_{xx}/\partial x\partial y$ and $\partial^2\sigma_{yy}/\partial x\partial y$ produce a torque, as illustrated in Figure 4 (a) and (b), respectively.

Now, it is necessary to collect all the contributions to the torque caused by the second derivatives of stress in order to require that their sum must be zero. In this context, the equilibrium conditions (2 a, b) must also be considered. In conclusion, one obtains

$$\frac{\partial^2\sigma_{xx}}{\partial x\partial y} = \frac{\partial^2\sigma_{yy}}{\partial x\partial y}, \tag{4}$$

or equivalently

$$\frac{\partial^2\tau_{xy}}{\partial x^2} = \frac{\partial^2\tau_{xy}}{\partial y^2}. \tag{5}$$

Here, it should be noticed that equation (4) implies equation (5) and vice versa, if the equations (2 a, b) are valid. Figure 5 demonstrates that the torques caused by $\partial^2\tau_{xy}/\partial x^2$ and $\partial^2\tau_{xy}/\partial y^2$ point along opposite directions. Consequently, the balance equations (4) and (5) are fully justified.

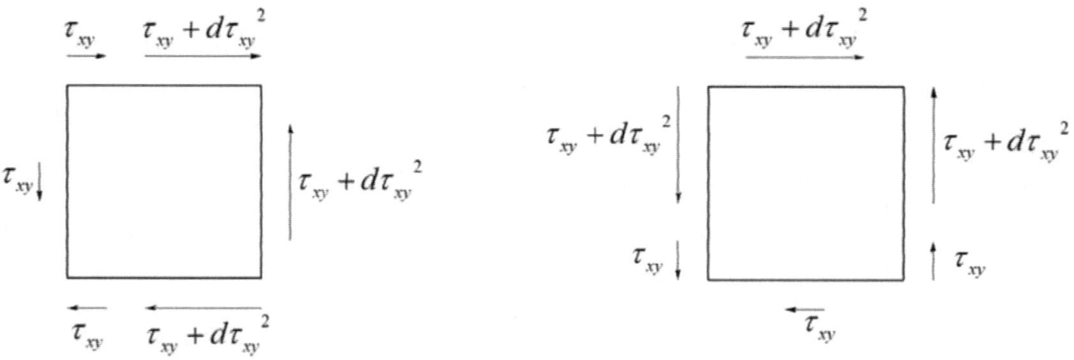

Figure 5. The torques caused by $\partial^2\tau_{xy}/\partial x^2$ and $\partial^2\tau_{xy}/\partial y^2$ have opposite sense of direction.

For symmetry reasons, $\partial^2\tau_{xy}/\partial x\partial y$ cannot introduce a torque, because both coordinate directions are equally loaded. This assessment completes the contributions related to the second derivatives of stress.

Thus, we come to the question whether consideration of higher order derivatives leads to further relations. In principle, higher order derivatives of stress may contribute to a torque. But the equilibrium equations obtained in this way are not independent from those fundamental equations (2 a, b) and (4), which have already been established. Finally, there occurs the question, whether a combination of equations (1) and (4) could lead to a relaxation of the conditions in the sense that only the sum of all torques must vanish. But the condition for the symmetry of the Cauchy stress cannot be ruled out by a condition for second derivatives of stress, because the former condition is derived from a smaller length scale. If the volume element is small enough, then a hypothetical asymmetry of Cauchy stress would outmatch the effect related to second order derivatives. On the other hand, the conditions for second order derivatives of stress are not unnecessary either, because conservation of angular momentum is generally valid for infinitesimal and for finite volumes. In consequence, the equations (1), (2 a, b) and (4) represent an irreducible and complete set of equilibrium conditions.

2.4. Three dimensional representation of balance conditions

For sake of simplicity, we have so far restricted ourselves to the two dimensional case. Hereafter, the results for the three dimensional representation are summarized: The symmetry of the Cauchy stress writes as

$$\tau_{xy} = \tau_{yx}, \quad (6\text{ a}) \qquad \tau_{yz} = \tau_{zy} \quad (6\text{ b}) \qquad \text{and} \qquad \tau_{xz} = \tau_{zx}. \quad (6\text{ c})$$

The ordinary equilibrium conditions read as

$$\frac{\partial\sigma_{xx}}{\partial x} + \frac{\partial\tau_{xy}}{\partial y} + \frac{\partial\tau_{xz}}{\partial z} + f_x = 0 \tag{7a}$$

$$\frac{\partial\sigma_{yy}}{\partial y} + \frac{\partial\tau_{xy}}{\partial x} + \frac{\partial\tau_{yz}}{\partial z} + f_y = 0 \tag{7b}$$

$$\frac{\partial\sigma_{zz}}{\partial z} + \frac{\partial\tau_{xz}}{\partial x} + \frac{\partial\tau_{yz}}{\partial y} + f_z = 0 \tag{7c}$$

Equivalently, one may express this condition in the form

$$div(\sigma) + \vec{f}_b = 0, \tag{8}$$

where $\vec{f}_b$ is the vector of body force density. Finally, the balance equations for the second order derivatives of the stress tensor are

$$\frac{\partial^2\sigma_{xx}}{\partial x\partial y} = \frac{\partial^2\sigma_{yy}}{\partial x\partial y} \tag{9a}$$

$$\frac{\partial^2\sigma_{yy}}{\partial y\partial z} = \frac{\partial^2\sigma_{zz}}{\partial y\partial z} \tag{9b}$$

$$\frac{\partial^2\sigma_{xx}}{\partial x\partial z} = \frac{\partial^2\sigma_{zz}}{\partial x\partial z} \tag{9c}$$

Further relations may be derived from combinations the equations given here.

3. Comparison with the linearized theory of elasticity

Even though the equations (4 a, b) and (9 a-c) were derived from a fundamental conservation law, one cannot combine these results with existing theories in arbitrary style. In particular, the linearized theory of elasticity incorporates relations for the second derivatives of stress, which do in general not agree with our new results. In order to clarify this issue, the Beltrami-Michell equations [13, 14] are briefly reemphasized here. Linear elasticity is based on the linearized strain tensor ε_{ij} defined as

$$\varepsilon_{xx} = \frac{\partial u}{\partial x} \qquad \varepsilon_{yy} = \frac{\partial v}{\partial y} \qquad \varepsilon_{zz} = \frac{\partial w}{\partial z}$$
$$\varepsilon_{xy} = \frac{1}{2}\left(\frac{\partial u}{\partial y} + \frac{\partial v}{\partial x}\right) \qquad \varepsilon_{yz} = \frac{1}{2}\left(\frac{\partial v}{\partial z} + \frac{\partial w}{\partial y}\right) \qquad \varepsilon_{xz} = \frac{1}{2}\left(\frac{\partial u}{\partial z} + \frac{\partial w}{\partial x}\right) \tag{10 a-f}$$

where u, v and w are the displacements in x-, y- and z-direction, respectively. Owing to the rules of partial derivatives, the second derivatives $\frac{\partial^2 f}{\partial x \partial y}$ and $\frac{\partial^2 f}{\partial y \partial x}$ of any differentiable function are equivalent. Consequently, one derives the Saint-Venant compatibility equations, among which we here mention

$$\frac{\partial^2 \varepsilon_{xx}}{\partial y^2} + \frac{\partial^2 \varepsilon_{yy}}{\partial x^2} = 2\frac{\partial^2 \varepsilon_{xy}}{\partial x \partial y} \tag{11 a}$$

$$\frac{\partial^2 \varepsilon_{yy}}{\partial z^2} + \frac{\partial^2 \varepsilon_{zz}}{\partial y^2} = 2\frac{\partial^2 \varepsilon_{yz}}{\partial y \partial z} \tag{11 b}$$

$$\frac{\partial^2 \varepsilon_{xx}}{\partial z^2} + \frac{\partial^2 \varepsilon_{zz}}{\partial x^2} = 2\frac{\partial^2 \varepsilon_{xz}}{\partial x \partial z} \tag{11 c}$$

Further, Hooke's law for isotropic linear elastic materials writes as

$$\begin{pmatrix} \sigma_{xx} \\ \sigma_{yy} \\ \sigma_{zz} \\ \tau_{xy} \\ \tau_{yz} \\ \tau_{xz} \end{pmatrix} = \frac{E}{(1+\nu)(1-2\nu)} \begin{pmatrix} 1-\nu & \nu & \nu & 0 & 0 & 0 \\ \nu & 1-\nu & \nu & 0 & 0 & 0 \\ \nu & \nu & 1-\nu & 0 & 0 & 0 \\ 0 & 0 & 0 & 1-2\nu & 0 & 0 \\ 0 & 0 & 0 & 0 & 1-2\nu & 0 \\ 0 & 0 & 0 & 0 & 0 & 1-2\nu \end{pmatrix} \cdot \begin{pmatrix} \varepsilon_{xx} \\ \varepsilon_{yy} \\ \varepsilon_{zz} \\ \varepsilon_{xy} \\ \varepsilon_{yz} \\ \varepsilon_{xz} \end{pmatrix}, \tag{12}$$

where E is the Young's modulus and ν is the Poisson ratio. Now, combination of (11 a-c) and (12) implies the Beltrami-Michell equations, which write as [14]

$$\sigma_{ij,kk} + \frac{1}{1+\nu}\sigma_{kk,ij} = 0 \tag{13}$$

in index notation. Equation (13) represents a set of six elementary equations. According to this notation, summation must be carried out over repeated indices. Obviously, these conditions are not identical with our equations (9 a-c), although they all comprise the second derivatives of stress. For the moment, it is unclear whether some equivalence between (9) and (13) can be found. Therefore, this fundamental question is here discussed for the case of an elementary example:

In this context, we choose the Kirsch problem [14, 15], which has analytically been solved within linearized elasticity. For this example of a hole in a uniaxial stress field, stress distributions are known, which fulfil the biharmonic equation [14]. Consequently, the Beltrami-Michell equations are also satisfied by this solution. However, a cross-check shows that these classical solutions from literature are not in agreement with the novel equations (4 a, b) proposed in the present article. Therefore, the equations (4) and (9) on the one hand, and equation (13) on the other are not equivalent. In the following section, we give an explanation for this discrepancy.

4. Material frame indifference

Material frame indifference means that a material law should be formulated in such a way where the response of the material is independent of an observer's position in space and time [16, 17]. In the case of static equilibrium this principle usually reduces to Euclidean objectivity. When material properties are characterized by tensors, then the tensors should transform like objective tensors during rotation of the coordinate system. This requirement is nowadays widely used as prerequisite for the objective formulation of material models. The linearized theory of elasticity is, however, not in perfect accordance with this principle. Unfortunately, the linearized strain tensor defined in equation (10) is not an objective tensor. Instead, objective strain tensors may be derived from the Lagrangian representation of continuum mechanics. This picture of continuum mechanics is based on the deformation gradient tensor

$$F_{ij} = \frac{\partial X_i}{\partial x_j} \tag{14}$$

where X_i and x_j are the coordinates of the deformed and the undeformed system, respectively. Owing to the theorem of polar decomposition, the deformation gradient

$$F = U \cdot R \tag{15}$$

may be decomposed into a symmetric stretch tensor U and a matrix R describing rigid body rotation. The stretch tensor may be obtained from the Cauchy-Green tensor

$$C = F^T \cdot F \tag{16}$$

through the equation

$$U^2 = C\,. \tag{17}$$

The matrices U and C are both symmetric, and they share their principal axes. Now, there are several possibilities how a finite strain tensor may be related to the stretch tensor U. In the present article, we like to choose that strain tensor, which shows the best affinity to the linearized strain. For this purpose, we use the Biot strain tensor [18, 19]

$$E^{Biot} = U - I\,. \tag{18}$$

Thus, one may reformulate generalized Hooke's law

$$\begin{pmatrix} \sigma_{xx} \\ \sigma_{yy} \\ \sigma_{zz} \\ \tau_{xy} \\ \tau_{yz} \\ \tau_{xz} \end{pmatrix} = \frac{E}{(1+\nu)(1-2\nu)} \begin{pmatrix} 1-\nu & \nu & \nu & 0 & 0 & 0 \\ \nu & 1-\nu & \nu & 0 & 0 & 0 \\ \nu & \nu & 1-\nu & 0 & 0 & 0 \\ 0 & 0 & 0 & 1-2\nu & 0 & 0 \\ 0 & 0 & 0 & 0 & 1-2\nu & 0 \\ 0 & 0 & 0 & 0 & 0 & 1-2\nu \end{pmatrix} \cdot \begin{pmatrix} E_{xx}^{Biot} \\ E_{yy}^{Biot} \\ E_{zz}^{Biot} \\ E_{xy}^{Biot} \\ E_{yz}^{Biot} \\ E_{xz}^{Biot} \end{pmatrix} \tag{19}$$

in material frame indifferent style. This objective version of Hooke's law may now be combined with the exact stress conditions (9a-c), whereas the formulation of equation (12) should be related to the condition (13).

5. Energy formulation

Instead of using a set of stress conditions, one may formulate the same theory with use of an expression for the internal elastic energy. With respect to the Nöther theorem [20, 21], a system of point masses will automatically preserve momentum and angular momentum, if the system can be described by a Lagrange function of the energy, which is invariant with respect to translation and rotation of the coordinate system. On the other hand, it has in numerous cases been demonstrated that continuum models may be described by discrete systems of material points [22]. In consequence, one may expect that a material model described through an expression for the internal energy will automatically satisfy the fundamental conservation laws of linear and angular momentum, if the required invariance conditions for the energy are satisfied. In this context, equilibrium states are defined as minima of energy. In conclusion, we formulate the internal energy of an elastic material

$$W = \frac{1}{2}\int_V C^*\left(E^{Biot}\right)^2 dV \tag{20}$$

as quadratic form of an objective strain tensor, where integration is carried out over the deformed volume. Here, C^* represents the elastic tensor of the material. Such energy formulations are well known in the field of nonlinear elasticity [23]. In the present article we claim that this description automatically satisfies the novel stress conditions (9a-c).

6. Summary and conclusions

The conservation laws of momentum and angular momentum are among the most fundamental principles of physical sciences. However, simplifications of mechanical theories can lead to unexpected inaccuracies. In particular, the linearized theory of elasticity is not in perfect accordance with the principle of material frame indifference, because the linearized strain is not an objective tensor. In consequence, one derives compatibility conditions, which are not essentially exact. Those compatibility conditions imply conditions for second derivatives of stress which show the same inaccuracy. On the other hand, precise conditions for the second spatial derivatives of stress in equilibrium states may be obtained directly from a postulated balance of angular moments. The results for the stress conditions may be combined with material laws satisfying the principle of material objectivity. In the present study, we have reformulated generalized Hooke's law in material frame indifferent style. Consequently, the novel stress conditions can be used in combination with this material model. However, the results obtained here are applicable in a much wider sense. The conditions leading to balance of angular moments may for instance be applied to elasto-plastic materials. But we have excluded materials like Cosserat solids, which are governed by elastic interactions deviating from the Cauchy stress. Nevertheless, the novel results are valid for a large class of materials. Finally, the energy formulation, which is in some sense related to the Nöther theorem, points the way how the novel stress conditions could easily be implemented by analogy to existing numerical algorithms.

Acknowledgments
The financial support of the Austrian Ministry of Science, Research and Economy and the National Foundation for Research, Technology and Development is gratefully acknowledged.

References

[1] Truesdell C 1968 Essays in the History of Mechanics, Whence the Law of Moment of Momentum? Berlin, Heidelberg, New York: *Springer*
[2] Poisson S D 1833 Traite de Mecanique, 2nd Ed, **2**, Paris: *Bachelier*
[3] Euler L 1776 Nova methodus motum corporumrigidorum determinandi, *Novi Commentarii Academiae Scientiarum Petropolitanae*, 20 1968 *Opera Omnia (2)* 99
[4] Casey J 1992 *Archive for history of exact sciences* **43** (4) 329
[5] Atanackovic T M and A Guran A 2000: Theory of elasticity, New York: *Springer Science and Business Media*, LLC
[6] Cosserat E and Cosserat F 1909 Théorie des corps déformables, *Librairie Scientifique* Paris: A. Hermann et Fils, English translation by Delphenich D 2009 Theory of deformable bodies
[7] Eringen A C 1967 Theory of micropolar elasticity, Princeton University
[8] Nowacki W 1986 Theory of asymmetric elasticity, Warzawa, *Polish scientific publishers*
[9] Mindlin R D 1963 *ExperimentalMechanics* **3** 1
[10] Toupin R 1962: *Arch. Rat. Mech. Anal.* **11** 385
[11] Cauchy A L 1827 Sur les relations qui existent, dansL'etat d' equilibre d'un corps solideoufluide, entre les pressionsou tensions et les forces acceleratrices, in *exercices de mathematiques* **2** (Paris, Bure Freres)
[12] Hutter K and Jöhnk K 2004 *Continuum Methods of Physical Modeling* Berlin, Heidelberg, New York: *Springer*
[13] Michell J H 1899 *Proc. London Mathematical Society* **31** 100
[14] Sadd M A 2014 Elasticity, theory, applications, and numerics, 3rd edition, *Academic Press*, Elsevier
[15] Kirsch 1898 *Zeitschrift des Vereinesdeutscher Ingenieure* **42** 797
[16] Liu I-S 2004 *Continuum Mech. Thermodyn.* **16** 177
[17] Noll W 1995 On material frame indifference, Technical Report, Carnegie Mellon University
[18] Biot M A 1965Mechanics of incremental deformations, New York, *Wiley*
[19] Bertram A 2012 Elasticity and plasticity of large deformations: An introduction, Berlin, Heidelberg, New York *Springer*
[20] Nöther E 1918 *Nachr. Akadem. Wiss. Göttingen, Math.-Phys. Klasse* II, pp. 235-257
[21] Neuenschwander D E 2017 Emmy Noether's wonderful theorem, Baltimore, *Johns Hopkins University Press*
[22] Bergman D J and Inan E, Editors 2004 Continuum models and discrete systems, NATO Science Series II. Mathematics, Physics and Chemistry **158** Dordrecht, Boston, London: *Kluwer Academic Publishers*
[23] Basar Y and Weichert D 2000 Nonlinear continuum mechanics of solids: Fundamental mathematical and physical concepts, Berlin, Heidelberg, New York *Springer*

IOP Conf. Series: Journal of Physics: Conf. Series **1045** (2018) 012005 doi:10.1088/1742-6596/1045/1/012005

Investigation of liquid crystal polymer structure-property relationships between crystal orientation and dielectric behavior

A Sullivan, A Saigal and M Zimmerman

Department of Mechanical Engineering, Tufts University, Medford MA 02155, USA

Email: anthony.sullivan@tufts.edu and Anil.Saigal@tufts.edu

Abstract. Liquid crystalline polymers (LCP's) make up a class of performance materials that derive favorable mechanical, chemical, and electrical characteristics from their long-range molecular ordering. This unique microstructure gives rise to anisotropic bulk behavior and an understanding of the driving forces behind this morphology is essential to the design of manufacturing processes for isotropic material production. In this investigation, the preferred crystal orientation in injection molded LCP plaque samples was measured using wide-angle x-ray scattering (WAXS) and the degree of order in the material was quantified using an order parameter and an anisotropy factor. Additionally, the dielectric constant of the samples was measured with respect to the mold direction (MD) and transverse direction (TD). Specimens were analyzed both as-injection-molded and with the skin layer removed, to investigate the effects of hierarchal structure on material properties. It is shown that preferred orientation along the shear direction in the LCP samples corresponds to dielectric anisotropy, and mechanically removing the shear-aligned layer results in a more isotropic dielectric response.

1. Introduction

Liquid crystalline polymers (LCP's) constitute a class of performance materials that exhibit high mechanical strength, chemical inertness, inherent flame retardancy and frequency-stable dielectric response [1, 2]. Due to these favorable properties, LCP's are ideal candidates for various engineering applications, such as high strength-to-weight ratio components, semiconductor packaging and high-frequency electronics. The unique LCP microstructure, which in contrast to conventional amorphous polymers consists of long-range molecular ordering to form crystalline domains, gives rise to anisotropic bulk properties [3-5]. This directionally-dependent behavior can be problematic from an industrial perspective, and thus an understanding of the driving forces behind this morphology, both inherent to the material and induced during manufacturing, is essential for isotropic material production.

Typical manufacturing processes for thermotropic LCP's include injection molding and cast film extrusion. Both processes expose the polymer to high shear and temperature gradients that ultimately drive the formation of the hierarchal LCP morphology [6]. During injection molding, shear gradients along the flow direction compete with extensional forces in the transverse direction as the material flows from the gate into the mold cavity [7]. Crystallization of the polymer is initiated at the cooler mold surface, forming the skin layer, and shear stresses arise from flow of the LCP bulk over this frozen skin, aligning the crystals along the flow direction [8, 9]. In the center of the mold cavity, transverse extensional gradients compete with the shear forces, producing a core with a more isotropic state, or a preferred alignment perpendicular to that of the skin [8-10]. A schematic illustration of the LCP flow during injection molding, and the resulting morphology is shown in Figure 1.

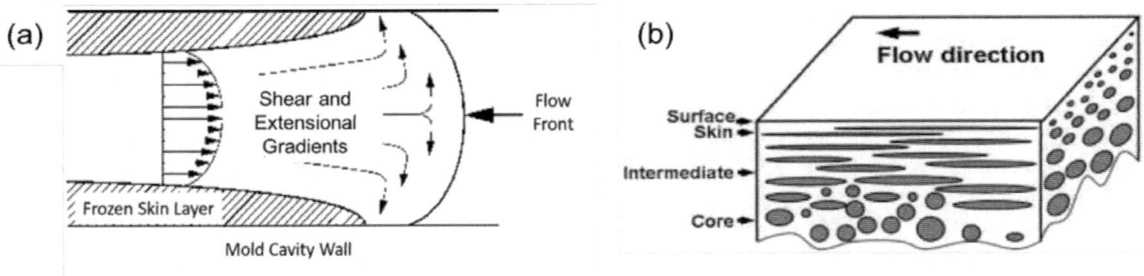

Figure 1. a) Injection molding flow behaviour [11], b) hierarchal structure from injection molding [6]

The hierarchal structure that results from the injection molding process ultimately drives the anisotropic LCP behavior. In this investigation, the structure-property relationship between LCP morphology and dielectric response was probed for an injection molded thermotropic LCP. The orientation state in the polymer was determined and compared to the measured dielectric constant along the mold direction (MD) and the transverse direction (TD). This analysis was conducted for LCP plaque samples, both as-injection-molded and with the skin layer mechanically removed, to understand the contributions of the hierarchal microstructure to the macroscopic polymer properties. It was hypothesized that preferred orientation along the shear direction in the LCP samples would correspond to dielectric anisotropy between MD and TD, and that removal of the shear-aligned layer would result in a more isotropic dielectric response.

The LCP orientation state was determined by wide-angle x-ray scattering (WAXS) in transmission mode. The thickness-averaged measurement of direction and degree of preferred orientation has been a common tool for investigating LCP morphology [12-16]. Due to the ordering of rod-like LCP molecules, shown in Figure 2a, typical scattering is manifested as intensity peaks perpendicular to the direction of average molecular orientation, shown in Figure 2b, where MD runs north to south in the representative 2D WAXS pattern (thickness direction normal to the plane shown) [3]. Because of the non-polar nature of the molecules, scattering is symmetric about the orientation axis.

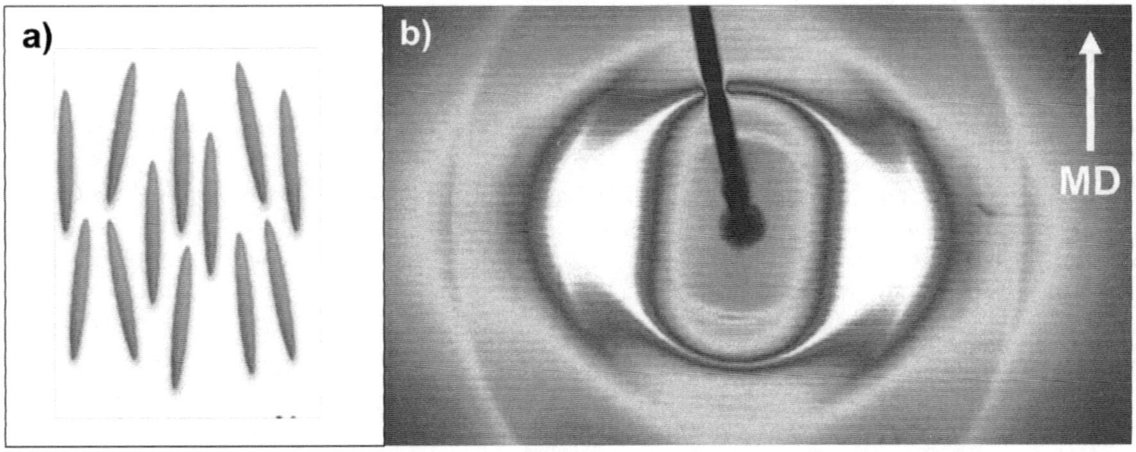

Figure 2. a) Schematic LCP orientation and b) resulting WAXS pattern

In addition to the direction of average orientation in the material, the degree of order in the material may be quantified by integration of the intensity data to calculate a scalar order parameter [17]:

$$S = \frac{1}{2}\langle(3\cos^2\theta - 1)\rangle \quad (1)$$

where θ is the angle between an individual molecular axis and the direction of preferred orientation (the director) and ⟨…⟩ denotes the average over all molecules [18]. Mitchell and Windle have provided a methodology for extracting the order parameter from the 2D WAXS intensity data [15].Under this convention, the degree of order is bounded between values of zero and one for a perfectly isotropic state and a state perfectly aligned with the director, respectively. It should be noted that the orientation convention shown above assumes a uniaxial preferred orientation in the material, which is not necessarily valid for the bulk injection molded LCP material [17]. The scalar order parameter may be used for plaque samples with a single mode of orientation, or to describe the degree of preferred alignment in a single regime of the hierarchal structure. An additional metric to quantify bulk LCP order that does not make an assumption about the orientation state in the material is calculated from the tensorial description of the material orientation. Following the work of Rendon et al., this anisotropy factor (AF) may be calculated as:

$$AF = \sqrt{(\langle\cos^2\theta\rangle - \langle\sin^2\theta\rangle)^2 - 4\langle\sin\theta\cos\theta\rangle^2} \quad (2)$$

where θ is the angle with respect to the flow direction, and ⟨…⟩ denotes the average over all molecules [9]. This quantity, like the scalar order parameter, ranges between 0 and 1 for the isotropic and perfectly aligned cases, respectively, making it an ideal metric for comparing overall degrees of orientation in different materials.

The dielectric behavior of the plaques was measured using the resonant cavity perturbation (RCP) method. This non-contact technique determines the complex permittivity of the material:

$$\varepsilon = \varepsilon_r' + i\varepsilon_r'' \quad (3)$$

where ε_r' is the dielectric constant (also noted as D_k), and describes the material's ability to elastically store energy in an applied electric field, and ε_r" is the dielectric loss, which describes the energy dissipated as heat in the applied field [19]. The non-polar LCP main chains, and reduced molecular mobility resulting from the crystalline order, result in both low D_k and low dielectric loss, making LCP's ideal insulators in high-frequency signal-carrying devices where they reduce crosstalk and power dissipation [19, 20]. The dielectric constant was measured with respect to MD and TD and compared to the orientation state in the material, both as-injection-molded and with the shear layer removed, to illuminate the structure-property relationship in the LCP.

2. Experimental

2.1. Sample Preparation

The material system utilized in this investigation was a commercially-available thermotropic LCP. Plaque samples were injection molded with a Sodick Plustech Model TR40EHV IM machine (Schaumburg, IL, USA). The specimens were prepared in a rectangular mold with a midpoint coat hanger gate, the cavity measuring 76.2 mm along MD, 50.8 mm in TD, and 0.5 mm along the thickness. Samples for WAXS and dielectric testing were $(27.1\ \text{mm})^2$ square samples, punched approximately two-thirds of the way from the gate end in the middle of the plaque to minimize entrance and wall effects. For the skin versus core analysis, the outer shear layer of the samples was identified visually by color and texture change apparent in micrographs of the plaque cross-sections, and was found to be approximately 65 μm thick. This region was mechanically removed using a Buehler Ecomet IV polisher (Lake Bluff, IL, USA).

2.2. Wide-Angle X-Ray Scattering

The 2D WAXS orientation measurements were performed on the samples using a D/MAX-RAPID II by Rigaku (The Woodlands, TX, USA). The analysis was performed in transmission mode at 40 kV and 20 mA with a Cu Kα x-ray generator. A 0.5-mm collimator was used in the analysis and the exposure time for each sample was 10 minutes. Azimuthal scans were performed to measure the degree of orientation with respect to an external frame of reference. For each sample, MD was oriented at 0 degrees and the scattering intensity was then measured as a function of angle in the plane of the sample (rotation axis parallel to the thickness direction), as shown in Figure 3.

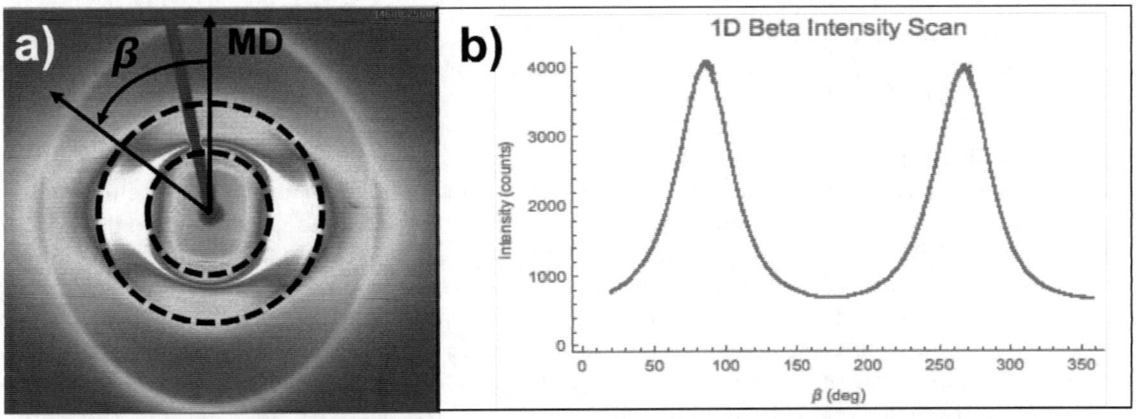

Figure 3. a) 2D WAXS pattern with angular convention, and b) resulting intensity versus angle used in the order parameter and anisotropy factor calculations below

Visual inspection of the 2D WAXS pattern demonstrates the direction of average molecular orientation, with an isotropic state depicted by a uniform ring, and an oriented state showing poles of higher intensity. The degree of order in the samples was quantified using a scalar order parameter by integrating the intensity data [15]:

$$S = \frac{\int_0^{\pi/2} I(\beta)\cdot P_2(\cos\beta)\cdot \sin\beta \, d\beta}{\int_0^{\pi/2} I(\beta)\cdot \sin\beta \, d\beta} \tag{4}$$

where S is the scalar order parameter, I is the intensity, and $P_2(\cos\beta)$ is the second Legendre polynomial of $\cos\beta$. Rendon et al. provide a detailed implementation of this methodology which was utilized in this investigation [9]. The resulting value provides a metric of the degree of isotropy (or anisotropy) from 0 to 1.

Following a similar method provided by Cinader and Burghardt, the degree of order in each sample was also found through the tensorial description of orientation [12]. Beginning with **u,** a unit vector representation of a point on the azimuthal intensity scan, where $u_1 = \cos\beta$ and $u_2 = \sin\beta$, the second moment tensor of **u** is given by:

$$\langle \mathbf{uu} \rangle = \begin{bmatrix} \langle u_1 u_1 \rangle & \langle u_1 u_2 \rangle \\ \langle u_1 u_2 \rangle & \langle u_2 u_2 \rangle \end{bmatrix} = \begin{bmatrix} \langle \cos^2\beta \rangle & \langle \sin\beta \cos\beta \rangle \\ \langle \sin\beta \cos\beta \rangle & \langle \sin^2\beta \rangle \end{bmatrix} \tag{5}$$

where $\langle \ldots \rangle$ is the weighted azimuthal intensity distribution $\mathbf{I}(\beta)$, with $\langle \cos^2\beta \rangle$ calculated as:

$$\langle \cos^2\beta \rangle = \frac{\int_0^{2\pi} \cos^2\beta \cdot I(\beta) \, d\beta}{\int_0^{2\pi} I(\beta) d\beta} \tag{6}$$

The anisotropy factor was finally calculated as described above in Equation 2.

2.3. Dielectric Constant Measurement

The dielectric constant was calculated for the injection-molded specimens using the RCP method, in which a sample is placed in a resonant cavity, perturbing the center frequency (*f*) from the empty cavity state. The shift in *f*, and the known sample and cavity volumes then can be used to calculate the dielectric constant of the material:

IOP Conf. Series: Journal of Physics: Conf. Series **1045** (2018) 012005 doi:10.1088/1742-6596/1045/1/012005

$$D_k = \frac{V_c(f_c - f_s)}{2V_s f_s} + 1 \tag{7}$$

where V_c is the volume of the resonant cavity, V_s is the volume of the sample, f_c is the center frequency of the empty cavity, and f_s is the center frequency with the sample in the cavity. The dielectric measurements were taken along MD and TD using an Agilent Technologies 8722ES vector network analyzer (VNA) and a rectangular waveguide operated at an empty cavity center peak of approximately 8.8 GHz [21].

3. Results and Discussion

The orientation results from this investigation are shown in Figure 4, with the as-injection-molded sample in (a) and the sample with the skin layer removed in (b). For both samples, a single mode of orientation along MD is seen, corresponding to a preferential alignment with the flow direction. Visual inspection of the 2D WAXS pattern alone does not conclusively show a change in scattering intensity between the two samples, however, calculation of the order parameter showed a value of 0.70 in the as-injection-molded sample and a value of 0.65 in the sample with the shear layer removed. A similar drop in the anisotropy factor between samples from 0.72 to 0.68, respectively, was also seen. Because a single mode of orientation is observed in both samples, it is expected that the scalar order parameter and the anisotropy factor would give similar results. The reduction in both order metrics with removal of the shear layer is indicative of a skin morphology that is aligned along the flow direction; by removing this highly-oriented region, the degree of order in the material is reduced. However, a uniaxial orientation along MD still exists in the sample with the skin layer removed, demonstrating that shear gradients dominated over transverse extensional gradients through the material thickness during the injection molding process. This is likely due to the thin plaque geometry.

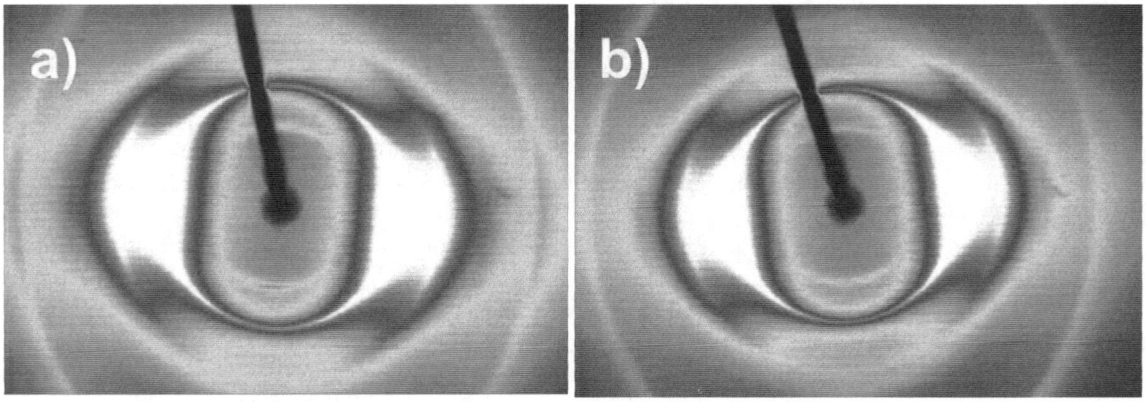

Figure 4. WAXS patterns for, a) as-injection-molded sample, and b) sample with skin layer removed

The corresponding sample dielectric behavior in MD and TD is presented below in Figure 5. In accordance with the orientation results, dielectric anisotropy is seen in both samples, with a greater spread between MD and TD in the as-injection-molded sample. Removal of the skin layer resulted in a decrease in dielectric constant along both MD and TD, however as illustrated in Figure 6, the percent change along MD was greater, thus the dielectric response became more isotropic. As with the reduction in order metrics, this again results from removal of the highly-aligned skin morphology, ultimately reducing the contribution of this region to the bulk material anisotropy.

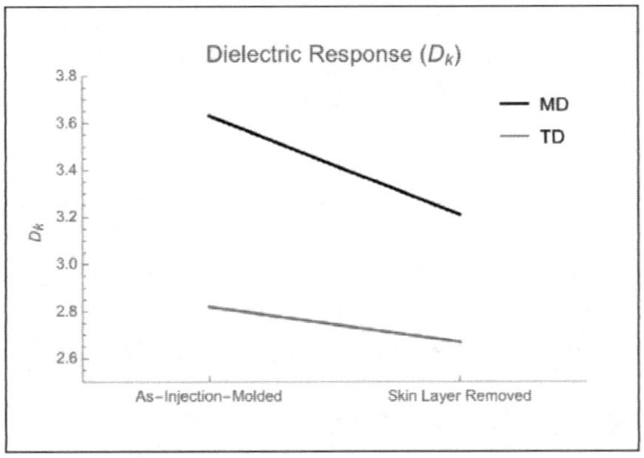

Figure 5. D_k of the LCP plaques

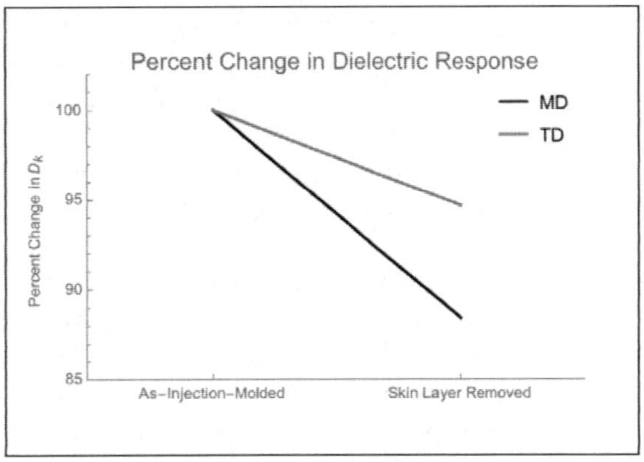

Figure 6. Percent change in D_k

4. Conclusions

The orientation state and dielectric constant of injection molded thermotropic LCP samples were presented in this investigation. The complex rheological and thermal conditions present during processing give rise to a hierarchal structure in the LCP morphology that results in macroscopic anisotropy of the material behavior. To better understand the structure-property relationships in LCP's, molecular orientation was analyzed using WAXS in transmission mode, and the dielectric constant, measured along MD and TD, was found using the non-contact RCP method. An order parameter and anisotropy factor were used to quantify the degree of order in the material, and the values were compared to the dielectric results. The LCP plaque samples were tested as-injection molded, and with the skin layer mechanically removed to illustrate the contributions of the hierarchal LCP morphology to the overall orientation state and macroscopic dielectric response.

It was found that the samples showed a single mode of orientation aligned with the flow direction, resulting from shear stresses through the material thickness during processing. Removal of the skin layer resulted in an expected reduction in the order parameter and anisotropy factor, as well as a decrease in the dielectric anisotropy between MD and TD. These results demonstrated the anisotropic contribution of the skin layer to the overall orientation state, and demonstrated the capability of the orientation metrics to predict approximate dielectric behavior. The observed structure-property relationship in the LCP injection molded plaques provides insight into the fundamental driving factors behind bulk material behavior. The subsequent understanding of how the structure develops as a function of processing presents an opportunity to predict the macroscopic LCP properties resulting from a given manufacturing process, and could eventually enable the design of improved manufacturing methods for the production of isotropic LCP materials.

Acknowledgements

The authors would like to acknowledge Rigaku, in The Woodlands, TX, for performing the wide-angle x-ray scattering measurements of the LCP samples, and assisting in the subsequent analysis.

References

[1] Cox M K 1987 *Molecular Crystals and Liquid Crystals Incorporating Nonlinear Optics* **153** (1) 415
[2] Thompson D C et al. 2004. *IEEE Transactions on Microwave Theory and Techniques* **52** (4) 1343
[3] Donald A M, Windle A H, Hanna S 2006 Liquid Crystalline Polymers, New York *Cambridge University Press.*
[4] Donald A M, Viney C and Windle A H 1983 *Polymer* **24**(2) 155
[5] Bedford S E and Windle A H 1990 *Polymer* **31** (4) 616
[6] Zhong G J and Li Z M 2005 *Engineering & Science* 2005 **45** (12) 1655

[7] Fang J et al. 2010*Polymer Engineering & Science* **50** (9) 1864
[8] Weng T, Hiltner A and Baer E 1986 *Journal of Materials Science* **21** (3) 744
[9] Rendon S et al. 2007 *Journal of Applied Polymer Science* **106** (4) 2502
[10] Rendon S et al. 2004 *Polymer* **45** (15) 5341
[11] Partlow W 2005 Injection Molding, *in Civil Engineering Handbook* 543
[12] Cinader D K and Burghardt W R 1999 *Journal of Polymer Science Part B: Polymer Physics* **37** (24) 3411
[13] Cinader D K and Burghardt W R 1998 *Macromolecules* **31** (25) 9099
[14] Rendon S, Burghardt W R, and Bubeck R A 2007 *Rheologica Acta* **46** (7) 945
[15] Mitchell G R and Windle A H 1988 Developments in Crystalline Polymers-2, Basset, D. C., Ed.; *Elsevier* London, 115
[16] Reyes-Mayer A et al. 2016 *Polymers for Advanced Technologies* **27** (6) 748
[17] Hermans P H P P 1939 *Kolloid-Zeitschrift* **88** (1) 68
[18] De Gennes P-G and Prost J 1995 The Physics of Liquid Crystals **83** 83. Oxford; New York Book: *Clarendon Press.*
[19] Ahmad Z 2012 *Polymer Dielectric Materials*, Dielectric Material, Dr. Marius Alexandru Silaghi (Ed.) InTech DOI: 10.5772/50638. Available from: https://www.intechopen.com/books/dielectric-material/polymer-dielectric-materials
[20] Tummala R R, 2001 Fundamentals of microsystems packaging, New York: *McGraw-Hill*
[21] Agilent 2006 Basics of Measuring Dielectric Properties of Materials, *Agilent Technologies*

IOP Conf. Series: Journal of Physics: Conf. Series **1045** (2018) 012006 doi:10.1088/1742-6596/1045/1/012006

Mechanical and optical properties of hybrid materials based on inorganic glass matrix and organic metal complex phosphors

I Ch Avetissov[1], O B Petrova[1], M O Anurova[1], A V Khomyakov[1], A A Akkuzina[1], I V Taidakov[2]

[1]Department of Chemistry and Technology of Crystals, Dmitry Mendeleev University of Chemical Technology of Russia, 125047 Moscow, Russia
[2]P.N. Lebedev Physical Institute of the Russian Academy of Sciences, 119991, Moscow, Russia

E-mail: igor_avetisov@mail.ru

Abstract. Hybrid materials (HM) consisting nanoparticles of an organic functional agent dispersed in an insulating matrix, are widely used in photonics as active and passive elements as well as integral optics components. Highly efficiently phosphors used in OLED technology are promising organic agent. One of the problems which can be solved by the creation of HM is protection of organic phosphors from environmental conditions (oxygen, water vapor, UV light, etc.) which lead to the destruction of metal-organic complexes and their loss of luminescence properties.

1. Introduction

Organo-inorganic hybrid materials (HM) are widely used in photonics as active and passive elements as well as integral optics components [1]. One of the problems which can be solved by the creation of HM is a creation of a material stable to environmental conditions and (oxygen, water vapor, UV light, etc.) which lead to the destruction of metal-organic complexes and their loss of luminescence properties [2]. The possibility of synthesis of photoluminescent HM based on 8-hydroxyquinoline [3] and phenantroline [4] metal organic phosphors and inorganic low melting glasses as a matrix using a non-aqueous technique has been demonstrated.

For the synthesis of HM by the melt technology, a glass matrix should fulfill a number of requirements on temperature, mechanical and optical characteristics. Glasses from lead borate and borosilicate systems meet the conditions mentioned above. These systems were previously studied for fabrication of low melting solvents for the growth of crystals from melt solution [5-6] and sealing glasses [7]. The substitution of lead oxide by lead fluoride in glass composition resulted to a decrease in melting point and viscosity [8-9], as well as the optical properties changing. In the 600–700 °C temperature range the above mentioned melts have viscosity in 3 orders of magnitude lower than the for B_2O_3 melt, which facilitates the process of mass transfer during synthesis with stirring. The glasses of these compositions have sufficiently high values of mechanical and chemical resistances.

One of the problems in the production of fluoro-borate glasses is pyrohydrolysis of lead fluoride, leading to significant losses of fluorine during the syntheses of glasses. This process is associated with the interaction of fluorides with water vapor at temperatures above 400°C [10]. The use of dried boron oxide and synthesis under dry atmosphere can significantly reduce fluoride losses.

The high content of OH-groups in the glass leads to quenching of the luminescence of many both inorganic activators [11-12] and organic metal complexes. Thus, the producing of HM based on glasses with a reduced content of OH-groups should lead to an increase in the luminescence intensity.

2. Experimental

2.1. Pyrification and drying of B_2O_3

To keep the fluorine vaporazation resulting from pyrohydrolisys which was controlled by residual water content in boric acid and commercial boron oxide preparations we prepared specially dried boron oxide (100 ppm residual water content) and used it for HM fabrication.

Boric acid purchased from KOMPONENTREAKTIV (Russia) (99.999 wt%) had been recrystallized several times from acidified hot solution. Impurity content of recrystallized boric acid was determined by the inductively coupled plasma technique (NexION 300D, Perkin Elmer Inc.). The final probe of boric acid was as pure as 99,9998wt% (metal based). Extra pure B_2O_3 was made by slow decomposition of the boric acid in a platinum crucible using the multy-step heating with one hour exposition at 130, 170, 300°C. This regime allowed preserving intense water vaporization. The heating of a preparation to 700°C and further exposition during 6 hours resulted to the B2O3 melting followed by the formation of homogeneous transparent mass. The final removal of the residual water was made by annealing of the amorphous B_2O_3 at 500°C during 3 hours under dynamic vacuum (10^{-1} Pa) in a home-made reactor in a platinum crucible. The residual water content was measured by a humidity analyzerAGS50 (AXIS, Poland) and it was less than 500 ppm.

2.2. Synthesis of glasses and HM

As raw materials we used boric acid (99.999 wt% from KOMPONENTREAKTIV, Russia), additionally purified and dried B_2O_3 (see above), commercial boron oxide (99.999 wt% from IREA2000 Ltd., Russia) and specially prepared dried boron oxide (99.9995 wt%).

The glasses of nominal compositions of $70PbO\text{-}30B_2O_3$ and $70PbF_2\text{-}30B_2O_3$ were synthesized using PbO and PbF_2 of at least high purity grade (LANHIT Ltd.).

All glasses were obtained in closed corundum crucibles (30 ml) under the same conditions: 50 g batch charge, 950°Ctemperature, 15 minutes exposure. The melt was poured into heated steel molds.

In the present research we have synthesized and studied hybrid materials based on 8-hydroxyquinoline metal complexes with metals of I, II, and III groups of the Periodic Table having general formula Mq_n (Liq, Kq, Rbq, Mgq_2, Znq_2, Alq_3). Purity of 8-oxiquinolin metal complexes measured by the ICP-MS technique was not less 99,99wt% (metal based).

Besides, HM's were synthesized in alumina crucibles. A pre-synthesized glass (5 g) was melted at 700°C then cooled up to 550°C and a powdered organometallic phosphor (0.1 wt%) was poured into the melt. Then a mixture was stirring during 10-15 s. The HM samples were obtained by rapid casting into steel molds.

2.3. Measurement techniques

Elemental analysis Scanning Electron microscopy (VEGA-3-LMU, Tescan) equipped with an energy-dispersive spectrometer (EDS Inca Energy X-MAX-50, Oxford Instruments) supported by AZTec software was used for data gathering and processing. The glass composition was determined by SEM-EDX analysis after carbonization, the beam energy was 30 KeV for glasses.

The micro-hardness was determined by the Vickers method using a PMT-3 micro-hardness tester, in the range of loads 50 - 200 g.

The absorption spectra of glasses and HM were recorded using a Varian Cary 5000 (UV/VIS/IR) spectrophotometer in 400–3000 nm wavelength range with 1 nm step and a Tensor 27 (Bruker) IR-Fourier spectrometer (2750-6250 nm).

A QE65000 (Ocean Optics, USA) spectrometer with a 785 nm excitation for Raman spectra recording was used.

IOP Conf. Series: Journal of Physics: Conf. Series **1045** (2018) 012006 doi:10.1088/1742-6596/1045/1/012006

A Fluorolog FL3-22 spectrofluorimeter (Horiba JobinYvon, USA) with double grating excitation and emission monochromators was used for luminescent measurements in a wavelength range of 400–700 nm with a 0.1 nm step.
All measurements were carried out at room temperature.

3. Results and discussion

As a result of the synthesis, transparent optically high-quality glass-matrices were obtained. Glasses obtained in the 70PbO - $30B_2O_3$system were yellowish while the $70PbF_2$-$30B_2O_3$system made it possible to get colorless preparations.

Analysis of the elemental composition of the glasses (Table 1) showed that the dried boron oxide reduces two times the losses of fluorine compared to the commercial boron oxide, and three times compared to the boric acid. We associate this result with the suppression of pyrohydrolysis in a drained system. The fluoro-borate melt is more aggressive than the borate melt, as a result of its interaction with the crucible, about 4 mol.% aluminum oxide is transferred to the glass, compared to 1.5 mol.% for the borate melt. However, the use of dried boron oxide reduces the proportion of dissolved aluminum oxide by half, both for borate and fluoro-borate systems. Thus, the use of dried boron oxide reduces the aggressiveness of borate melts, which can be used in the production of crystals and glasses based on high purity borate.

Table 1. The property of glass-matrixes

Batch composition (mol%)	Volatilization of fluorine, $(C_F^{batch} - C_F^{glass})/C_F^{batch} \cdot 100\%$	Al content (at%) ±0,5	Micro-hardness (kg/mm^2)±10	Cutoff wavelength (nm) ±1
$70PbF_2$-$30(H_3BO_3)_2$	41	8.7	144	398
$70PbF_2$-$30B_2O_3$	26	9.6	139	397
70PbF-$30B_2O_3$ (dried)	13	4.0	144	365
70PbO-$30(H_3BO_3)_2$	-	2.7	143	461
70PbO-$30B_2O_3$	-	3.2	147	482
70PbO-$30B_2O_3$ (dried)	-	1.5	157	449

This comparatively small transmission (less than 75%) of pure, transparent glasses could be explained by the large losses in the reflection (Fresnel losses) arising from the large refractive index, inherent in all high-lead glasses (Figure 1).

The cutoff of the transmission of glasses depends on the fluorine content: in glasses of nominal composition $70PbF_2$-$30B_2O_3$, obtained from non-dried components, it lies about 400 nm, in a glass of the same nominal composition, obtained from dried boron oxide - 365 nm. For glasses in the 70PbO-$30B_2O_3$system, the transmission edge is in the region of 450 - 480 nm. Photoluminescence (PL) of the used organic phosphors is achieved by the excitation of ligands with the energy more than 3 eV [13]. Thus the glasses with the cutoff wave length longer that 400 nm will absorb the excitation energy and we will fail to get the luminescence in hybrid materials.

The long-wavelength edge of the transparency window lies at around 3500 nm. A broad band in the 2950 nm (3400 cm^{-1}) region is associated with the absorption of OH-groups. As the water content in the initial reagents decreases, this band naturally decreases. Moreover, in the fluoro-borate glasses, the band is initially smaller, which may be due to a decrease in the hydroxyl group content as a result of volatilization upon interaction with fluorides, and also with a lower viscosity of fluoro-borate melts, which contributes to the volatilization of water vapor.

IOP Conf. Series: Journal of Physics: Conf. Series **1045** (2018) 012006 doi:10.1088/1742-6596/1045/1/012006

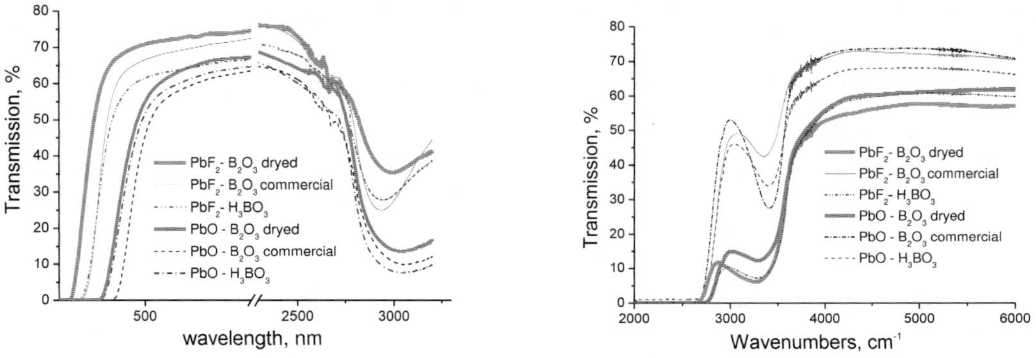

Figure 1. Transmission spectra of glass-matrices. The thickness of the samples is 3 mm.

Several broad bands are visible on the Raman spectra (Figure 2). The spectra of borate and fluoro-borate glasses differ evidently. The band intensity corresponding to the vibrations of tetrahedral units of PbO_4 (306 cm^{-1}) is noticeably lower in fluoro-borate systems than in borate systems, which is due to the fluoride surroundings of lead in fluoro-borate glasses and the predominance of Pb-F bonds, whose vibrational energy is much lower.

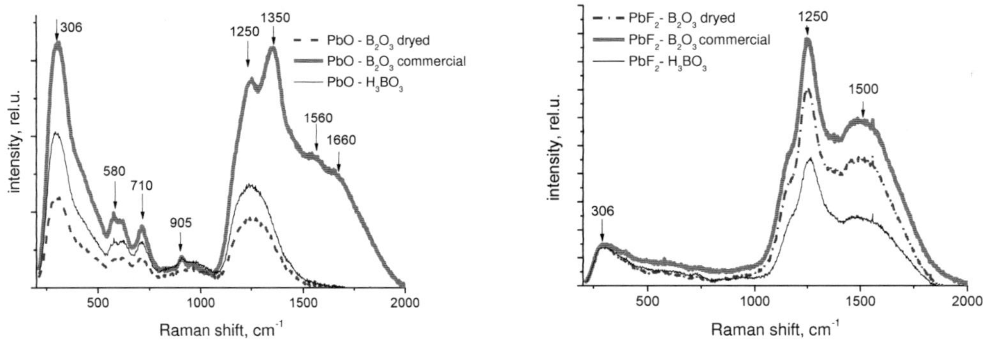

Figure 2. Raman spectra of glass-matrices

As it was demonstrated in Ref. [14], the replacement of PbO by PbF_2 led to an increase in the intensities of the combinational lines associated with the oscillation of the BO_4 units (Table 2). This was explained by the fact that the bonds with PbO_4 stabilized the triple coordination of boron in the glass. The energy of the phonons of the BO_4 units is lower than that of BO_3, which is favorable for the production of luminescent materials. The reduced energy of the phonon matrix reduces the probability of non-radiative relaxation in the activator ions and promotes an increase in the luminescence intensity.

IOP Conf. Series: Journal of Physics: Conf. Series **1045** (2018) 012006 doi:10.1088/1742-6596/1045/1/012006

Table 2. Raman peaks in glass-matrices

Peaks (cm^{-1})	Oscillations [14-15]
306	Deformation oscillations of tetrahedral units PbO_4
580	Deformation oscillations of B-O-B in various borate groups with a predominance of BO_4- units
710	Symmetric stretching vibrations of a six-membered borate ring with one unitsBO_4.
905	Valence oscillations of B-O in BO_4 with the participation of bridging oxygen
1250	Deformation oscillations of B-O in BO_3-triangles and boroxsol rings with non-bridging oxygen
1350	Deformation oscillations of B-O in BO_4 with one or two non-bridging oxygen
1500	Deformation oscillations of B-O in BO_4 with three non-bridging oxygen
1560	Valente oscillations of B-O-B in triangles BO_3
1660	Valente oscillations of B-O-B in triangles BO_3

The micro-hardness of all the glasses obtained was similar (Figure 3, Table 1). The values of the hardness of the glass are affected by a number of factors. In this case, some factors are oppositely directed: so changing the oxide to fluoride usually reduces hardness, but at the same time the melt aggressiveness and concentration of aluminum oxide increase, which contributes to the hardness increase [16]. The micro-hardness of the synthesized glasses varied from 140 to 150 kg/mm^2 (Figure 3), while for HM's we obtained the values up to 210 kg/mm^2 due to the fluoride loss and alumina content increase (from crucible interaction) of the melt during HM' synthesis. The micro-hardness both glasses and HM's fabricated using dried B_2O_3 was higher comparing to the synthesis with boric acid and commercial B_2O_3.

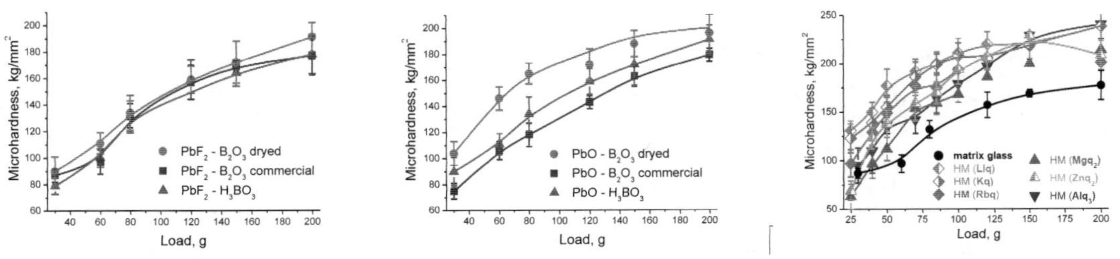

a b c

Figure 3. Micro-hardness a) glass-matrices in system 70PbO-30B_2O_3; b) glass-matrices in system 70PbF_2-30B_2O_3; c) glass-matrices in system 70PbF_2-30B_2O_3 and HM based its

All HM showed a wide PL band in the range 400-700 nm (Figure 4).

The use of dried B_2O_3 led to an increase in the PL intensity (Figure 5). We assumed that the PL intensity growth for matrix made with the dried B_2O_3 could be resulted for reducing of hydroxyl groups' content in the HM. The latter could result either to the reduction of luminescence quenching at a non-radiative transitions on OH-groups' oscillations or to the increasing of an organic ligands content in the case of the less aggressive glass melt.

4. Conclusions

The use of dried boron oxide in the synthesis of fluoro-borate glasses and organo-inorganic hybrid materials on their basis resulted to the reducing the pyrohydrolysis of the fluoride components, to the reducing of glass melt aggressiveness, and to the increasing of the glass transparency and the PL intensity in the case of the hybrid materials.

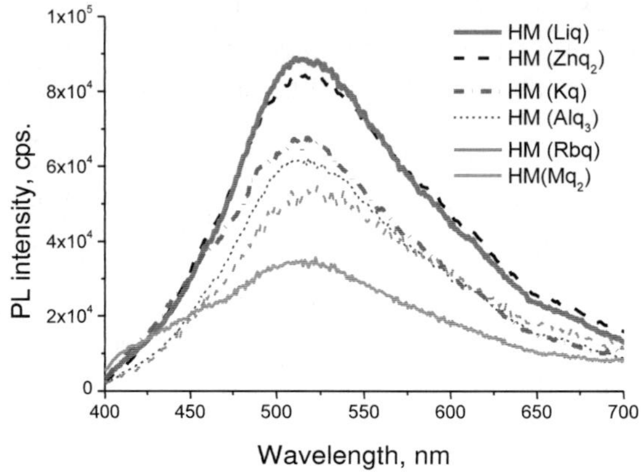

Figure 4. The PL spectra of HM's based glass-matrices in system $70PbF_2$-$30B_2O_3$

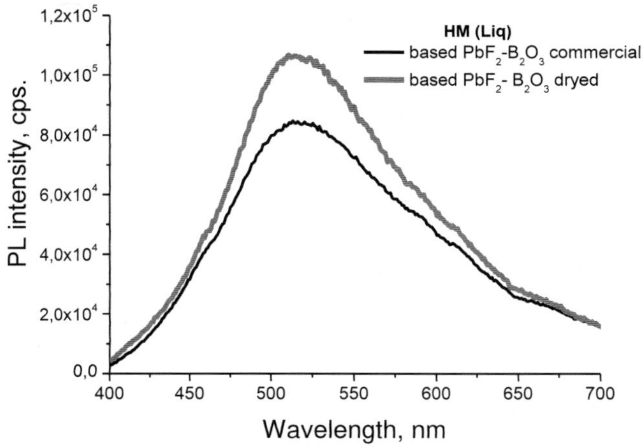

Figure 5. The PL spectra of HM (Liq) based glass-matrices $70PbF_2$-$30B_2O_3$

Acknowledgments
The research was financially supported by the Ministry of Education and Science of the Russian Federation RFMEFI57716X0218.

References

[1] Lebeau B and Innocenzi P 2011 *Chem. Soc. Rev.* **40** 886
[2] Papadimitrakopoulos F, Zhang X-M 1997 *Synt. Met.***85** 1221
[3] Avetisov R, Petrova O, Khomyakov A, Mushkalo O, Akkuzina A, Cherednichenko A, Avetissov I 2014 *J. Cryst. Grow.* **401** 449
[4] Petrova O, Taydakov I, Anurova M, Akkuzina A, Avetisov R, Khomyakov A, Mozhevitina E, Avetissov I 2015 *J. Non-Cryst. Sol.* **429** 213
[5] Coe I M, Elwell D 1974 *J. Cryst.Grow.* **23** 345
[6] Geller R F, Bunting E N 1939 *J. Res. Natl. Bur. Stand.* **23** 275
[7] Goleus V, Dimitrov Ts, Nosenko A, Amelina A, Gordeev J 2015 *Scientific Research at the University of Rousse* **54** 93
[8] Korjakova Z, Bitt V 2004 *Komponenty i tehnologii* **5** 126 (in Russian)
[9] Gressler C A, Shelby J E 1988 *J. Appl. Phys.* **64** 450

[10] Buchinskaya I I, Fedorov P P 2004 *Russ. Chem. Rev.* **73** (4)., 371
[11] Lunter SG, Fyodorov YuK. 1994 *Proc. F. Simp. Light Materials, Laser Technology Material for Optic Telecommun.* **2** 327
[12] Karlsson G, Laurell F, Tellefsen J., Denker B., Galagan B., Osiko V., Sverchkov S 2002 *Appl.Phys. B* **75** 41
[13] Taidakov I V, Lobanov A N, Lepnev L S, Vitukhnovskii A G 2014 *Rus. J. Coord. Chem.* **40** (1) 16
[14] Pisarska J 2010 *Optica Applicata* **XL** (2) 205
[15] Cheng Y, Xiao H, Guo W, Guo W 2007 *Ceramics International* **33** 1341
[16] Galagan BI, Glushchenko IN, Denker BI, Kalachev YuL, Mikhailov VA, Sverchkov SE 2011 *Glass Phys. Chem.* **37** (3) 258.

IOP Conf. Series: Journal of Physics: Conf. Series **1045** (2018) 012007 doi:10.1088/1742-6596/1045/1/012007

Melt shear viscosity of original and recycled PET in wide range shear rate

Attila Bata, Gergely Toth, Dorottya Nagy, Karoly Belina

Department of Materials Technology, Faculty of Mechanical Engineering and Automation,
John von Neumann University, Hungary

E-mail: bata.attila@gamf.uni-neumann.hu

Abstract. Among the plastic materials, the PET is the most often used one. The collecting of the material to be recycled is partially solved. This way we have great quantity of reground PET. During the regrinding the structure of molecules changes and these properties can be measured by rheology methods. Polymer melts are -Newtonian fluids and the flow curves of the materials can be used to characterize the molecular weight and the molecular weight distribution. This research is focused on the characterization of original and ground PET materials. Inherent viscosities were determined to characterize the materials. Flow curves of pure and blended materials were measured at different temperatures.

1. Introduction

Nowadays polyethylene terephthalate (PET) is one of the most common plastics, also because of its wide spread use. For the PET bottles there are certain places where recycled PET is already used and produced. In this case, about 10-25 % of the recycled PET is used. Our goal is to increase the quantity of the recycled PET. From the recycled materials the new bottles, products can be made. This requires a development path and a more sophisticated technology since it is not easy to determine for once used bottle what was stored in it, how it was used previously [1]. PET is a hygroscopic thermoplastic polymer, therefore drying process is very important, since it affects the rheological properties of the material, the subsequent processing, and the product characteristics [2].

The rheological analysis mentioned in this paper is of polyethylene terephthalate (PET) base material, in which a specific raw material - a melt viscosity of structural, elastic and viscoelastic behavior was investigated. The study was carried out by ARES-G2 type oscillational rheometer, and capillary rheometer with which the material was examined at the low and the high shear rate. The Goettfert Rheograph 25 original granulate was analyzed between 270-290 °C. Then, the original raw material was processed with injection molding machine and then it was granulated again. There was a full investigation at the same temperatures. The measuring values are presented on diagrams and from these the conclusions were drawn.

2. Experimental

2.1 Material

RAMAPET N180 is a general purpose, non-reheat PET resin for bottles, film, thermoformed containers and other transparent applications. It has good optical properties, the mechanical properties are representative for a PET resin.

The PET is hygroscopic, and since moisture adversely affects the IV during the melt processing of the polymer, it must be dried prior to molding. Undried PET pellets can contain up to 0.2 % (2000 ppm) of moisture. After drying, the moisture content is typically below 50 ppm [3].

The PET has good surface hardness, stiffness, and dimensional stability that absorbs low volume of moisture. The PET material is heat-resistant until 180 °C, it does not decrease in strength, its melting temperature is around 260 °C. Above 250 °C the crystallite section is continuously disappearing, so the melting process of the material could be observed. Usually, the material is processed in the molten state, so it is worth to study the rheological properties only above this temperature point [4]. The measurements were performed at 270 °C, 280 °C and 290 °C.

3. Method

3.1 The rotational viscometry

Our measurements were performed by a shear viscosity investigation method; that means a low range of shear rate and temperature steps near the processing temperature. This was the reason why the rotational viscometry was measured in the low shear rate range (0.05 – 20 1/s). With the rotational viscometry we could measure torsion flow in a polymer melt between parallel plates. The measuring technique conceptual layout is shown in Fig. 1. The CMT means combined motor and transducer concept [5].

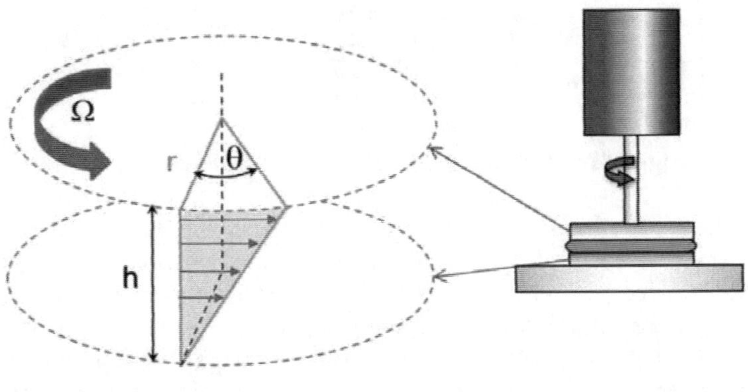

Figure 1. Rotation viscometry CMT conceptual layout and marking system [5]

From the marking system the stress, deformation and deformation rate components can be calculated with the equations described below:

$$\sigma = \frac{2}{\pi r^3} \times \mathrm{M} \quad (1)$$

$$\gamma = \frac{r}{h} \times \theta \quad (2)$$

$$\dot{\gamma} = \frac{r}{h} \times \Omega \quad (3)$$

where r is the circle plate radius; h is the gap between the plates; is the driving engine rotation (rad); is the driving engine angular speed (rad/s) and M is the torque (Nm) [5].

The stress and deformation rate can be calculated from expressions (1) to (3) and the shear viscosity can be defined. It is a frequent measuring technique to determine shear viscosity for small molecular liquids, or oils and thermoplasts at low shear rate. With this method we can measure at higher temperature (to 600 °C) and monitor the degradation process. For our testing we used the SMT (Separate motor and transducer) conceptual TA Ares G2 type rotation viscometer with 25 mm parallel circle plates and 0.4 mm gap. In this type of rotation viscometer the drive engine and the transductor are separated [6].

3.2 The capillary rheometer

The capillary rheometer is a measuring device in which we are able to measure the shear viscosity of thermoplastics. The shear rate ranges between 100 and 10 000 1/s near to the processing temperature [7].

The GoettfertRheograph 25 capillary rheometer was used. The measuring method is based on continuous melt flow in known geometry capillary(ies), while the pressure drops is measured between the capillary inlet side and the atmospheric pressure (Fig. 2.).

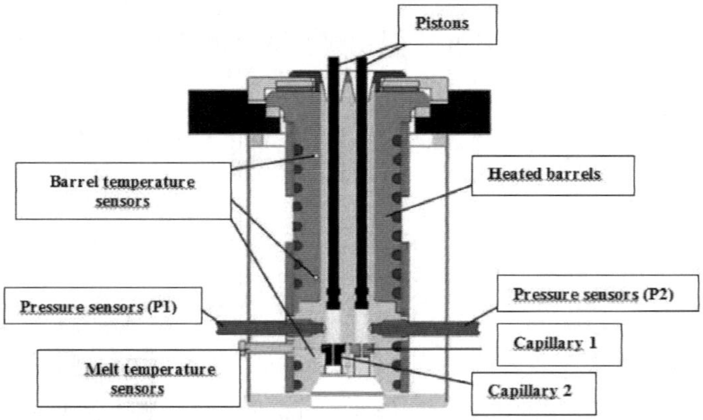

Figure 2. Capillary rheometer conceptual layout [7]

The variable parameters are the barrel temperature and the piston speed (parallel to the deformation rate). In this case, the shear viscosity depend on temperature and deformation (shear) rate. In our tests, the shear rate changed between 100-10.000 1/s and the barrel temperatures changed in three steps around the material supplier recommended process temperature.

The measuring method uses two other geometry capillaries because of the entrance and exit effects (pressure drop). With this equipment in one step we can do the Bagley correction (correct the pressure drop) which gives the real shear viscosity [8].

4. Results

4.1 The capillary rheometer

The viscosity curves were determined as functions of the shear rate. The measurements were performed with original material and with recycled material. The viscosity of the original material is shown in Fig. 3.

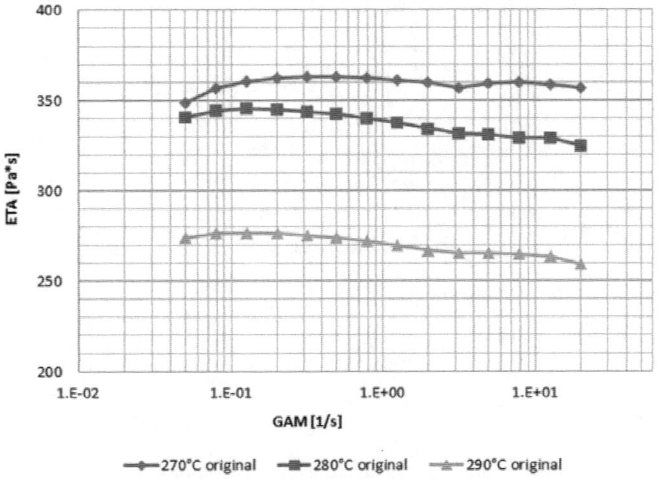

Figure 3. Original PET shear viscosity curves

The curves at different temperatures are not the same. It can be observed that the viscosity is the highest at 270 °C. The higher is the temperature, the lower is the viscosity. From the differences of the curves, the transformation of the molecular structure can be determined. The viscosity of the recycled materials is shown in Fig. 4.

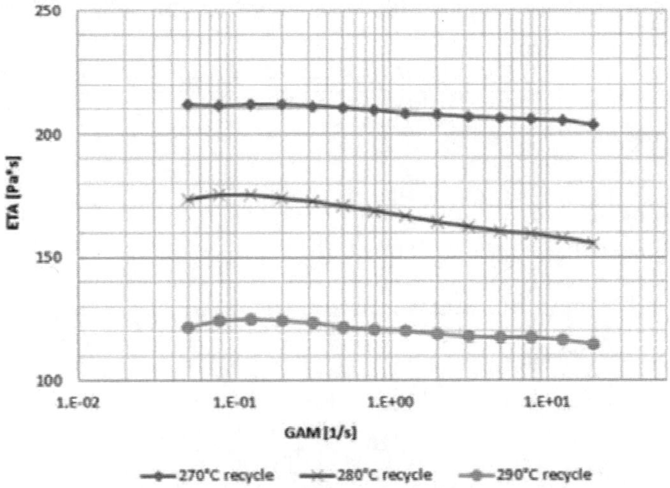

Figure 4. Recycled PET shear viscosity curves

The recycled material shows the same tendency (Fig. 4). Increasing the temperature the viscosity decreases. During the recycling and grinding processes, the molecules in the materials can easily degrade. This is caused by the rising thermal and mechanical strain during the processes. The degradation makes the molecular weight lower, so the molecular density dispersion widens. The polymer with shorter chains can move easier.Above the glass transition temperature the segments of the polymer move increasingly.The viscosities of the original and recycled materials are shown in Fig. 5.

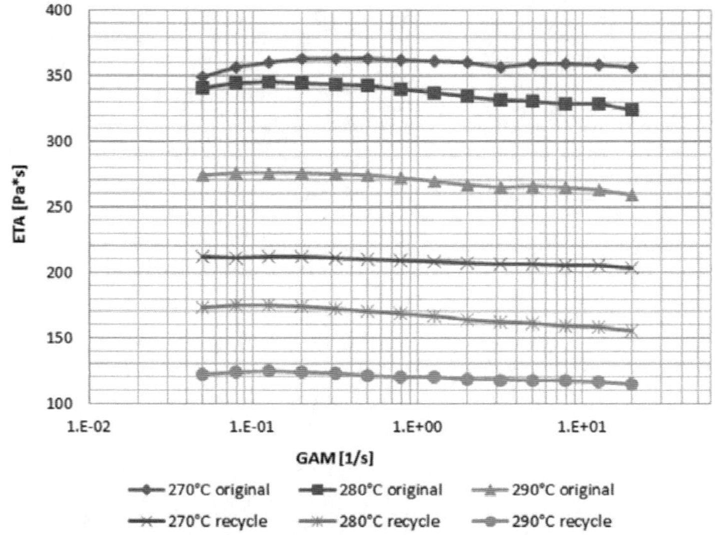

Figure 5. Original and recycled PET shear viscosity curves

In each diagram the viscosity curves show a typical form (Fig. 5). In the curves of newtonian fluids it could be recognised that the curves are almost linear and not dependent on the shear rate. In our measurements, the viscosity curves approach the horizontal line, but they do decrease within a small range. It was confirmed that these materials behave like non newtonian fluid. At a constant temperature, the difference between the original and the recycled materials approaches 150 Pa*s. The impact of the recycling is the continuously decreasing molecular weight. The lower molecular weight means easier molecular movement.

4.2The capillary rheometer
The measurements were performed with original material and with recycled material. The viscosities of the original and recycled materials are shown in Fig. 6.

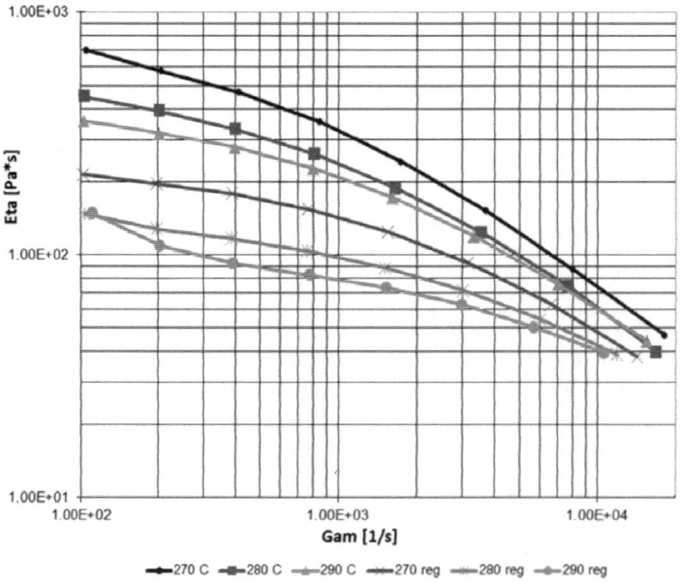

Figure 6. Original and recycled PET shear viscosity curves

Comparing the data to the 270 °C viscosity curve in the lower shear range: There is a ~30 – 35 % reduction at 280 °C, and at 290 °C we can see a 40 – 50 % reduction. As we approach the larger shear ranges (10000 - 18000 1 / s) the differences are reduced: at 280 °C ~15 – 18 %, while at 290 °C ~7 – 13 %. In the initial shear speed range, the recycled material shows a 70 % reduction at 270 °C, while in the higher shear range this value shows a 15 % - 30 % reduction. We could observe the following viscosity reductions in the results: at 280 °C, ~ 65 % and at 290 °C ~ 50 %. Although the initial viscosity values at 290 °C show a different tendency compared to the curve we measured at the previous two temperatures, this can be attributed to a measurement error with a great certainty. From these values we can conclude that we have to consider the viscosity change of the material during the recycling, because with such a degree of deviation we could assume that the material will not be able to withstand the requirements, which are set by design.

5. Conclusion

In our investigation we measured one type of material, but the temperature of the process was different. An original and a recycled material were examined. The measurements were performed by an ARES G2 rotational rheometer and Goettfert Rheograph 25 capillary rheometer at 270 °C, 280 °C and 290 °C. The shear rate was changed from 0.05 to 20 1/s and 100-10 000 1/s. It was determined that the viscosity of the recycled material is lower than the value of the original material at each temperature. The molecules of the recycled material were supposed to shorten during the grinding process. If the temperature increases, the shear viscosity decreases, but there was no linear correlation between these two variables. The structure viscous behavior of the polymer melt was revealed. This area offers lot of possibilities for future research, so we hope to continue with further investigations.

6. Acknowledgement

This research is supported by the EFOP-3.6.1-16-2016-00014 project. The Project is supported by the Hungarian Government and co-financed by the European Social Fund.

References

[1] Ronay F 2006 Physical Recycling of the Material of PET Bottles (in Hungarian) PhD Thesis, Budapest University of Technology and Economics

[2] Czvikovszky T, Nagy P, Gaal J 2000 The basics of polymer technology (in Hungarian) *University publisher* Budapest. http://www.tankonyvtar.hu/hu/tartalom/tkt/polimertechnika-alapjai/adatok.html

[3] http://plasticker.de/recybase/docs/27055_1463559873.pdf

[4] Ronkay F, Czigany T 2004 Properties Modifying of Recycled PET with Additives. Proceedings of the Fourth Conference on Mechanical Engineering Hungary, 170-174

[5] General Rheo Training_2015_v1 with Questions https://www.scribd.com/document/340563607/General-Rheo-Training-2015-v1

[6] TA Instruments Factory Ares G2 Rotational Viscometer Equipment Education Material – Study of Rheology Theory and Applications http://www.tainstruments.com

[7] Szucs A 2010 Investigation Polimer Melt Properties High Shear Rate (in Hungarian) PhD Thesis, University of Miskolc www.kerpely.uni-miskolc.hu/downloads/docs/cikkek/szucsandras-ertekezes.pdf

[8] Toth G, Bata A, Belina K 2017 *IOP Conf. Ser.: Mater. Sci. Eng.* **175** 012056 https://doi.org/10.1088/1757-899X/175/1/012056

IOP Conf. Series: Journal of Physics: Conf. Series **1045** (2018) 012008 doi:10.1088/1742-6596/1045/1/012008

Rheological behavior of Hydroxyethylcellulose (HEC) Solutions

K Benyounes[1], S Remli[2] and A Benmounah[2]

[1] Laboratoire Génie Physique des Hydrocarbures, University M'Hamed Bougara de Boumerdes, Avenue de l'indépendance, 3500 Boumerdes, Algeria
[2] Unité de recherche Matériaux, Procédés et Environnement, University M'Hamed Bougara de Boumerdes, Avenue de l'indépendance, 3500, Boumerdes, Algeria

Email: khaled_benyounes@yahoo.fr

Abstract. The polymers have a wide applicability in several domains particularly in the food, cosmetics, pharmaceuticals industry and petroleum industry. They guarantee a high degree of mix-ability, pump-ability, and pour-ability during their treatment. For this, the flow properties of aqueous non ionic derivative cellulose solutions at different concentrations were measured over a wide range of shear rates (0-1000 s^{-1}), and the temperature was kept at 20°C. The studied concentrations of the polymer solutions show a marked non-Newtonian shear-thinning behavior. No yield stress and thixotropy were detected. In order to define the type of flow behavior, several rheological models such as Power law, Williamson, and Mendes-Dutra were used to fit the rheogramms based on the correlation coefficient (R^2). These models are all applicable and have an equivalent ability to describe the shear flow behavior of concentrated solutions. Rheological parameters such as consistency index, flow index, derived from these models were then subjected to further analysis.

1. Introduction

Natural polymers have a large range of applications such as food industry, detergents, cosmetics, coating technology, pharmaceuticals, and drilling muds [1-6]. The HEC polymer has a very high affinity for water, when the dry polymer is introduced into water, HEC powders swell and hydrate, which means the polymer molecules become associated with many water molecules. The HEC powder in aqueous solution can exhibit high viscosity, which varies depending on the shear rate and temperature, showing a pseudoplastic behavior. For flow in the matrix, a fluid-rheology function was used with shear thickening at high shear rates and shear thinning at medium and low shear rates. For flow in fractures, a shear-thinning constitutive law with a zero shear-rate plateau was used [7]. A clay dispersions aging study in aqueous solutions of CMC showed the effect of time, in particular the existence of a yield stress and its evolution. From a structural point of view, HEC is very similar to CMC, grace to its non-ionicity, it is effective for reducing filtration and for thickening solutions in presence of salt. However, HEC is used in drilling and completion fluids. Magnesium oxide stabilizes the thickening action of HEC in salt waters at high temperatures [3]. The thickening medium is caused by chain entanglements between the long polymer molecules, which are solvated and extended in an aqueous medium. Rheometric measurements have shown the effect of the polymer and the KCl on the rheophysical parameters of bentonite- polymer system [8], the addition of KCl to the clay suspension causes an increase in viscosity and the yield stress is explained by aggregate of clay particles due to the compression of the electrical double layer, moreover, such characteristics are interest in many other industrial areas and a further development of the utilization of this polymer may be expected from these studies. Particularly Cellulose derivative polymers are intensively used in pharmaceutical applications to modify the release of drugs in tablets and capsule formulations and to improve their dissolution in the gastrointestinal fluids. This process is known as hydrophilization [9] showed that both low and high viscosity grade cellulose ether polymers can be mixed uniformly, in different proportions, in order to produce matrices with modulated drug release properties. At higher concentrations, the swelling behavior changes drastically, suggesting a gradual degradation of the matrices and the dissolution of ibuprofen from mixtures of MC/HPMC or HPC/HPMC matrices was found to be more effective when either the MC or HPC content was increased. [10] studied the flow behavior of two types of cellulose derivatives (CMC and HEC) in aqueous solution, he showed that the viscous behavior is roughly similar with two critical concentrations c^* and c^{**}, delimiting three flow regions, and a third critical concentration c_e, intermediate between c^* and c^{**}, determined from the variation of the relaxation time with concentrations. There have been several studies of interaction between cellulose derivatives and surfactants, [11] showed that some homogeneous mixtures have higher viscosity and elasticity than predicted by the log-additivity rule from partial concentration, the relative polymer proportion at which the maximum synergy is obtained and the degree of synergy vary depending on the substituent content. Rheological studies of cellulose derivative are of importance for estimating and understanding on a molecular basis, the function of these components in fracturing fluids and in drilling fluid. Several works were carried to improve the stability of emulsions by adding natural cellulose derivatives to increase the viscosity of the medium [12], [13], to perform and control the texture of foodstuffs, they are also used as thickener, binder, stabilizer suspending and water retaining agent [14], [15]. In [16] study, oscillatory test was used to study, the viscoelastic properties of aqueous and milk systems. Both the type of dispersing media and the CMC concentration clearly affected the viscoelastic behavior of samples, which ranged from liquid-like to weak gel.

2. Materials and Methods

2.1. Materials

The HEC used in this study is a commercial product, which is used by the oil companies. Its moisture content and the pH value of aqueous 0.5 wt.% solution (25°C) were of 8.0% and 7.0 respectively. HEC can be produced by reacting ethylene oxide with cotton. The structure of the repeating unit of HEC is shown in Figure 1. Polymer solutions were prepared in different concentrations ranged from 0.2 to 0.5 wt. %. After mixing each polymer dispersion was poured in a covered vessel and left for one day at room temperature.

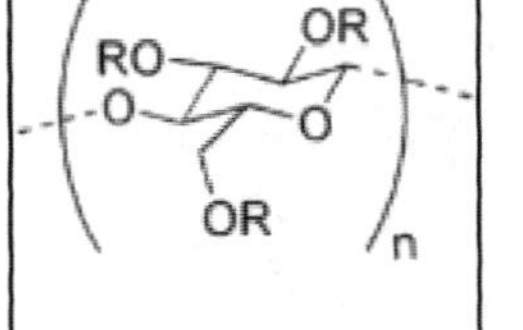

R = H or CH_2CH_2OH

Figure 1. Molecular structure of hydroxyethylcellulose

2.2. Preparation of solutions

The cellulose derivative is soluble in both cold and hot water; it needs an intensive agitation upon introduction into an aqueous medium in order to avoid a formation of lumps.

In this work, water solutions of an anionic cellulose derivative with 0.2, 0.3, 0.4 and 0.5 % (w/w) concentration were prepared, the polymer was dissolved in distilled water, while a magnetic stirrer was dispersing the solution at ambient temperature for 24 hours until it was perfectly dissolved.

2.3. Rheological measurements
The rheological measurements were performed at 20°C using a controlled stress rheometer AR 2000 (Fig.2) from TA instruments using the concentric cylinder measuring system under various conditions. Furthermore, to prevent water evaporation, a cover was placed around the geometry.

The sample was pre-sheared during 2 min at shear rate of 100 s^{-1} followed of 1 min of rest. The flow behavior was investigated over a wide range of concentrations (0.2 –0.5%). Rheological parameters (shear stress, apparent viscosity and shear rate) were obtained from the software.

Figure 2. Rheometer AR 2000

3. Rheological Models
The flow curve of fluid will not only depend on its composition, but also on the protocol used in the flow test. Therefore, in the literature, there are many rheological models, the flow curves were fitted to the following equations.

3.1. Power law model
Power law is a two-parameter model indicating a non linear relationship between shear rate and shear stress (Ostwald, 1925), which is given as follows:

$$\tau = k.\dot{\gamma}^n \quad (1)$$

Where k is the consistency and n is the flow behavior index. The different values of n indicate the fluid behavior. For a Newtonian fluid, n = 1. If n < 1, the fluid is called pseudoplastic; if n > 1, the fluid is dilatants.

3.2. Williamson Model
Williamson law is three-parameter model. The equation is:

$$\eta = \frac{\eta_0}{1 + k.\dot{\gamma}^n} \quad (2)$$

3.3. Mendes-Dutra model
Mendes-Dutra rheological law is defined by four parameters, which is given by the following relation:

$$\tau = \left(1 - \exp\left(\frac{-\eta_0.\dot{\gamma}}{\tau_0}\right)\right).(\tau_0 + k.\dot{\gamma}^n) \quad (3)$$

Where η_0, refers to the asymptotic values of viscosity at very low shear rates respectively, k is the consistency,τ_0 is the yield stress, and n is the flow behavior index.

The different models were fitted to measured rheograms using rheological data analysis software, which also estimated the correlation coefficient. In a general way, an acceptable fit gives a correlation coefficient > 0.90.

4. Results and Discussion

4.1. Flow behavior

Both of the shear rate and concentration dependencies of shear flow behavior were reported from experimentally obtained data. Figures 3 and 4 show the flow curves of the polymer solutions at different concentrations. It can be clearly seen that the apparent viscosity increases with increasing concentrations, it exhibit a non Newtonian shear-thinning behavior. This rise in the apparent viscosity is due to the increase in the intermolecular interactions between the cellulose derivative polymer molecules [10], [17]; These properties could be attributed to polymer's long chain molecules that tended to orient themselves in the direction of flow. As the shear stress was increased, the more chains rearranged themselves along the shear direction, and then the shear resistance to flow (viscosity) decreased. When a lower stress was imposed on the same solution, the viscosity was higher because random molecular orientations exhibited increased resistance to flow [18], there was no evidence of a yield stress (Figure 5).

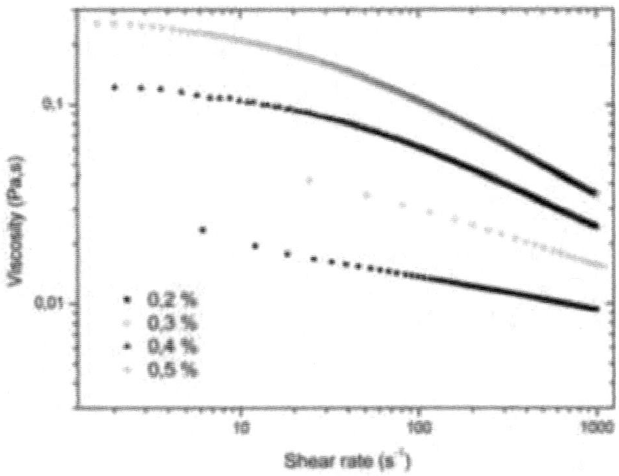

Figure 3. Shear rate dependence of viscosity for different concentrations

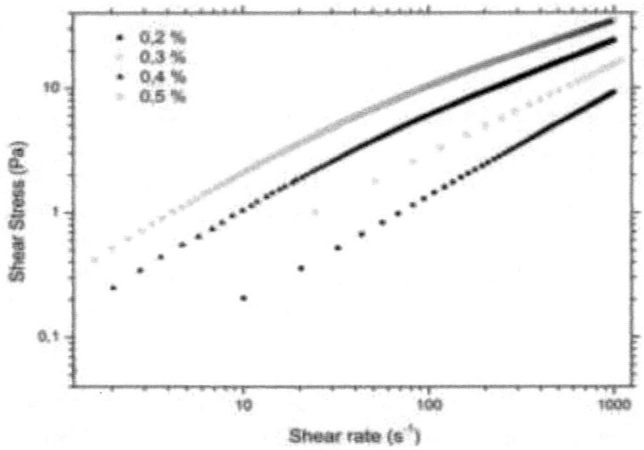

Figure 4. Flow curves for different concentrations

The curves exhibit a shear rate dependant viscosity above a critical shear rate $\dot{\gamma}_{crit}$ representing transition from Newtonian behavior to non Newtonian behavior; as the concentration increases the value of $\dot{\gamma}_{crit}$ is shifted to lower shear rates, below γ_{crit} the solution viscosity is independent of shear rate and is called the zero-shear viscosity $\eta_{0,}$ this result is in good agreement with [10] and [13].

Thixotropy of a solution is quantified by the solution ability to regain its gel structure when the liquid is allowed to rest for a longer period time [4]. Regarding the time-depending rheological behavior of HEC polymer, the obtained results show that there is no thixotropy for all the studied solutions(Figure 6).

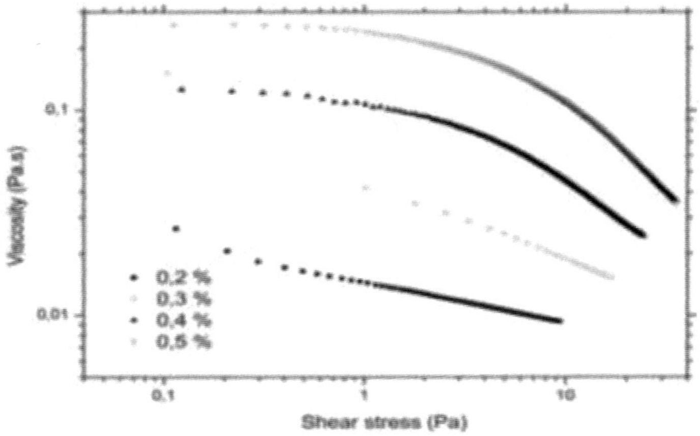

Figure 5. Viscosity as a function of shear stress for different concentrations

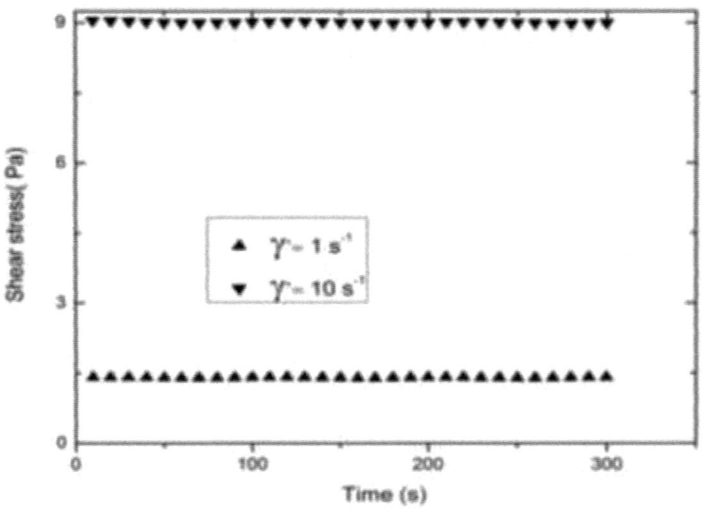

Figure 6. Shear rate response as a function of time for a concentration of 4%.

4.2. Applicability of rheological flow models

The Power law model shows a large agreement between the obtained results and the experimental measured data over a great range of shear rates tested, it gives a good ability for predicting the flow behavior of the polymer solution. The Williamson and Mendes-Dutra models are in good agreement

IOP Conf. Series: Journal of Physics: Conf. Series **1045** (2018) 012008 doi:10.1088/1742-6596/1045/1/012008

with the experimentally data, over a whole range of shear rates tested and for concentrations over 0.4% (Fig.7), they have the value of the correlation coefficient greater than 0,980. Furthermore, no discrepancy is observed between these two models, indicating that they have the same ability to predict the flow behavior of concentrated polymer derivative solutions [4] .

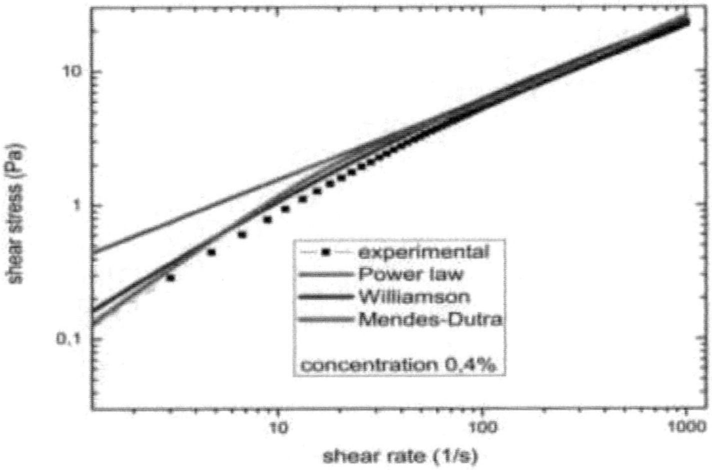

Figure 7. Applicability of some flow models

5. Conclusion

The present work analyses the effect of concentration on the rheological behavior of cellulose derivative solutions. Viscosity was strongly influenced by concentration. The results suggested that solutions behaved as non Newtonian shear thinning at the test conditions. The consistency index decreased with increasing concentration. The time-dependent viscosity has a constant value corresponding to an equilibrium state. From these results, and for the concentrations used, the solutions were found to be not thixotropic. The power law, Williamson and Mendes-Dutra models are all applicable and have an equivalent ability to describe the shear flow behavior of concentrated solutions.

References

[1] Alemdar A Oztekin N Gungor N Ece O I Erim F B 2005 *Colloid.Surface. A* **252** 95
[2] Bakshi M S Kaur R Kaur I Kumar Mahajan R Sehgal P and Doe H 2003 *Colloid. Polym. Sci.* **281** 716
[3] Darley H C and Gray G R 1988 Composition and properties of drilling and completion fluids, *Gulf Professional Publishing*, Texas
[4] Ghannam M T and Esmail M N 1997 *J. Appl. Polym. Sci.* **64** 289
[5] Gómez-Díaz D and Navaza J M 2002 *Electron. J. Environ. Agric. Food Chem* **1** 12
[6] Holme K R Hall L D Speers R A and Tung M A 1988 *Food Hydrocolloid.* **2** 159
[7] Ma Y and McClure M W 2017 *SPE Reserv. Eval. Eng.* **20** 394
[8] Benyounes K and Benmounah A 2012 *Int. J. Phys. Sci.* **7** 1790
[9] Kamel S Ali N Jahangir K Shah S and El-Gendy A 2008 *Express. Polym. Lett.* **2** 758
[10] Castelain C Doublier J L and Lefebvre J 1987 *Carbohyd. Polym.* **7** 1
[11] Alvarez-Lorenzo C Duro R Gomez-Amoza J Martinez-Pacheco R Souto C. and Concheiro A 2001 *Colloid. Polym. Sci.* **279** 1045
[12] Radi M and Amiri S 2013 *J. Disper. Sci. Technol.* **34** 582
[13] Yaseen E Herald T Aramouni F and Alavi S 2005 *Food. Res. Int.* **38** 111
[14] Mirhosseini H Tan C.P Aghlara A Hamid N S Yusof S and Chern B H 2008 *Carbohyd. Polym.* **73** 83

[15] Menezes R Marques L Campos L Ferreira H Santana L and Neves G 2010 *Appl. Clay. Sci.* **49** 13
[16] Bayarri S and Costell E 2011 *J. Food. Process. Eng.* **34** 1903.
[17] Li J Xie W Cheng H Nickol R G and Wang P G 1999 *Macromolecules* **32** 2789
[18] Yang X H and Zhu W L 2007 *Cellulose* **14** 409

IOP Conf. Series: Journal of Physics: Conf. Series **1045** (2018) 012009 doi:10.1088/1742-6596/1045/1/012009

Characteristic features of the rheological behavior of dielectric layered-structure suspensions of hydrated metal oxides in an electric field

L Eshchenko[1], E Korobko[2], Z Novikova[2]

[1]The Belorussian State Technological University, 13a Sverdlova Street, Minsk 220006, Belarus

[2] A.V. Luikov Heat and Mass Transfer Institute, National Academy of Science of Belarus, 15 P. Brovka Street, Minsk 220072, Belarus

E-mail: evkorobko@gmail.com

Abstract. In this paper suspensions containing hydrated oxides of magnesium and iron were objects of rheological studies. The different content of the “stoichiometric” water in fillers is attributable to the conditions of their precipitation and aging. According to the results of TGA - DTA “super-stoichiometric” water of all the samples was removed in the temperature range 130 - 150°C, indicating its strong bond with the surface. Filler ER activity is determined by the change in the shear stress of the ER-suspensions containing them under the external electric field. The important role in the manifestation of the ER effect is played not only by the quantity of “super-stoichiometric” water in the interlayer space of the disperse phase particles, but also by the steric conditions.

1. Introduction

The uprising of the rheological indices of electrorheological fluids (ERF) under the action of an external electric field depends on the properties of a disperse phase whose role may be played by both water-containing and water-free compounds [1-7]. Some researchers [8-10] attribute the manifestation of the electrorheological (ER) effect to the presence of the adsorbed activator (the substance capable of forming hydrogen bonds such as hydroxyl-carboxyl-or amino-containing compounds) on the surface of the disperse phase particles. The most common activator is water entering into the composition of the water-containing disperse phase. The preparation of such disperse phase (ERF filler) is possible by means of both water adsorption in water-free compounds, more often oxide ones, and by chemical precipitation of hydrated compounds from aqueous solutions. In this case, the energy state of H_2O molecules and the form of their bonding in the hydrated filler may differ substantially and differently influence the ERF fluidity (viscosity, yield point, shear modulus, etc.). Water-containing ERF fillers, silicas (silica gel, aerosil, diatomite), metal oxides activated by H_2O adsorption have been studied most extensively. Occurring in this case is the filling of the surface of the particles with water and formation of polymolecular H_2O layers bound by the H bonds. The presence of film polarization ensures the transfer of charge carriers only along the particle surface. In an ERF based on particles with adsorbed water, the weakening or complete degeneration of the ER effect in time is possible especially on heating because of the irreversible desorption of weakly bound H_2O molecules.

In fillers based on hydrated compounds obtained by chemical precipitation, in addition to the structural and crystalline-hydrate H_2O there may also exists other O_mH_n groups capable of being preserved up to relatively high temperatures (200-250°C). They can potentially be used for creating thermally stable ERFs. The relationship between the nature, structure of O_mH_n groups in such type of fillers, and their rheological activity in an electric field is scarcely discussed in scientific publications. In the last 30 years, scientists of various countries have carried out numerous rheological investigations of ERF based on a water-containing disperse phase [1, 2, 4, 5 and 8]. In this connection, of particular interest is the investigation of the fillers of the ERFs, which are based on lamellar hydrated oxides of polyvalence metals, which, according to [11-13], consist of positively charged alternating two-dimensional metal-hydroxide layers, in the interlayer space of which the number of water molecules needed for ensuring the electrorheological (ER) activity of suspensions can be localized. The form and type of bonding as well as the quantity of H_2O molecules depend in the first place on the metal cation nature, structure of lamellar compounds, and on the conditions in which they were prepared. Moreover, in chemical compounds, the O_mH_n groups, which represent the interlayer water, can possess specific mobility, which exerts its influence on proton transfer and, correspondingly, on the magnitude of the conduction current.

Earlier [14, 15] we have studied the ER activity of the modifications of hydrated aluminum oxides (boehmite, pseudoboehmite, AlOOH, bayerity, $Al(OH)_3$), obtained by chemical precipitation under different conditions and differing in the type and quantity of H_2O molecules in them. It was shown [14] that fillers based on boehmite, $Al_2O_3{\cdot}H_2O$, bayerite, $Al_2O_3{\cdot}3H_2O$, and containing only structural H_2O in the form of hydroxyl groups have a low ER activity. At the same time, the ERF based on pseudoboehmite which contains "super-stoichiometric" or unstructured water manifest a considerable ER effect. Such water molecules whose quantity is 0.25-0.8 mole per mole of Al_2O_3 are mainly localized in the interlayer space of the structure. It is shown that they are bound by strong hydrogen bridges in the boehmite lattice, are somewhat deformed, possess thermal stability, and are efficient activators of the disperse phase particles. The appearance of the proton conductivity in this case can be explained by the dissociation of H_2O molecules in an electric field with formation of protons and OH groups with strong hydrogen bonds. This ensures the appearance of strong structural bridges from disperse phase particles in an ERF.

Taking into account the role of the interlayer H_2O molecules in the initiation of the ER effect, as it is displayed in [14, 15], it is of scientific and practical interest to reveal the dependence of the ER effect on the quantity of active O_mH_n groups entering into the composition of lamellar hydrated compounds obtained by chemical precipitation. This will allow one to obtain new experimental data needed for elucidating the nature of proton-containing groups, their role in proton processes in a disperse phase for purposeful creation of electrically active and thermally stable fillers for ERFs.

The aim of the present work is investigation of the influence of the composition and structure of chemically precipitated hydrated magnesium and iron oxides on the ER activity of suspensions based on them.

2. Experimental

Electrosensitive particles of the filler of an ERF based on hydrated magnesium and iron oxides were obtained by the method of chemical precipitation in the metal salt-base-water system. As metal-containing reagents we used magnesium and iron chlorides. The precipitation of hydrate magnesium and iron oxides was effectuated in an alkaline medium, after which the sediment was kept for 6-8 hours under the liquid phase formed as a result of the chemical precipitation. Thereafter, it was separated from the liquid phase by filtering, washed and dried at a temperature of 20-30°C and then at 90°C up to a constant mass. The different content of the "stoichiometric" water in fillers is attributable to the conditions of their precipitation and aging. For the purposes of comparison we obtained hydrated aluminum oxide, the electrorheological properties of which were studied earlier in [14].

The compounds obtained were identified by the results of x-ray phase analysis with the use of a 08 Advance diffractometer of Bruker AXS Company (Germany). The content of magnesium and iron on conversion into the mass of oxide, as well as the quantity of water were determined by the well-known methods. The dispersity of sediments was determined by a scanning electronic microscope "Jeol JSM-5610LV". The IR spectra of the compounds were obtained with the aid of a NEXUS IR-Fourier spectrometer of THERMO NICOLET Company. The thermal stability of the compounds was evaluated by the results of a differential-thermal analysis recorded by a thermoanalytical system TGA/DSC-1/1600 HF (METTLER TOLEDO Instruments, Switzerland).

The ER activity of the synthesized hydrated metal oxides was evaluated by the change in the shear stress of ERFs containing 10.0 wt.% of the filler in an dielectric oil under the action of electric field of strength from 0 to 3 kV/mm at the shear rate $\gamma = 17.24\ s^{-1}$. The shear stress of ERF was measured by "Rheotest 2.1" rotational viscometer with an axially cylindrical cell at room temperature.

3. Results and discussion

The influence of the conditions of precipitation of hydrated metal oxides and their composition are presented in Table 1.

Table 1. Condition of precipitation of hydrated metal oxides and their composition

No. of sample	Nature and concentration of metal salt, moles/liter	pH of precipitation	Chemical composition	Phase composition
1	$MgCl_2$ 1.0	9.5-10.0	$MgO{\cdot}[H_2O]{\cdot}\{0.16H_2O\}$	
2	$MgCl_2$ 0.5	9.8-10.0	$MgO{\cdot}[H_2O]{\cdot}\{0.28H_2O\}$	$Mg(OH)_2$ (brucite)
3	$MgCl_2$ 0.05	9.9-10.0	$MgO{\cdot}[H_2O]{\cdot}\{0.45H_2O\}$	
4	$FeCl_3$ 0.5	12.0-12.5	$Fe_2O_3{\cdot}[H_2O]{\cdot}\{0.3H_2O\}$	α-FeOOH (goethite)
5	$Al_2(SO_4)_3$ 0.5	9.5-10.0	$Al_2O_3{\cdot}[H_2O]{\cdot}\{0.35H_2O\}$	AlOOH (pseudoboehmite)

[ ] –quantity of structured water
{ } – quantity of "superstoichiometric" water

According to the x-ray-phase analysis, the synthesized samples of fillers Nos.1-3 are crystalline hydrated magnesium oxides. They have a typical lamellar structure of brucite in which each Mg^{2+} ion is located between six OH groups and is bound with three OH ions of one layer and three OH ions of another one. Sample No. 4 is iron oxyhydroxide having the structure of goethite, which is a transitional one from a chain structure to a lamellar one. Its base is the densest hexagonal packing of O^{2-} ions, since the H^+ protons are located between pairs of O^{2-} ions not occupying particular space.

It has been established from the results of electronic-microscopic investigations that the synthesized samples of fillers are highly dispersed ones. Figure 1 presents microphotographs of samples No.2 and No.4, from which it is seen that they are polydisperse in composition. Sample No.2 (brucite) has particles of spherical shape of size 1-3 μm and their agglomerates of size 10-50 μm, pointing to the ability of primary particles to agglomerate. The prevailing size of the goethite particles (samples No.4) is 4.0-8.0 μm.

IOP Conf. Series: Journal of Physics: Conf. Series **1045** (2018) 012009 doi:10.1088/1742-6596/1045/1/012009

Figure 1. Microphotographs of a fillers sample No.2 (a), sample No.4 (b).

Proceeding from the stoichiometric composition of the indicated compounds, in particular, of $Mg(OH)_2(MgO \cdot H_2O)$, $FeOOH(Fe_2O_3 \cdot H_2O)$, in them, corresponding to 1 mole of metal oxide there is 1 mole of chemically bound or structural water, which in the form of OH groups participates in the construction of the crystal lattice. In addition to this type of water, the samples also contain "superstoichiometric" water. As shown in Table 1, its quantity for magnesium oxide depends on the precipitation conditions, in particular, on the concentration of the magnesium salt solution. Just as in the case of pseudoboehmite, the molecules of "superstoichiometric" H_2O can be both localized in the interlayer space of the lamellar structure of the compounds and adsorbed on the surface of crystals with which they are strongly bound by hydrogen bonds.

For the "superstoichiometric" H_2O molecules of brucite and goethite just as of pseudoboehmite, typical is the relatively high thermal stability, which is seen from the thermograms presented in Figure 2. The curves of heating the tested compounds have two endothermal peaks. The first, small peak is fixed in the region of T=100-250°C for brucite and T=100-230°C for goethite. The second, deep peak is located in the region of higher temperatures.

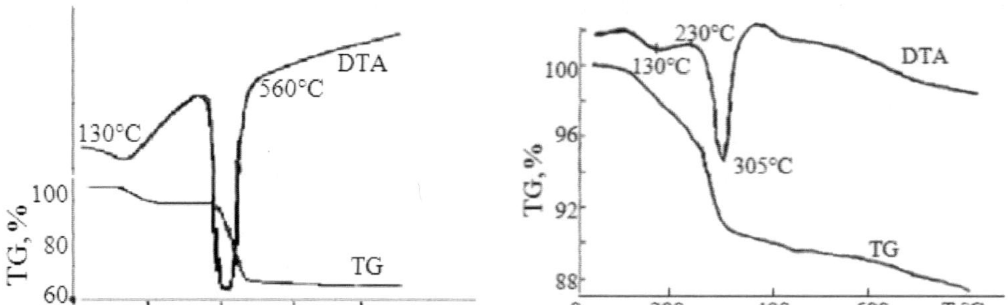

Figure 2. Thermograms of a fillers No.2 (a), No.4 (b).

Table 2 presents the results of the thermogravimetric analysis of samples No.2 and No.4 with indication of the amount of the removed H_2O in the region of the first and second endoeffects that was calculated from the thermogravimetric curves (TG) of samples shown in Figure 2.

Table 2. Results of thermogravimetric analysis

No. of sample	Amount of water removed, mole				
	Temperature interval, °C				
	90–130	130–200	200–360	360–560	560–800
2	0.23	0.05	0.0	1.0	0.0
4	0.10	0.19	0.74	0.09	0.19

As is seen from Figure 2, for sample No.2 the splitting out of 1 mole of chemically bound H_2O in the form of OH groups is fixed in the region of the second endoeffect at temperatures 360-560°C. In sample No.4, the chemically bound water is removed both in the region of the second endoeffect and on subsequent increase in the temperature, up to 800°C. In the temperature range 90-200°C, 0.28 and 0.29 moles of H_2O are removed from samples No.2 and No.4, respectively, which agrees with the amount of "superstoichiometric" water determined by gravimetrical analysis. Consequently, the "superstoichiometric" H_2O is removed from the region of the first endoeffect from both brucite and goethite. A comparatively high temperature of the dehydration of the "superstoichiometric" water molecules indicated the existence of strong H bond in them.

The energy state of H_2O molecules in brucite and goethite was estimated by the IR spectra in comparison with the results obtained for pseudoboehmite. The vibrational spectra of the compounds studied contain all basic features of the spectral display of hydrogen bonds in the O_mH_n groups (Figure 3).

The presence of the absorption bands (Figure 3) in the range 3700-3000 cm^{-1} indicates the formation of hydrogen bonds of different strength by H_2O molecules with the OH groups that were coordinated by the metal cation. For brucite, there is a narrow intense absorption band at 3700 cm^{-1} and a wide low-intensity band at 3436 cm^{-1}, as well as the absence of the bands of deformation vibrations of H_2O molecules in the spectrum. This may point to a specific energy state of the "superstoichiometric" H_2O in brucite initiated by the strong electrostatic field of the magnesium cation. Under its effect the OH bonds in water molecules can become weaker. Proceeding from the position of the band of valence vibrations of the OH groups of H_2O molecules an assumption can be made on the formation of strong hydrogen bonds in goethite, just as in the case of pseudoboehmite AlOOH.

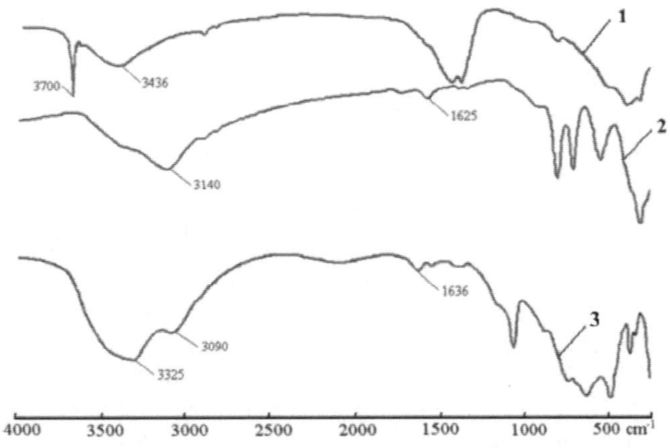

Figure 3. IR spectra of hydrated oxides magnesium, $Mg(OH)_2{\cdot}0.28H_2O$ (1); iron, $FeOOH{\cdot}0.3H_2O$ (2); aluminum, $AlOOH{\cdot}0.35H_2O$ (3).

The goethite, FeOOH, and pseudoboehmite, AlOOH, have the most similar IR spectra in the region of valence and deformation vibrations. However, the displacement of the bands of valence vibrations of the OOH groups of water molecules to the side of smaller frequencies is most typical of the IR spectra

IOP Conf. Series: Journal of Physics: Conf. Series **1045** (2018) 012009 doi:10.1088/1742-6596/1045/1/012009

of pseudoboehmite. This may mean that due to its small radius the Al^{3+} cation (0.51Å) is capable of very closely approaching the unshared pair of water molecules and attract them by its strong electrostatic field.

It was shown earlier [14, 15] that the energy state of H_2O molecules determined by steric conditions, polarizing action of the cation field, and by other factors is closely connected with the manifestation of the ER effect.

The dependences of the shear stress (τ) and current density (j) of ERF based on the studied hydrated magnesium, iron, and aluminum oxides on the electric field strength (E) are presented in Figures 4 and 5.

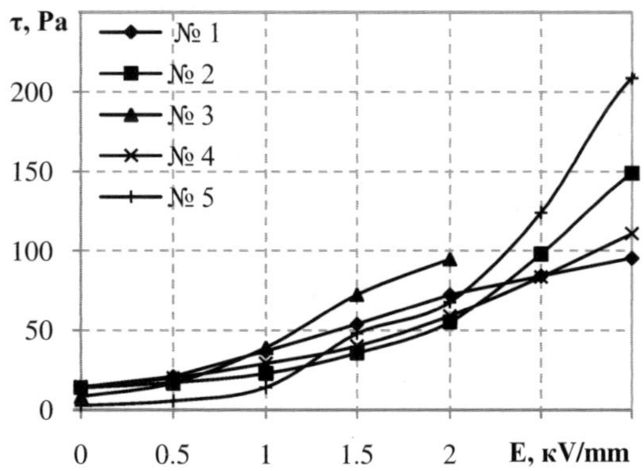

Figure 4. Dependence of shear stress on electric field strength for samples Nos. 1-5.

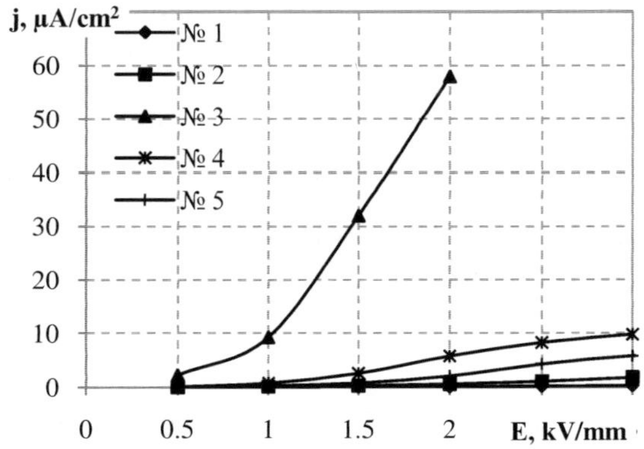

Figure 5. Dependence of current density on electric field strength for samples Nos. 1-5.

It should be noted that for the ERF based on brucite, which contains 0.45 moles of "superstoichiometric" water (sample No.3) a high current density (> 50 μA/cm² at E=1.5 kV/mm) is fixed. A further increase in the electric field strength leads to the overestimation of the threshold values of current density and to the break-down occurrence. On reduction of the amount of water in brucite to 0.28 moles (sample No.2) the current density also decreases, which allows one to increase the electric field strength up to 3.0 kV/mm and to attain the values of up to about 150 Pa for the shear stress of the ERF. A decrease in the number of moles of H_2O in brucite to 0.16 moles decreases the shear stress to 100 Pa (sample No.1). Thus, the intensity of the rise of the τ(E) curves for the ERF based on brusite is determined by the quantity of the "superstoichiometric" water. This seems to be associated with the concentration of mobile protons which depends on both the place of localization of water molecules and their energy state as well as on the steric conditions determined by the filler structure. It is seen form Figure 4 that at the content of the number of moles of "superstoichiometric" unstructured water close to about 0.3 moles in samples of brucite, goethite, pseudoboehmite (samples No.2,4,5, respectively) their ER activity does not differ substantially in an electric field of strength of up to 2.0 kV/mm. An increase in the electric field strength up to 3.0 kV/mm leads to a sharper increase in the shear stress of the ERF based on brucite, $Mg(OH)_2{\cdot}0.28H_2O$ (sample No.2) and pseudoboehmite, $AlOOH{\cdot}0.35H_2O$ (sample No.5), which is less typical of the ERF based on goethite, $FeOOH{\cdot}0.3H_2O$ (sample No.4). The strong shift of the band of valence vibrations of the OH groups of water molecules to the low-frequency region on the IR spectrum of goethite (3140 cm^{-1}, Figure 2) and their thermal stability make it possible to assume the presence of strong hydrogen bonds of H_2O molecules and the possibility of their protolytic dissociation in an electric field. It seems that by virtue of the specific features of the crystal structure of goethite, in which there are twinned layers of the densest packing of O^{2-} ions, the transfer of the proton over a great distance to oxygen atoms can be hampered. As a result, a continuous chain of hydroxyl groups that interacts via the hydrogen bonding and ensures the transfer of the charge by the proton mechanism is not formed. In the structure of brucite, the Mg^{2+} cations occupy all octahedron voids between two layers of OH groups located

IOP Conf. Series: Journal of Physics: Conf. Series **1045** (2018) 012009 doi:10.1088/1742-6596/1045/1/012009

according to the principle of the densest hexagonal packing, which ensures additional paths for the charge transfer in the interlayer space. In this case, for the transfer of the proton by the relay-race mechanism a small amount of H_2O is required, which is localized in the interlayer space. The increase in the shear stress in the ERF based on pseudoboehmite is larger than in the ERFs based on brucite and goethite, which contain 0.28-0.35 moles of "superstoichiometric" water at an electric field strength higher than 2 kV/mm. This is explained by the fact that, according to the IR spectra, stronger polarization is typical of the "superstoichiometric" water molecules in pseudoboehmite. It is determined by the donor-acceptor properties of aluminum and by the characteristic features of the structure of boehmite, in which the charge transfer may take place via the continuous network of the H bonds in the bulk of a substance, which plays the decisive role in the manifestation of the ER effect.

4. Conclusions

Thus, as follows from experimental data, despite the lamellar structure of all the ERF fillers studied, in the interlayer space of which approximately the same quantity of H_2O molecules is localized that form strong H bonds the ER activity of ERF differs substantially. This fact makes it possible to assume that the important role in the manifestation of the ER effect is played not only by the quantity of "superstoichiometric" water in the interlayer space of the disperse phase particles, but also by the steric conditions, i.e., by the characteristic features of the lamellar structure of fillers.

References

[1] Block H, Kelly J P, Qin A, Watson T 1990 *Langmuir* **6** 6

[2] Klingenberg D J, Zukoski IV C F 1990 *Langmuir* **6** 15

[3] Lemaire E, Bossis G, Grasselli Y Rheological Behavior 1992 *Langmuir* **8** 2957

[4] Whittle M, Bullough W A, Peel D J, Firoozian R 1994 *Physical Review E* **49** 5249

[5] Wen W, Ma H, Tam W Y, Sheng P 1997 *Physical Review E* **55** R1294

[6] Ikazaki F, Kawai A, Uchida K, Kawakami T, Edamura K, Sakurai K, Anzai H, Asako Y 1998 *J. of Physics D: Applied Physics* **31** 336

[7] Hao T, Kawai A, Ikazaki F 1998 *Langmuir* **14** 1256

[8] Shen C, Wen W, Yang S, Sheng P 2006 *Journal of Applied Physics*. **99** 106104.

[9] Belza T 2008 *Colloids and Surfaces A: Physicochemical and Engineering Aspects* **316** 89

[10] Kim Y D 2004 *J. of Colloid and Interface Science* **269** 205

[11] Moroz E M, Shefer K I, Zyuzin D A 2006 *React. Kinet.Catal.Lett* **87** 367

[12] Vorontsova O A, Lebedeva O E, Ressner F 2009 *Kinetics and Catalysis* **50** 899

[13] Isupov V P 1999 *J. Strukt. Khim.* **40** 832

[14] Korobko E V, Eshchenko L S, Bedik N A, Zhuk G M 2007 *Kolloid. Journal* **69** 201

[15] Korobko E V, Bedik N A, Eshenko L S 2011 *Electro-Rheological Fluids and Magneto-Rheological Suspensions* (*Proc. of the 12th International Conference on Electro-rheological Fluids and Magnetorheological Suspensions, Philadelphia*) ed R Tao (Singapore: World Scientific) pp 293–299

IOP Conf. Series: Journal of Physics: Conf. Series **1045** (2018) 012010 doi:10.1088/1742-6596/1045/1/012010

Significance of rheology characterisation in mineral flotation process performance

S Farrokhpay[1,2]

[1]School of Chemical Engineering, The University of Queensland, St Lucia, Brisbane 4068, Australia
[2]GeoRessources, University of Lorraine, 2 rue du Doyen Marcel Roubault, 54518 Vandoeuvre-lès-Nancy, France

E-mail: s.farrokhpay@uq.edu.au

Abstract. Mineral and metal extractive industries play an important role in the current global economics, and mineral processing is one of the key areas to achieve and maintain sustainability. It is known that the rheological behaviour of mineral slurries affects their processing. However, in addition to controlling the transportation of slurries around processing circuits, rheological behaviour also influences separation processes such as flotation. Although the impact of rheology in unit operations such as grinding and slurry transport has received considerable attention, this has not been the case for flotation. This paper explains the relationship between rheology and froth flotation performance. The results revealed that the mineral flotation performance can be improved by controlling the rheology in flotation process.

1. Introduction

Froth flotation is a process for selectively separating hydrophobic materials from hydrophilic gangues. This process is widely used in mineral processing, paper recycling, de-inking recycled newsprint, and waste-water treatment industries. Historically, flotation was first used in the mining industry as a great enabling technology of the 20th century. Flotation process is schematically illustrated in Figure 1. The particles with attached air bubbles are carried to the surface and removed, while the particles that remain completely wet remain in the liquid phase. Froth flotation can be adapted to a broad range of mineral systems to selectively separate valuable mineral. For example, flotation is used in separating sulfide minerals from silica gangue, potassium chloride from sodium chloride, coal from ash, phosphate minerals from silicates.

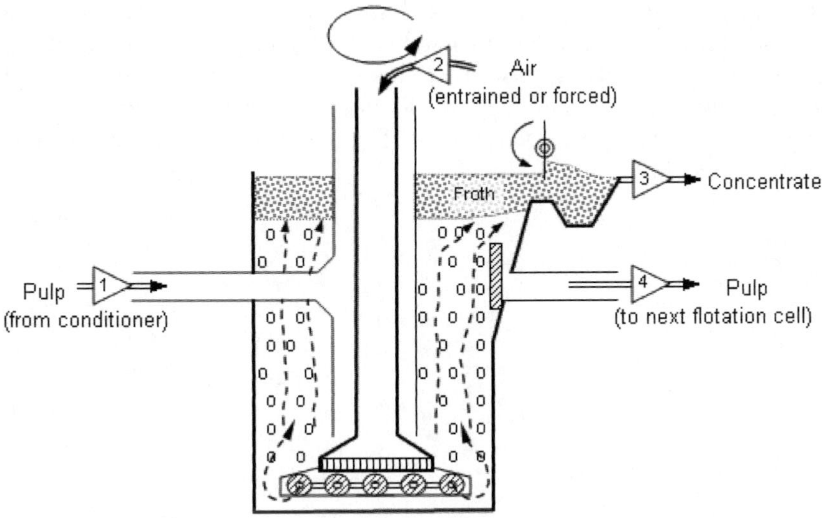

Figure 1. A schematic diagram of froth flotation process

2. Rheology in Froth Flotation

2.1 Rheology in pulp phase

The rheological behaviour of mineral slurries indicates the level of inter-particle interaction or aggregation, and therefore, it can be used as a useful processing control parameter. The flotation separation of minerals may be influenced by inter-particular interactions, and rheological data can offer a direct approach to further understand these interactions. For example, rheology has been used to investigate particle interactions in oxide minerals slurries [1, 2], clays [3, 4], coal [5], and sulfide minerals [6]. It should be noted that although the impact of rheology in unit operations such as grinding and slurry transport has received attention in the past, this has not been the case for flotation. In fact, the pathways by which the pulp or froth rheology influences flotation performance is not yet fully understood [7].

It is worth mentioning that accurately measuring the rheological properties of mineral slurries is difficult, since the solid particles tend to settle during measurement. Researchers have used methods such as pumping slurry during measurement [8], using capillary viscometers [9] or stirring the suspension while measuring with Couette-type viscometer [10], to overcome this problem. However, finding a proper practical way to measure the rheology of such slurries remains a challenge for research in this area. Turbulence is also another parameter which may affect the rheology measurements and therefore turbulence correction procedures have been suggested to process the measured data [11].

The significance of rheology in minerals processing has been reviewed previously [12, 13]. A direct measurement technique (Vane method) at a constant low shear rate mode, was applied to measure the yield stress of concentrated fine particles [14]. However, due to the practical limitations, it is often difficult to conduct measurements at such a low shear rate. In such cases, the yield stress can be estimated using mathematical models involving the extrapolation of the data to a zero shear rate. It is also worth mentioning that Nguyen and Boger have concluded that the Vane technique is the most suitable method for a material having yield stress of greater than approximately 10 Pa [15].

As mentioned, rheological investigations can be used to understand the interactions occurring in the flotation process [16, 17]. The viscosity of flotation medium may also affect the flotation recovery. A higher recovery of coarse composite copper bearing particles was reported by increasing the viscosity of the flotation medium (when a glycerol water mixture was used) [18].

The relationship between change in rheology (viscosity or yield stress), and flotation performance is shown in Figures 2 to 4 [6, 17, 19]. In Figure 2, the concentration of phyllosilicates or fibres has direct

IOP Conf. Series: Journal of Physics: Conf. Series **1045** (2018) 012010 doi:10.1088/1742-6596/1045/1/012010

relationship with the slurry yield stress, but it does increase the copper recovery in flotation. In Figure 3, the viscosity of the Carlin ore slurry was changed by adding dispersant. At the same time, the gold flotation recovery was considerably improved by about 5%. The data presented in Figure 4 were obtained by mixing montmorinite and copper ore. It shows that the best copper flotation recovery is indeed obtained when the ore slurry has the lowest yield stress. For detail, please see the original paper published by Farrokhpay et al.[20].

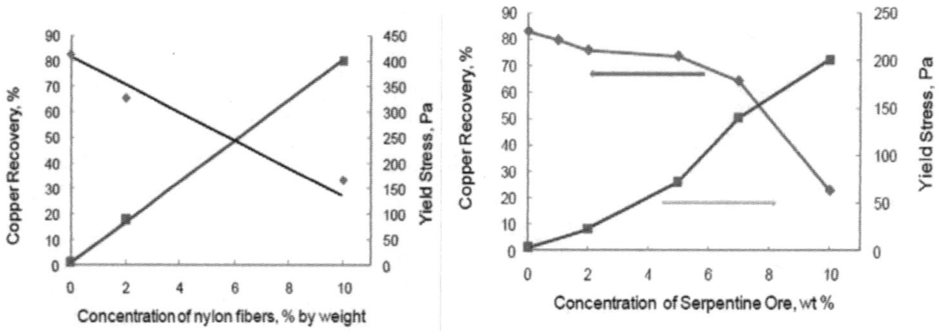

Figure 2. Relationship between rheology and copper recovery in the presence of serpentine and nylon fibres

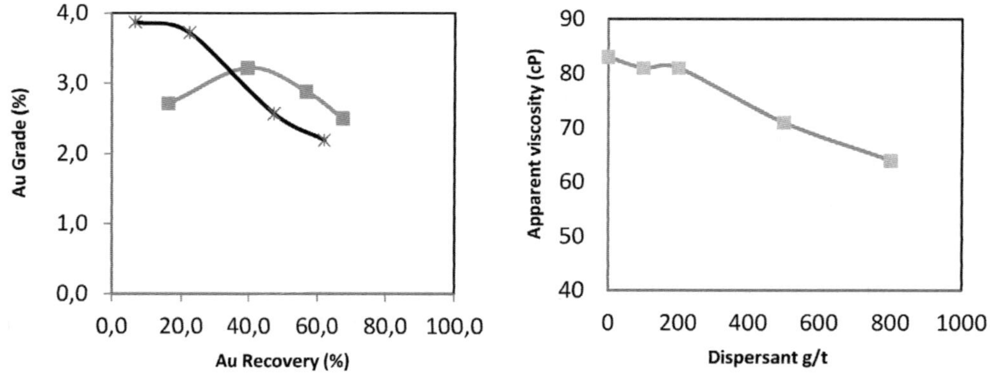

Figure 3. The effect of adding dispersant to gold flotation recovery and grade

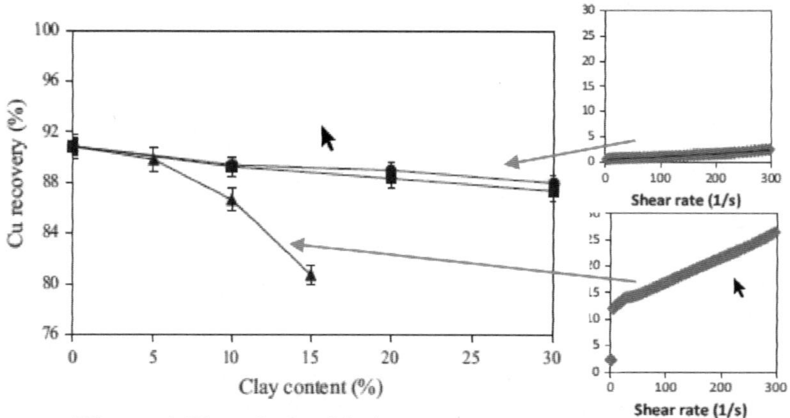

Figure 4. The relationship between copper recovery and ore slurry yield stress (by adding clays)

IOP Conf. Series: Journal of Physics: Conf. Series **1045** (2018) 012010 doi:10.1088/1742-6596/1045/1/012010

2.2 Rheology in froth phase

As showed in Figure 1, flotation contains two phases of slurry (pulp), and froth. Froth has three phases of water, air, and solid. Therefore, it bursts and collapses if left for any length of time as water and solids drain from the bubble surfaces. Therefore, it was not possible to accurately measure the froth rheology until Li et al. introduced a reliable method to do that [21].

Rheology has been mentioned as an important froth property. In flotation plants, experienced operators use their fingers to test if the froth is viscous to adjust the operating variables [22]. Until recently, avery limited number of studies were available on froth rheology [23, 24]. These studies were performed in either batch flotation cells or in industrial cells using complex ore mixtures. The results showed that rheological measurements are strongly correlated with certain froth properties, such as solid or water content and froth grade. However, due to the complexity of the applied system, they were not able to identify the mechanisms that govern the rheological behaviour of froth. A correlation between froth rheology and the grade of hydrophobic and hydrophilic minerals (chalcopyrite and quartz, respectively) is shown in Figure 5 [22]. Shi and Zheng [22] used froth 'torque' as a proxy for shear stress. The torque values can be converted to shear stress values using equation provided by Nguyen and Boger [25]. A correlation has also been observed between froth viscosity and mineral recovery (and grade) in phosphate flotation [26]. However, the reason for the froth rheology change was not yet clear [27]. A more viscous pulp may result in a more stable froth [28]. Li et al. has exclusively investigated the froth rheology and its effect on flotation performance in recent years [21, 29-31].They have observed correlation between froth viscosity and mineral recovery in froth flotation process.

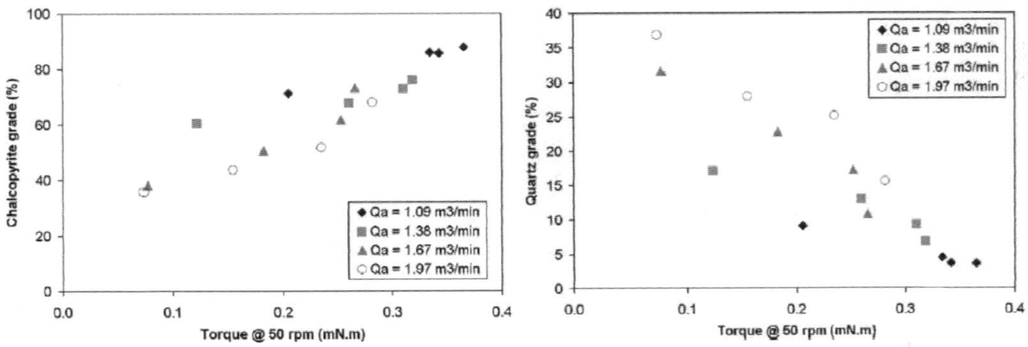

Figure 5. Froth rheology as a function of froth flotation concentrate grade for (left) chalcopyrite and (right) quartz [22]

3. Conclusion

The relationship between rheology measurements, and froth flotation performance was discussed. It was shown that the mineral flotation recovery and/or grade can be improved by controlling the rheology in flotation process. Flotation consists of two separate phases of pulp (slurry) and froth, and it has been revealed that rheology of both phases is related to the overall flotation performance. However, it should be emphasized that one should choose the measurement method carefully. The particles in the pulp phase intend to settle (due to gravity), so rheology measurements should be done in a way to overcome this problem. Froth is also a three phase system (air, water and solid). Therefore, it bursts and collapses if left for any length of time during the rheology measurements. Reliable methods to measure the rheology of both phases are now available. Therefore, it is suggested to use the rheological measurements as a controlling tool in flotation process development.

4. References

[1] Rubio-Hernández, F J, Ayúcar-Rubio, M F, Velázquez-Navarro, J F, Galindo-Rosales, F J, 2006 *J. Colloid Interface Sci.* **298** 967

[2] Farrokhpay, S, Morris, G E, Fornasiero, D, Self, P, 2010 *Powder Technol.* **202** 143
[3] Ndlovu, B, Farrokhpay, S, Bradshaw, D, 2013 *Int. J. Miner. Process.* **125** 149
[4] Zhou, Z A, Worku, O, Wang, L X, Xu, Z, Masliyah, J H, 2012 *CIM Journal.* **3** 54
[5] Dincer, H, Boylu, F, Sirkeci, A A, Atesok, G, 2003 *Int. J. Miner. Process.* **70** 41
[6] Farrokhpay, S, Ndlovu, B, Bradshaw, D, 2016 *Miner. Eng.* **96-97** 59
[7] Patra, P, Nagaraj, D R, Somasundaran, P. *The XI International Seminar on Mineral Processing Technology.* 2010: NML Jamshedpur, India
[8] Akroyd, T J, Nguyen, Q D, 2003 *Exp. Therm Fluid Sci.* **27** 507
[9] Kawatra, S K, Bakshi, A K, 1996 *Int. J. Miner. Process* **.47** 275
[10] Gustafsson, J, Toivakka, M, Koskinen, K K, 2005 *Ann. T. Nord. Rheol. Soc.* **13** 277
[11] Shi, F N, Napier-Munn, T J, 1996 *Int. J. Miner. Process.* **47** 153
[12] Boger, D V, 2000 *Miner. Process. Extr. Metall. Rev.* **20** 1
[13] Farrokhpay, S, 2012 *Miner. Eng.* **36–38** 272
[14] Nguyen, Q D, Boger, D V, 1985 *J. Rheol.* **29** 335
[15] Nguyen, Q D, Boger, D V, 1992 *Annu. Rev. Fluid. Mech.* **24** 47
[16] Genc, A M, Kilickaplan, I, Laskowski, J S. *8th UBC-McGill-UA International Symposium on the Fundamentals of Mineral Processing: Rheology and Processing of Fine Particles.* 2010: CIM, Vancouver
[17] Patra, P, Bhambhani, T, Nagaraj, D R, Somasundaran, P. *8th UBC-McGill-UA International Symposium on the Fundamentals of Mineral Processing: Rheology and Processing of Fine Particles.* 2010: CIM, Vancouver
[18] Farrokhpay, S, Ametov, I, Grano, S, 2011 *Adv. Powder Technol.* **22** 464
[19] Farrokhpay, S, Bradshaw, D, Dunne, R. *World Gold.* 2013 Brisbane: AusIMM
[20] Farrokhpay, S, Ndlovu, B, Bradshaw, D, 2016 *Miner. Eng.* **96-97** 59
[21] Li, C, Farrokhpay, S, Shi, F, Runge, K, 2015 *Miner. Eng.* **71** 89
[22] Shi, F N, Zheng, X F, 2003 *Int. J. Miner. Process.* **69** 115
[23] Moudgil, B M, 1993 *Miner. Metall. Proc.* **10** 100
[24] Rubinstein, J B, Melik-Gaikazyan, V I, *Characterisation of flotation froths*, in *Frothing in Flotation II*, J.S. Laskowski and E.T. Woodburn, Editors. 1998 *Gordon & Breach.* 145
[25] Nguyen, Q D, Boger, D V, 1983 *J. Rheol.* **27** 321
[26] Moolman, D W, Eksteen, J J, Aldrich, C, van Deventer, J S J, 1996 *Int. J. Miner. Process.* **48** 135
[27] Farrokhpay, S, 2011 *Adv. Colloid Interface Sci.* **166** 1
[28] Farrokhpay, S, Zanin, M, 2012 *Adv. Powder Technol.* **23** 493
[29] Li, C, Runge, K, Shi, F, Farrokhpay, S, 2016 *Powder Technol.* **29** 55
[30] Li, C, Runge, K, Shi, F, Farrokhpay, S, 2018 *Miner. Eng.* **115** 4
[31] Li, C, Runge, K, Shi, F, Farrokhpay, S, 2016 *Powder Technol.* **287** 216

IOP Conf. Series: Journal of Physics: Conf. Series **1045** (2018) 012011 doi:10.1088/1742-6596/1045/1/012011

Influence of compacting times and pressures on rheological properties of alumina and quartz ceramic powder mixtures

László A. Gömze[1, 2, 3] and Emese Kurovics[1]

[1]Intsitute of Ceramics and Polymer Engineering, University of Miskolc (Hungary)
[2] IGREX Engineering Service Ltd, Igrici (Hungary)
[3]Tomsk State University, Tomsk (Russia)

E-mail: femgomze@uni-miskolc.hu and mse205@gmail.com

Abstract. In this research the compacting and rheological properties of alumina and quartz powder mixtures were examined depending on forming pressures and times. On the basis of the achieved relative compaction curves as function of forming pressure and time the authors could successfully describe the changes of rheological behavior and rheological model of these powder mixtures. With increasing the compression pressure of the alumina and quartz powder mixtures considerable changes of rheological properties could be observed.
In this work the authors have shown how the compacting deformations of cylindrical specimens are depending on compaction pressures and times. On the basis of their research the authors could successfully illustrate also how the rheological behaviors and models of alumina-quartz ceramic powder mixtures are changing in the compacting die cavities at low values of forming pressures.

1. Introduction

In our days the ceramic materials and ceramic matrix composites [1-25] are used in very wide range of industry and human life thanking to their excellent physical, mechanical, chemical and biological properties. Because of the variety of their chemical, physical and material compositions, several techniques and technologies are applied to develop ceramics and ceramic based composites, including sol-gel technology, spark plasma sintering, spray drying, injection molding, hot pressing and many other technological processes [26-49]. In spite of the wide variety of the production technologies generally the ceramic materials and ceramic based composites have hard crystal grains with high values of elastic modulus depending on their material compositions and structures [17, 50 and 51].

Today application of rheology is playing very important role both in materials science [52-72] and in technology processes [73-81] especially in forming technology. To prepare products from alumina and quartz ceramic powder mixtures there are several forming methods and technologies can be used including slip casting [82], injection molding [83] and pressing [45, 48, 84-87].

2. Materials and experiments

For the experiment were used alumina powder Nabalox 315 and silica powder from Fehérvárcsurgó. Before compacting the specimens the alumina powders were mixed with quartz powders and milled together in a laboratory planetary ball mill Retch PM 400 trough 20 minutes of each at 200 rotation/minute. The microstructure and size distribution of the used powders are shown in Figure 1 and Figure 2. The so prepared powder mixtures were compacted through one-sided pressing to rods in a cylindrical die cavity with 10mm diameter and filling depth of 30mm on uniaxial hydraulic press.

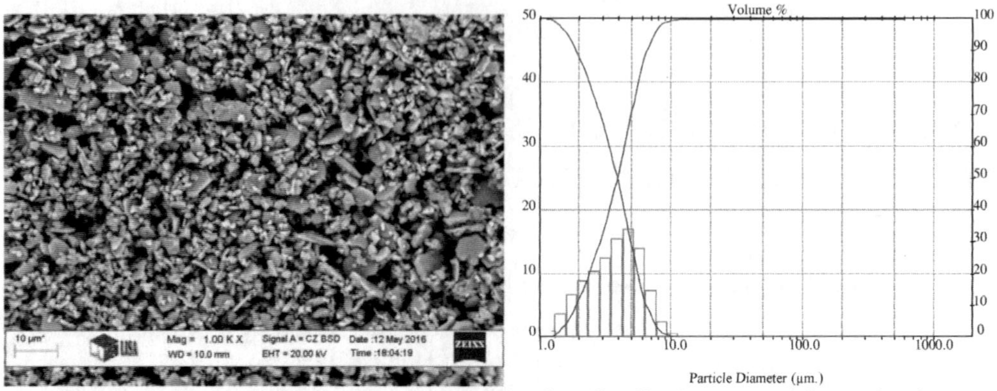

Figure 1. Surface micrograph and particle size distribution curve of the alumina

Figure 2. Surface micrograph and particle size distribution curve of the quartz

During the experiments the compactions (displacements of pressing punch) were recorded by a computer as a function of compression pressures. In first stage the compaction deformations were studied at low compacting pressures and further the so pre-compacted powders in the die cavity were final compacted at 30bars, 60bars, 90bars, 120bars and 150bars in the hydraulic cylinder at final pressure values as shown in Table 1.

Table 1. Final pressure values of compacting

Pressure in the hydraulic cylinder	Pressure forces	Compacting pressures
bar	N	MPa
30	5966	76
60	11932	152
90	17898	228
120	23864	304
150	29830	380

IOP Conf. Series: Journal of Physics: Conf. Series **1045** (2018) 012011 doi:10.1088/1742-6596/1045/1/012011

3. Results and discussions

In first the compactions were studied at low compacting pressures and a typical such curve is shown in Figure 3. The compression chart shows the deformation and rheological behavior of the mixed alumina quartz powders under small values of forming pressures. In the first stage under a low compaction pressure the deformation is growing very quickly and can be regarded as almost ideally viscous material and characterized by Newton's model. In the second stage the evolving stress causes the grains to roll on each other consequently filling the particle gaps, there by deformation occurs and the material can be modeled as row-linked viscous-elastic body. In the third stage the particles are continuing to fill the voids, meanwhile the compaction pressure is growing more quickly. In this stage the compaction is non-linear and can be modeled as parallel-linked viscous-elastic body, due to the frictions between the particles during their movement in the die cavity.

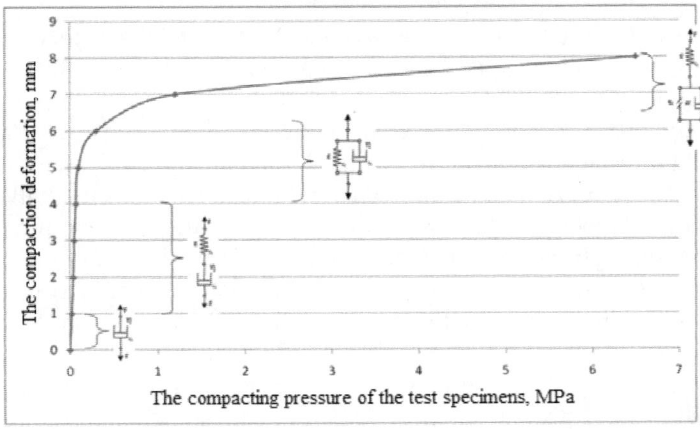

Figure 3. The compression chart with the determined rheological behavior

Finally in the fourth stage the particles of the ceramic powder are getting closer and closer to each other causing a significant increase in pressure meanwhile the compaction is relatively small. In this stage the ceramic powder mix behaves as a "quasi-plastic" material and can be modeled as it is shown above in Figure 3.

After the study the compression properties of alumina-quartz mix powders at low pressure the pre-compacted materials were final compacted without moving out from the die cavity at compacting pressures from 76MPa to 380MPa as it is shown in Table 1. The compaction deformation at different final compaction pressures are shown in Figure 4. At the first test when the final compression pressure was 76MPa the measurement has started before the pressing punch would have achieved in the die cavity the surface of the pre-compacted powder. Because of this the measured "compaction deformation" was higher than in reality.

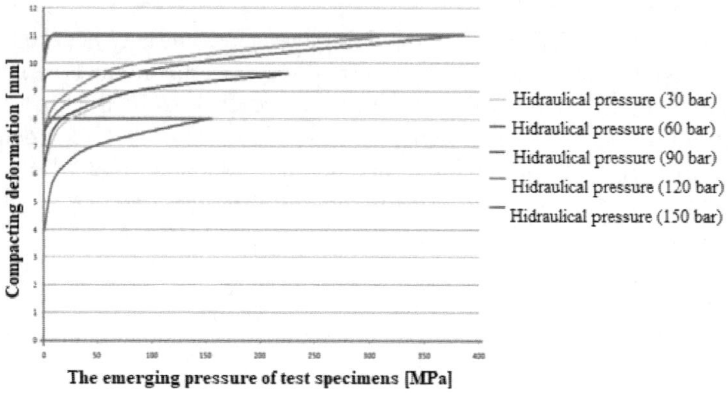

Figure 4. The compaction deformations of cylindrical specimens are depended on forming pressure

It is obvious in the Figure 4 that increasing the compaction pressures from 152MPa up to 304MPa the compaction deformation of pre-compacted rod specimens have increased from 8mm up to 11mm. The further increase of compression pressure up to 380MPa didn't influence on compaction deformation. These curves show that it is impossible to develop the density of the used ceramic powder mixtures in the cylindrical die cavity through further increment of compacting pressure.

In the next experiments (Figure 5) the pre-compacted ceramic rods were further compacted without moving out from the die cavity at compacting pressures from 76MPa to 380MPa, but the pressing punch was kept at the final compression pressure by 10-15sec. It is very well seen in Figure 5 that at the achieved compaction deformations the compacting pressures have decreased with increasing the time thanking to the mechanical stress relaxation inside of the powder mixtures in the die cavity. This phenomena informs us also that the particles of the powder mix have continued their movements and orientations in the die cavity without changing their total geometrical volume.

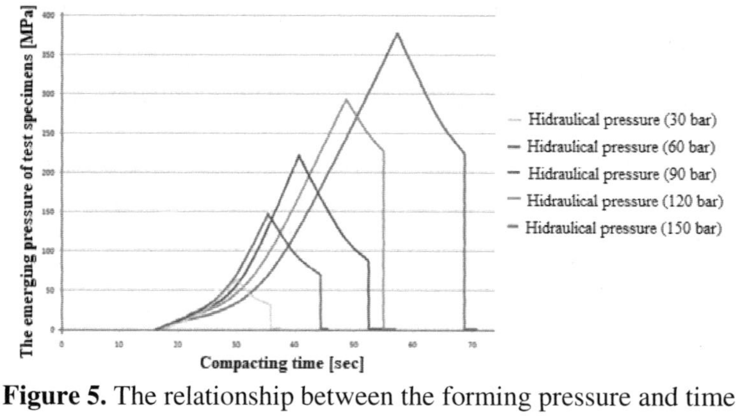

Figure 5. The relationship between the forming pressure and time

4. Conclusions

From the realized by authors experiments it is obvious that at low compaction pressures in the cylindrical die cavity the ceramic mixtures can be characterized with different rheological models depending on forming pressure.

The achieved results of the realized by authors examination show that it is impossible to develop the density of the used ceramic powder mixtures in the cylindrical die cavity through further increment of compacting pressure higher than 304MPa.

During uniaxial compaction of ceramic powders the mechanical stress relaxation is starting already inside of the die cavity.

There are further experiments are required to find the rheological model of ceramic powder mixtures in the cylindrical die cavities at final (high) compaction pressures.

Acknowledgement
The authors would like to acknowledge to IGREX Engineering Service Ltd for their support and help during this research work.
The described article was carried out as part of the EFOP-3.6.1-16-2016-00011 "Younger and Renewing University – Innovative Knowledge City – institutional development of the University of Miskolc aiming at intelligent specialization" project implemented in the framework of the Szechenyi 2020 program. The realization of this project is supported by the European Union, co-financed by the European Social Fund.

References
[1] Kuznyecova T V, Ljutyikova T A, Szamcsenko Sz V, Vorobjov A I 2005 *Építőanyag-JSBCM* **57** (1) 16 http://dx.doi.org/10.14382/epitoanyag-jsbcm.2005.4
[2] Boymanns G, Gebhhardt F, Libal L G, Schlacht D 2006 *Építőanyag-JSBCM* **58** (1) 19 http://dx.doi.org/10.14382/epitoanyag-jsbcm.2006.5
[3] Rádóczy T, Kovács K 2006 *Építőanyag-JSBCM* **58** (2) 34 http://dx.doi.org/10.14382/epitoanyag-jsbcm.2006.8
[4] Laczkó L, Eniszné Bódogh M 2007 *Építőanyag-JSBCM* **59** (1) 2 http://dx.doi.org/10.14382/epitoanyag-jsbcm.2007.1
[5] Varga G 2007 *Építőanyag-JSBCM* **59** (1) 6 http://dx.doi.org/10.14382/epitoanyag-jsbcm.2007.2
[6] Szilágyi K, Gömze A L, Polacsek G 2008 *Építőanyag-JSBCM* **60** (2) 44 http://dx.doi.org/10.14382/epitoanyag-jsbcm.2008.9
[7] Kulkov S N, Savchenko N 2008 *Építőanyag-JSBCM* **60** (3) 62 http://dx.doi.org/10.14382/epitoanyag-jsbcm.2008.10
[8] Morikawa A, Suzuki T, Kikuta K, Suda A, Shinjo H 2009 *Építőanyag-JSBCM* **61** (1) 2 http://dx.doi.org/10.14382/epitoanyag-jsbcm.2009.1
[9] Gömze A L, Gömze N L 2009 *Építőanyag-JSBCM* **61** (2) 38 http://dx.doi.org/10.14382/epitoanyag-jsbcm.2009.7
[10] Keita A-S, Naciri A En, Delachat F, Carrada M, Ferblantier G and Slaoui A 2009 *IOP Conf. Ser.: Mater. Sci. Eng.* **6** 012016 https://doi.org/10.1088/1757-899X/6/1/012016
[11] Ershova N I, Kelina I Yu 2009 *Építőanyag-JSBCM* **61** (2) 34 http://dx.doi.org/10.14382/epitoanyag-jsbcm.2009.6
[12] Fündling S, Li S, Postels B, Al-Suleiman M, Wehmann H H, Bakin A and Waag A 2009 IOP Conf. Ser.: Mater. Sci. Eng. 6 012003 https://doi.org/10.1088/1757-899X/6/1/012003
[13] Katahira K, Ohmori H 2009 *Építőanyag-JSBCM* **61** (4) 114 http://dx.doi.org/10.14382/epitoanyag-jsbcm.2009.21
[14] Kuribayashi K, Ohmiya D 2010 *Építőanyag-JSBCM* **62** (1) 11 http://dx.doi.org/10.14382/epitoanyag-jsbcm.2010.3
[15] Uchida T, Nakamichi S, Sunaoshi T, Morimoto K, Kato M and Kihara Y 2010 *IOP Conf. Ser.: Mater. Sci. Eng.* **9** 012004 https://doi.org/10.1088/1757-899X/9/1/012004
[16] Kulkov S N, Grigoriev M V 2010 *Építőanyag-JSBCM* **62** (3) 66 http://dx.doi.org/10.14382/epitoanyag-jsbcm.2010.13
[17] Gömze A L and Gömze N L 2011 IOP Conf. Ser.: Mater. Sci. Eng. 18 082001 https://doi.org/10.1088/1757-899X/18/8/082001
[18] Naslain R 2011 *IOP Conf. Ser.: Mater. Sci. Eng.* **18** 082002 https://doi.org/10.1088/1757-899X/18/8/082002

[19] Zhu Q, Hu S, Jin W, Wen H, Chen W and Zakharova G 2011 *IOP Conf. Ser.: Mater. Sci. Eng.* **18** 082013 https://doi.org/10.1088/1757-899X/18/8/082013
[20] Zharikov E V, Zaramenskikh K S, Popova N A, Faikov P P, Iskhakova L D, Gerke M N, Kutrovskaya S V and Nogtev D S 2011 *Glass and Ceramics* **68** 80
[21] Štubna I, Podoba R, Bačík P, Podobník L 2013 *Építőanyag-JSBCM* **65** (2) 48 http://dx.doi.org/10.14382/epitoanyag-jsbcm.2013.11
[22] Lisachuk G V, Kryvobok R V, Fedorenko E Y, Zakharov A V 2015 *Építőanyag-JSBCM* **67** (2) 20 http://dx.doi.org/10.14382/epitoanyag-jsbcm.2015.4
[23] Kandeel A M 2015 *Építőanyag-JSBCM* **67** (3) 98 http://dx.doi.org/10.14382/epitoanyag-jsbcm.2015.16
[24] Ayub N, Rafique U 2017 *Építőanyag-JSBCM* **69** (3) 98 http://dx.doi.org/10.14382/epitoanyag-jsbcm.2017.17
[25] Sadowski T, Craciun E, Marsavina L 2018 *AIP Conference Proceedings* **1932** 030036 http://dx.doi.org/10.1063/1.5024186
[26] Hegman N, Szűcs P, Lakatos J 2003 *Építőanyag-JSBCM* **55** (3) 101 http://dx.doi.org/10.14382/epitoanyag-jsbcm.2003.18
[27] Veinot J G C, Henderson E J and Hessel C M 2009 IOP Conf. Ser.: Mater. Sci. Eng. 6 012017 https://doi.org/10.1088/1757-899X/6/1/012017
[28] Kaygorodov A, Khrustov V 2010 *Építőanyag-JSBCM* **62** (4) 116 http://dx.doi.org/10.14382/epitoanyag-jsbcm.2010.21
[29] Zhu X H, Rhun G Le, Defaÿ E and Aïd M 2010 *IOP Conf. Ser.: Mater. Sci. Eng.* **8** 012002 https://doi.org/10.1088/1757-899X/8/1/012002
[30] Çopuroğlu M, O'Brien S, Malič B, Kužnik B, Kosec M, Zhu X, Defaÿ E and Winfield R 2010 *IOP Conf. Ser.: Mater. Sci. Eng.* **8** 012007 https://doi.org/10.1088/1757-899X/8/1/012007
[31] Lee C G, Nakamura Y, Nakamura H, Uehara M, Maeda H 2011 *Építőanyag-JSBCM* **63** (3-4) 52 http://dx.doi.org/10.14382/epitoanyag-jsbcm.2010.9
[32] Reschke V, Laskowsky A, Kappa M, Wang K, Bordia R K, Scheffler M 2011 *Építőanyag-JSBCM* **63** (3-4) 57 http://dx.doi.org/10.14382/epitoanyag-jsbcm.2010.10
[33] Ruzimuradov O N 2011 *IOP Conf. Ser.: Mater. Sci. Eng.* **18** 032004 https://doi.org/10.1088/1757-899X/18/3/032004
[34] Devaux X, Tsareva S Yu, Kovalenko A N, Zaramenskih K S, McRae E and Zharikov E V 2012 *Physica E* **44** 1028
[35] Borosnyói A 2013 *Építőanyag-JSBCM* **65** (1) 26 http://dx.doi.org/10.14382/epitoanyag-jsbcm.2013.6
[36] Egész Á, Buyakova S P, Kulkov S N, Gömze L A 2014 *Építőanyag-JSBCM* **66** (2) 52 http://dx.doi.org/10.14382/epitoanyag -jsbcm.2014.11
[37] Rodríguez Torres G M, Zarate Medina J, Contreras García M E 2016 *Építőanyag-JSBCM* **68** (4) 120 http://dx.doi.org/10.14382/epitoanyag-jsbcm.2016.21
[38] Zharikov E V, Duong T T T, Faikov P P, Popova N A and Sovyk D M 2016 *Inorganic Materials: Applied Research* **7** 163
[39] Torres-Romero A, Cajero-Juárez M, Contreras-García M E 2017 *Építőanyag-JSBCM* **69** (1) 8 http://dx.doi.org/10.14382/epitoanyag-jsbcm.2017.2
[40] Zharikov E V, Kapustin V V, Faikov P P, Popova N A, Barmin A A, Ivanov A V and Rizakhanov R N 2017 *IOP Conf. Ser.: Mater. Sci. Eng.* **175** 012065 https://doi.org/10.1088/1757-899X/175/1/012065
[41] Wang Y, Shabalin I L, Zhang L and Zhdanov V B 2011 *IOP Conf. Ser.: Mater. Sci. Eng.* **18** 082009 https://doi.org/10.1088/1757-899X/18/8/082009
[42] Liu J, Fu Z, Wang W, Zhang J, Wang H, Wang Y, Lee S and Niihara K 2014 *J. Eur. Ceram. Soc.* **34** 3095 http://dx.doi.org/10.1016/j.jeurceramsoc.2014.04.004
[43] Brodnikovska I V, Deriy A I, Petrovsky V Ya 2014 *Építőanyag-JSBCM* **66** (3) 58 http://dx.doi.org/10.14382/epitoanyag-jsbcm.2014.12

[44] Cho H B, Nakayama T, Suematsu H, Suzuki T, Jiang W, Niihara K, Song E, Eom N S A, Kim S and Choa Y H 2016 *Composites Science and Technology* **129** 205 http://dx.doi.org/10.1016/j.compscitech.2016.04.033
[45] Shinoda S, Aoba T, Suzuki T, Nakayama T, Suematsu H and Niihara K 2017 *Jpn. J. Appl. Phys.***56** 019201 https://doi.org/10.7567/JJAP.56.019201
[46] Isaev A, Grechishnikov V, Pivkin P, Ilyukhin Y, Kozochkin M, Peretyagin P 2016 *Építőanyag-JSBCM* **68** (2) 46 http://dx.doi.org/10.14382/epitoanyag-jsbcm.2016.8
[47] Buyakov S P, Kalatur E S, Buyakov A S, Kulkov S N 2016 *Építőanyag-JSBCM* **68** (3) 70 http://dx.doi.org/10.14382/epitoanyag-jsbcm.2016.12
[48] Zhukov I, Buyakov S P, Kulkov S N 2016 *Építőanyag-JSBCM* **68** (3) 74 http://dx.doi.org/10.14382/epitoanyag-jsbcm.2016.13
[49] Kinelovskii S A, Maevskii K K 2017 *Építőanyag-JSBCM* **69** (1) 29 http://dx.doi.org/10.14382/epitoanyag-jsbcm.2017.6
[50] Gömze A L and Gömze N L 2010 *Építőanyag-JSBCM* **62** (4) 98 http://dx.doi.org/10.14382/epitoanyag-jsbcm.2010.18
[51] Gömze A L and Gömze N L 2017 *IOP Conf. Ser.: Mater. Sci. Eng.* **175** 012001 https://doi.org/10.1088/1757-899X/175/1/012001
[52] Gömze A L 1983 *Építőanyag* **35** (5) 173
[53] Sadowski T and Golewski P 2012 *Comp. Mater. Sci.* **64** 285 https://doi.org/10.1016/j.commatsci.2012.02.048
[54] Sadowski T, Kneé M and Golewski P 2014 *Key Engineering Materials* **601** 25 https://doi.org/10.4028/www.scientific.net/KEM.601.25
[55] Burlayenko V and Sadowski T 2014 *Meccanica* **49** 2617 https://doi.org/10.1007/s11012-014-9924-y
[56] Dusenkova I, Malers J and Berzina-Cimdina L 2015 *J. Phys.: Conf. Ser.* **602** 012003 https://doi.org/10.1088/1742-6596/602/1/012003
[57] Khramchenkov E and Khramchenkov M 2015 *J. Phys.: Conf. Ser.* **602** 012005 https://doi.org/10.1088/1742-6596/602/1/012005
[58] Umerova S O, Dulina I O, Ragulya A V 2015 *Építőanyag-JSBCM* **67** (4) 119 http://dx.doi.org/10.14382/epitoanyag-jsbcm.2015.19
[59] Bansil R, Hardcastle J, Constantino M 2015 *Építőanyag-JSBCM* **67** (4) 150 http://dx.doi.org/10.14382/epitoanyag-jsbcm.2015.25
[60] Kulkov S N, Buyakova S P, Chatzinikolaidou M, Kocserha I 2015 *Építőanyag-JSBCM* **67** (4) 155 http://dx.doi.org/10.14382/epitoanyag-jsbcm.2015.26
[61] Sperberga I, Rundans M, Cimmers A, Krage L and Sidraba I 2015 *J. Phys.: Conf. Ser.* **602** 012007 https://doi.org/10.1088/1742-6596/602/1/012007
[62] Klovas A and Daukšys M 2015 *J. Phys.: Conf. Ser.* **602** 012009 https://doi.org/10.1088/1742-6596/602/1/012009
[63] Nagy R, Elekes A, Bartha L, Vágó Á 2016 *Építőanyag-JSBCM* **68** (4) 98 http://dx.doi.org/10.14382/epitoanyag-jsbcm.2016.17
[64] Vereschaka A, Kutin A, Sitnikov N, Oganyan G, Sharipov O 2016 *Építőanyag-JSBCM* **68** (4) 114 http://dx.doi.org/10.14382/epitoanyag-jsbcm.2016.20
[65] Khramchenkov M, Khramchenkov E 2016 *Építőanyag-JSBCM* **68** (2) 34 http://dx.doi.org/10.14382/epitoanyag-jsbcm.2016.6
[66] Khramchenkov M G, Usmanov R M 2017 *Építőanyag-JSBCM* **69** (4) 110 http://dx.doi.org/10.14382/epitoanyag-jsbcm.2017.19
[67] Kracalik M 2017 *Építőanyag-JSBCM* **69** (4) 116 http://dx.doi.org/10.14382/epitoanyag-jsbcm.2017.21
[68] Mastalska-Popiawska J and Izak P 2017 *J. Phys.: Conf. Ser.* **790** 012021 https://doi.org/10.1088/1742-6596/790/1/012021

[69] Reyes-Ocampo I, Aguayo-Vallejo J P, Ascanio G and Córdova-Aguilar M S 2017 *J. Phys.: Conf. Ser.* **790** 012028 https://doi.org/10.1088/1742-6596/790/1/012028
[70] Palaz S, Unver H, Ugur G, Mamedov A M and Ozbay E 2017 *IOP Conf. Ser.: Mater. Sci. Eng.* **175** 012014 https://doi.org/10.1088/1757-899X/175/1/012014
[71] Zsoldos G, Szoda K and Marossy K 2017 *IOP Conf. Ser.: Mater. Sci. Eng.* **175** 012028 https://doi.org/10.1088/1757-899X/175/1/012028
[72] Kinelovskii S A and Maevskii K K 2017 *IOP Conf. Ser.: Mater. Sci. Eng.* **175** 012030 https://doi.org/10.1088/1757-899X/175/1/012030
[73] Gonzalez-Gutierrez J, Stringari G B, Megen Z M, Oblak P, von Bernstorff B S and Emri I 2015 *J. Phys.: Conf. Ser.* **602** 012001 https://doi.org/10.1088/1742-6596/602/1/012001
[74] Tóth C, Soós Z 2015 *Építőanyag-JSBCM* **67** (4) 126 http://dx.doi.org/10.14382/epitoanyag-jsbcm.2015.20
[75] Wang Y, Lakho D A, Yao D 2015 *Építőanyag-JSBCM* **67** (4) 132 http://dx.doi.org/10.14382/epitoanyag-jsbcm.2015.21
[76] Kudaikulova G 2015 *J. Phys.: Conf. Ser.* **602** 012008 https://doi.org/10.1088/1742-6596/602/1/012008
[77] Maksarov V, Khalimonenko A, Olt J 2017 *Építőanyag-JSBCM* **69** (2) 66 http://dx.doi.org/10.14382/epitoanyag-jsbcm.2017.12
[78] Laufer N, Hansmann H and Koch M 2017 *J. Phys.: Conf. Ser.* **790** 012017 https://doi.org/10.1088/1742-6596/790/1/012017
[79] Noirez L and Baroni P 2017 *J. Phys.: Conf. Ser.* **790** 012024 https://doi.org/10.1088/1742-6596/790/1/012024
[80] Nagasawa Sh 2017 *IOP Conf. Ser.: Mater. Sci. Eng.* **175** 012003 https://doi.org/10.1088/1757-899X/175/1/012003
[81] Adamne Major A 2017 *IOP Conf. Ser.: Mater. Sci. Eng.* **175** 012027 https://doi.org/10.1088/1757-899X/175/1/012027
[82] Gömze A L 2015 Rheology Compilation of Scientific Papers I. Published by IGREX Ltd. ISBN 978-963-12-3088-8
[83] Egész Á, Gömze A L 2013 *Építőanyag-JSBCM* **65** (4) 107 http://dx.doi.org/10.14382/epitoanyag-jsbcm.2013.20
[84] Csányi J, Gömze A L, Kövér Zs I 2004 *Építőanyag-JSBCM* **56** (3) 103 http://dx.doi.org/10.14382/epitoanyag-jsbcm.2004.13
[85] Kurovics E, Buzimov A Y and Gömze A L 2016 *IOP Conf. Ser.: Mater. Sci. Eng.* **123** 012058 https://doi.org/10.1088/1757-899X/123/1/012058
[86] Kurovics E, Gömze A L 2017 Influence of sintering atmosphere and IG-017 bio-original additives on porosity of ceramics made from high purity Al_2O_3 and SiO_2 powders, *MultiScience - XXXI. microCAD International Multidisciplinary Scientific Conference* University of Miskolc https://www.researchgate.net/publication/316275265
[87] Kurovics E, Shmakova A, Kanev B and Gömze A L 2017 *IOP Conf. Ser.: Mater. Sci. Eng.* **175** 012013 https://doi.org/10.1088/1757-899X/175/1/012013

IOP Conf. Series: Journal of Physics: Conf. Series **1045** (2018) 012012 doi:10.1088/1742-6596/1045/1/012012

Kinetics of protopect in decomposition in vegetable raw material under high temperatures and pressures

R Gorshkova[1,2], D Khalikov[3], D Slobodova[2], A Slobodov[2]

[1]Institute of Macromolecular Compounds Russian Academy of Science, St. Petersburg, Russia
[2]ITMO University, St. Petersburg, Russia
[3]V. I. Nikitin Institute of Chemistry, Academy of Sciences of the Republic of Tajikistan, Dushanbe, Tajikistan

Email:gorshkova.raisa@gmail.com

Abstract. The kinetics of the catalytic decomposition of protopectin (PP) of pomace, peach, pumpkin, citrus, rhubarb, sunflower heads (SH) and beet pulp in HCl solution at pH = 2.0 and T = 120 °C and pressure 1.5 atm. The decay products of PP are separated according to the previously developed method into fractions: microgel (MG), pectin substances (PS) and oligosaccharides (OS). The experimental data were processed on the basis of the idea of a parallel and successive irreversible chemical reaction of the first order, the rate constants for the decay of the PP into the corresponding components-MG, PS, and OS, were evaluated, conclusions were made about the advantages and disadvantages of the high-temperature process.

1. Formulation of the problem

The effect of external parameters on the kinetics of chemical reactions in the liquid and solid states is mainly due to the influence of neighboring molecules on the change in the shape and dimensions of the activated complex, which at the macroscopic level leads to the dependence of the reaction rate constant (k) on the pressure and the aggregate state of the substance [1]. We have demonstrated in a number of works, on the example of the catalytic decomposition of protopectin (PP) of various sources, a dramatic increase in the rate of the hydrolysis reaction with increasing pressure and temperature, and, most importantly, simultaneous improvement in the quality of the target product, expressed in an increase in the content of galacturonic acid (GA) residues in pectin macromolecules and an increase in their molecular weight [2-6].

The complexity of raw materials composition and its interactions with reagents significantly complicates the problem about influence of the state parameters (temperature, pressure, composition) on physico-chemical properties of products. The available experimental and computational-model studies [7-13] don't solve this problem - due to the insufficient complexity of the studied systems, incompleteness of the required physico-chemical parameters, etc.

In connection with this, the purpose of this work was to study the decay of PP of different plant raw materials under the influence of high temperature and pressure, and also to estimate the kinetic parameters of the corresponding process.

IOP Conf. Series: Journal of Physics: Conf. Series **1045** (2018) 012012 doi:10.1088/1742-6596/1045/1/012012

2. Methodology

Inthis paper, squeezes of apples, pumpkins, stems and leaves of rhubarb, peach, beet pulp, mandarin crusts and sunflower heads (SH) were used as raw materials. Hydrolysis-extraction of PP was carried out with steam at a temperature of 120 ° C, pH = 2.0, a hydro module of 1:20, varying the pressure from 1.5 to 3.0 atm. and the duration of the process is from 3 to 10 minutes [2-6]. Hydrolyzate solutions were separated from the cellulose (CL), and the extract obtained was divided into three fractions conventionally named: microgel (MG), pectin substances (PS) and oligosaccharides (OS) [14, 15].

The content of the elements of galacturonic acid (GA) in pectin substances and the degree of esterification of carboxyl groups of GA (DE) were determined by the methods described in [16, 17], respectively.

3. Results and Discussion

Figure 1 shows the kinetics of yield of MG for different types of plant raw materials. Attention is drawn to the fact that, at a high temperature (120 ° C) and a pressure of 1.5 atm, the extreme kinetics of the yield of MH, which is so characteristic of the hydrolysis-extraction process at atmospheric pressure and temperature of 85 ° C (low-temperature process) [14, 15] except for KP and pumpkin squeezes. For them, the regularity of the yield of MH remains similar in form to the low-temperature process, but decreases in magnitude of the yield. Another feature of high-temperature hydrolysis of PP is a sharp decrease in the duration of the process, which terminates in almost 10 minutes (three hours in the low-temperature process), and as the pressure and temperature increase, the process accelerates further.

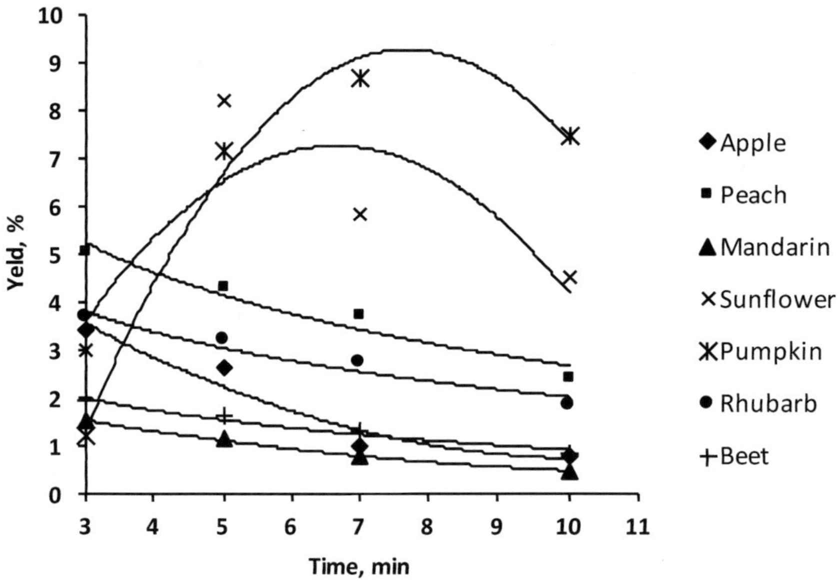

Figure 1. The kinetics of the yield of MG at T = 120 °C and P = 1.5 atm. for various raw materials. The output of the MG is calculated with respect to the initial mass of the object under study.

It is also important that the quality of the obtained PS, judging by the content of GA in the composition of the links of PS macromolecules, under chosen reaction conditions for all types of raw materials is sufficiently high [2-6].

The absence of a maximum on the graph of the kinetics of the MG reaction may be due not to the change in the mechanism of the sequential reaction characteristic of the low-temperature process at high temperature and pressure [14, 15], but with the acceleration of the reaction and the shift of the maximum to the initial region of the yield of MG. Proceeding from this, the data presented in Fig. 1 do not indicate a lack of a maximum in the kinetics of the yield of MG, but, on the contrary, confirm its presence, in the form of its right krill. This means that the rate constant of the successive decay of the MG into the PS and OS at a high temperature and pressure increases sharply, in comparison with the rate of decay of the PP by MG. In addition, the low yield of MH (in the 0-5% range, without taking into account KP and pumpkin) gives reason to consider the decay process of PP, not in the form of a sequential PP-MG-PV reaction, but in the form of a parallel decay reaction of PP to PS and OS.

Indeed, as can be seen in Fig. 2, the specific rate of decomposition of the protopectin of the studied plant materials on the MG (k (MG)) is much smaller than the corresponding decay parameter of the PP on the PS (k (PS)) and the OS (k (OS)). Among the studied objects, the greatest value of k (MH) is observed for a sunflower basket. This is natural, since, due to the high concentration of calcium ions in the composition of MG CL, the pectin chains are more crosslinked than other objects, and the subsequent disintegration of MG SH into the corresponding components becomes difficult.

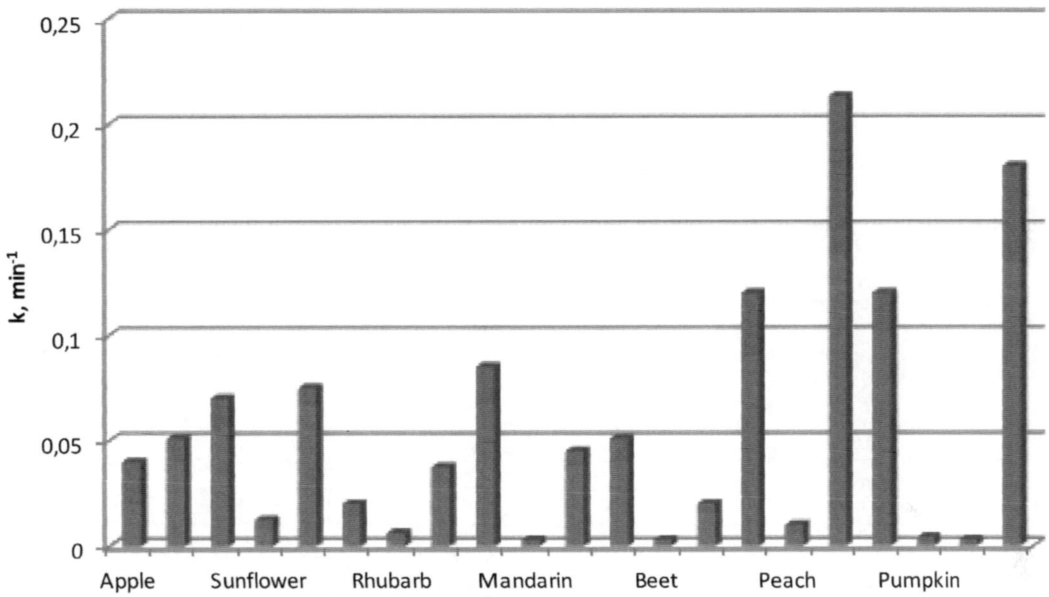

Figure 2. Kinetics of the decay of PP into fractions of MG, PS and OS for various types of plant raw materials. (The figure shows successively the value of k and the measurement error Δk, corresponding to the decay of the PP into MG, SH and OS).

Figure 2 also shows that the value of k in the decay of PP to PS (k (PS)) is not the same for different types of plant material. The greatest value of this parameter is achieved with the decay of the peach and PP, followed by pomace, citrus, rhubarb and beet. The value of k (PS) for a pumpkin is practically zero.

Finally, the corresponding objects of investigation differ significantly from each other in terms of the value of the specific rate of decay of the PP on the OS (k(OS)). The greatest value of k (OS) is observed for pumpkin, then, in descending order, beet, peach, citrus, rhubarb, pomace and CL.

Comparing the values of k (MG), k (PS) and k (OS) with each other, it can be concluded that under the conditions of high-temperature hydrolysis of various types of plant raw materials, the most optimal component ratio is achieved for KP (the maximum yield of PV with minimum MG and OS content) and the worst - in the decay of the pumpkin PP (the highest number of OS in the almost complete

IOP Conf. Series: Journal of Physics: Conf. Series **1045** (2018) 012012 doi:10.1088/1742-6596/1045/1/012012

absence of fractions of MG and PS).

Proceeding from this, the kinetic parameters of the decay of the PP SH with a change in pressure and a fixed temperature (120 °C) can be analyzed on the basis of the kinetics of the sequential PP-MG-PS reaction, in the manner described by us in [14]. The Table 1 shows the dependence of the constants k_1 and k_2 on the pressure.

Table 1. Dependence of the successive decay constants of PP on MG and PS

P, °K	k_1, мин$^{-1}$	k_2, мин$^{-1}$
1.5	0,0534	0,4967
2	0,0417	0,5385
3	0,0295	0,5530

The dependence of the rate of the chemical reaction on the pressure in the liquid state, which can be attributed to the particle swells swelling in the water, can be described on the basis of the Evans-Polanyi theory [16], within the framework of the free-volume model as a bulk-activated process. According to this idea, when the activated complex is formed, the initial volume of the molecule Vo varies by the activation volume ΔV </s>. For the reaction in the liquid state, in solutions [18]:

$$\frac{\partial(lnK)}{\partial p} = -\frac{\Delta V^{\neq}}{RT}, \qquad (1)$$

where p is the external pressure, R is the universal gas constant.

Based on the data in the table, we calculated the values of Lnk1 and its derivatives from the pressure (Fig. 3), and using the equation (1), the values of $\Delta V \neq$ at different pressures were calculated.

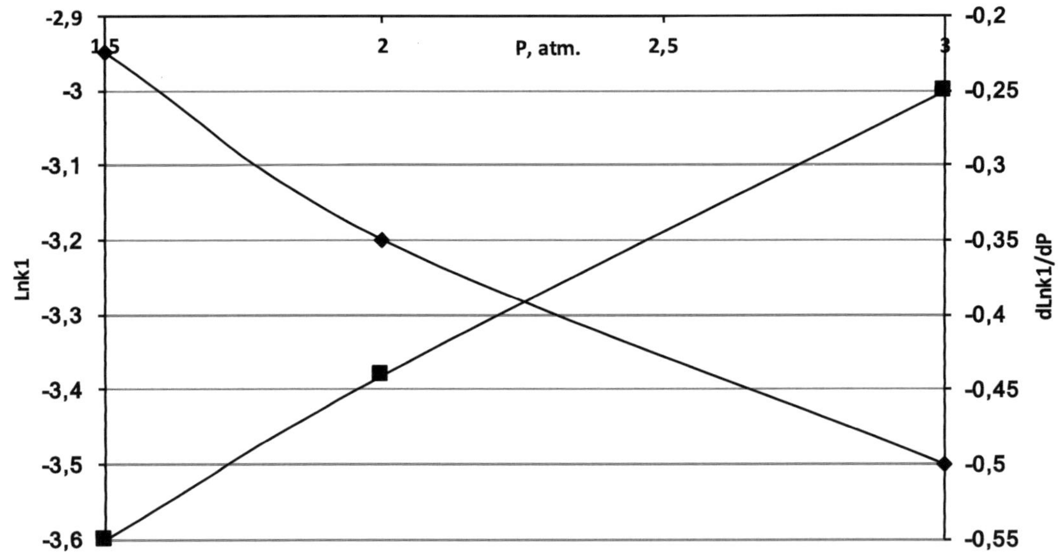

Figure 3. Dependence of Lnk1 and its first derivative on pressure at T = 120 ° C.

Figure 3 shows that the change in Lnk1 from pressure, contrary to the predictions of the Evans-Polanyi theory (equation (1)), is not linear, but is described by the second-order equation y = 0.1002x2 - 0.8475x - 1.8836, with a correlation coefficient equal to unit. The nonlinearity of this relationship is also found for numerous monomolecular reactions in the liquid and solid state, and the magnitude of the effect itself depends on the nature of the solvent, temperature and pressure [19, 20]. In this case, it is important that the pressure dependence of dLnk1/dP is linear and is described by the equation y =

0.2004x – 0.8475 with the correlation coefficient equal to unity (Fig. 3). This made it possible to calculate the amount of activation volume ($\Delta V \neq$) equal to -22.34, -19.40 and -13.51 cm^3/g-feed at a pressure of 1.5, 2.0 and 3.0 atm, respectively. The values obtained are high and, most likely, indicate the presence of a sufficiently large activation volume during the course of the decomposition reaction of PP in water vapor.

4. Conclusion

Thus, experimental data on protopectin decay of a plant cell at high temperature and pressure indicate a significant acceleration of the process of obtaining high-quality pectin substances in comparison with traditional methods and the possibility of optimizing the process to increase the yield of the target product for the selected plant material.

References

[1] Burov Yu M 2004 *Journal of Physical Chemistry* **78** (4) 682
[2] Gorshkova R.M, Nasriddinov A.S., Mukhidinov Z.K, Khalikov D.Kh., Khalikova S., Teshaev Kh.I. 2010 *News of Russian: "Chemistry and chemical technology" series* **53** (6) 87
[3] Gorshkova RM, Mukhidinov ZK, Makhkamov KhK, Johnmurodov AS, Khalikova S, Khalikov D Kh 2010 *Bulletin of the Tajik Technical University* **3** (11) 55
[4] Gorshkova RM, Mukhidinov ZK, Nasriddinov AS, Valiev MV, Khalikova S, Khalikov DKh, Liu LSh 2010 *News of the Academy of Sciences of the Republic of Tajikistan* **4** (141) 45
[5] Mukhidinov ZK, Gorshkova RM, Valiev MV, Khalikova S, Nasriddinov AS, Khalikov DKh, Fishman ML 2010 *Reports of the Academy of Sciences of the Republic of Tajikistan.* **53** (8) 617
[6] Mukhidinov Z.K, Gorshkova R.M, Valiev M.V, Khalikova S., Nasriddinov A.S, Khalikov D.Kh., Fishman M.L. 2010 *Reports of the Academy of Sciences of the Republic of Tajikistan.* **53** (9) 701
[7] Gömze L A, Lipusz D, Gömze L N 2013 *Materials Science Forum* **729** 350
[8] Savchenko N, Sevostyanova I, Sablina T, Gömze L A, Kulkov S 2014 *AIP Conference Proceedings* **1623** 547 http://dx.doi.org/10.1063/1.4899003
[9] Slobodov A, Uspenskiy Al, Yavshits S, Mischenko A 2015 *Epitoanyag – Journal of Silicate Based and Composite Materials* **67** (4) 164 http://dx.doi.org/10.14382/epitoanyag-jsbcm.2015.28
[10] Ralys R V, Uspenskiy A A, Slobodov A A 2016 *J. Non-Equilibrium Thermodynamics* **41** (1) 3
[11] Uspensky A A, Yavshits S, Lipin V, Zhigalov P, Slobodov A 2017 *IOP Conf. Ser.: Mater. Sci. Eng.* **175** 012018 https://doi.org/10.1088/1757-899X/175/1/012018
[12] Suvorov S A, Slobodov A A, Bocharov S V, Matuzenko M Yu 2003 *Refractories and Industrial Ceramics* **44** (2) 103
[13] Slobodov A A, Uspenskiy A B, Yavshits S G, Lipin V A, Kritskiy V G 2017 *WIT Transactions on Engineering Sciences* **115** 35
[14] Khalikov D Kh, Gorshkova R M, Mukhiddinov Z K, Avloev Kh Kh, Teshaev Kh I 2002 *Chemistry of natural compounds* **2** 118
[15] Khalikov D Kh , Gorshkova R M, Khalikova S, Avloev Kh Kh, Mukhiddinov Z K 2007 *Reports of the Academy of Sciences of the Republic of Tajikistan* **50** (4) 340
[16] Filizetti-Cozzi T M C C, Carpite N C. 1991 *Journal Anal. Biochem.* **197** 157
[17] Kelko C P 2001 Determination of DE. *Control metod.* March 7 3
[18] Evans MG, Polanyi M 1935 Some applications of the transition state method to the calculation of the reaction velocities, especially in solution - Trans. Faraday Soc. **31** 875.
[19] Eckert CA 1972 *Annual Review of Physical Chemistry*, H. Eyring - Editor for University of Utah **23** 239
[20] Asano T, le Noble WJ 1978 Activation Volumes and Reactions Volumes - *Chem. Rev.* **78** (4) 407

IOP Conf. Series: Journal of Physics: Conf. Series **1045** (2018) 012013 doi:10.1088/1742-6596/1045/1/012013

Mathematical modeling of protopectin decomposition under high temperatures and pressures

R Gorshkova[1,2], D Khalikov[3], D Slobodova[2], A Slobodov[2]

[1]Institute of Macromolecular Compounds Russian Academy of Science, St. Petersburg, Russia
[2]ITMO University, St. Petersburg, Russia
[3]V. I. Nikitin Institute of Chemistry, Academy of Sciences of the Republic of Tajikistan, Dushanbe, Tajikistan

Email:gorshkova.raisa@gmail.com

Abstract. On the basis of experimental data on the influence of the parameters of the process of hydrolysis-extraction of plant raw materials under the influence of high temperature and pressure, a full-factor mathematical protopectin decomposition model connecting the input (type of source of raw materials, process and parameters of the hydrolysis-extraction process) and output (output and physico-chemical parameters of each decomposition product of the PP) parameters. As a model of the effect of any hydrolysis-extraction parameter, a polynomial of the form is chosen: $y = \sum_{k=0}^{5}\left(\sum_{m=0}^{5}\left(\sum_{n=0}^{5} \beta_{mn}^{(k)} T_i^n\right) P^m\right)t^k$. To determine the coefficients, systems of linear algebraic equations are applied. For each pair of variables by the least squares method, using the accuracy functions, a number of coefficients were found, the total number of which was 385. This mathematical model formed the basis for the software "PEKTINI", which allows to identify optimal conditions for hydrolysis-extraction of protopectin at the design stage and direct the process of obtaining pectic polysaccharides in the direction of increasing yield and optimizing the quality.

1. Formulation of the Problem

In general, hydrolysis-extraction is a key stage determining the yield and quality of pectin. At the same time, the theoretical foundations of this process have not been adequately studied at present. There is no integrated model of the process, taking into account both the characteristics of plant raw materials, and the possibility of controlled extraction. In order to assess the influence of various factors on the individual stages of the hydrolysis-extraction process and their combined effect as a whole, it is necessary to express these relationships using mathematical models that will allow us to calculate the parameters of the process and find the optimal regimes for its.

A study of the extraction process in the solid-liquid system in the food industry is devoted to the work of many researchers. The main drawback of these works is that many of them are empirical and experimental, they consider separately hydrodynamic, thermal and mass-exchange processes and do not take into account external effects [1-6].

The more general computational, theoretical and model methods for investigating of interaction processes in multicomponent systems of different nature [7-13] don’t allow to solve this problem too – because of the complexity of account (for modeling) of all the main processes, lack of parameters (thermodynamic, kinetic and other characteristics) required for the application of these methods.

Therefore, it is necessary to solve a complex problem, allowing to take into account all the processes and factors that are operating.

2. Methodology

Thebasis of the mathematical model is the results of experimental studies of the decomposition of protopectin from various sources of raw materials. Hydrolysis-extraction was carried out by the developed method under the influence of high temperature and pressure. The duration of the process varied from 3 to 10 minutes, pressure from 1.5 to 3.0 atm., Temperature from 100 to 140°C [14-18]. The hydrolyzate solutions were separated from the cellulose, the extract obtained was divided into three fractions, conventionally named: microgel, pectin substances and oligosaccharides [19, 20]. For the input parameters were taken: the type of raw materials, the degree of maturity, the parameters of the hydrolysis-extraction process (pressure, temperature, hydromodule, hydrolyzing agent pH, duration of hydrolysis-extraction). Over the weekend, the yield of the target products, the content of galacturonic acid units in them, the molecular weight parameters, and the degree of esterification.

3. Results and Discussion

Theeffectiveness of the hydrolysis-extraction method developed depends on the temperature, pressure, and duration of the process. On the basis of experimental data, a full-factor model "PECTIN" was constructed [21, 22]. As a model of any hydrolysis parameter, a polynomial of the form is chosen:

$$y = \sum_{k=0}^{5}\left(\sum_{m=0}^{5}\left(\sum_{n=0}^{5}\beta_{mn}^{(k)}T_i^n\right)P^m\right)t^k, \tag{1}$$

where $m = \overline{0,5}$.

To determine the coefficients$\left\{\beta_{mn}^{(k)}\right\}$ applied, the procedure described below.

1. At the first stage, we represent the parameter y in the form:

$$y_{ij} = \sum_{k=0}^{5} a_{jk}^{(i)}\, t^k, \tag{2}$$

where: $i = \overline{1,4}$ – temperature option number, $T = (100,\ 120,\ 130, 140)$;
$j = \overline{1,3}$ - pressure variant number, $P = (1{,}5;\ 2;\ 3)$;
$k = \overline{0,5}$ - indices of the model coefficients (1);
the variable t – is the duration of the hydrolysis process, takes values from a given set options - $t \in (3,\ 5,\ 7,\ 10)$.

For each pair (i, j) by the method of least squares coefficients $\left\{a_{jk}^{(i)}\right\}$, using the accuracy functional:

$$J_{ij} = \sum_{q=1}^{4}\left(y_{ij}^{(q)*} - \sum_{k=0}^{5} a_{jk}^{(i)} t_q^k\right)^2 \to min \text{ by coefficients } a_{jk}^{(i)}, \tag{3}$$

where the index q – is the number of the variant of the duration of hydrolysis $q = \overline{1,4}$.

Coefficients $\left\{a_{jk}^{(i)}\right\}$ are determined from the solution$i \cdot j = 12$ systems of linear algebraic equations (SLAE), resulting in the following set of coefficients (Table 1):

Table 1. Coefficients $\left\{a_{jk}^{(i)}\right\}$, defined by the SLAE solution

	P_1	P_1	P_1	Coefficients$\left\{a_{jk}^{(i)}\right\}$		
T_1	y_{11}	y_{12}	y_{13}	$\left\{a_{1k}^{(1)}\right\}$	$\left\{a_{2k}^{(1)}\right\}$	$\left\{a_{3k}^{(1)}\right\}$
T_2	y_{21}	y_{22}	y_{23}	$\left\{a_{jk}^{(2)}\right\}$	$\left\{a_{jk}^{(2)}\right\}$	$\left\{a_{jk}^{(2)}\right\}$
T_3	y_{31}	y_{32}	y_{33}	$\left\{a_{1k}^{(3)}\right\}$	$\left\{a_{1k}^{(3)}\right\}$	$\left\{a_{1k}^{(3)}\right\}$
T_4	y_{41}	y_{42}	y_{43}	$\left\{a_{1k}^{(4)}\right\}$	$\left\{a_{1k}^{(4)}\right\}$	$\left\{a_{1k}^{(4)}\right\}$

For each pair (*i*, *j*) SLAE (1) are obtained by equating the partial derivative of the functional (3) with respect$a_{jk}^{(i)}$to zero:

$$\begin{cases} \frac{\partial J_{ij}}{\partial a_{j0}^{(i)}} = \sum_{q=1}^{4} F_{1q} \cdot \frac{\partial F_q}{\partial a_{j0}^{(i)}} = \sum_{q=1}^{4} F_q = 0 \\ \frac{\partial J_{ij}}{\partial a_{j1}^{(i)}} = \sum_{q=1}^{4} F_{1q} \cdot \frac{\partial F_q}{\partial a_{j1}^{(i)}} = \sum_{q=1}^{4} F_q \cdot t_q = 0 \\ \frac{\partial J_{ij}}{\partial a_{j2}^{(i)}} = \sum_{q=1}^{4} F_{1q} \cdot \frac{\partial F_q}{\partial a_{j2}^{(i)}} = \sum_{q=1}^{4} F_q \cdot t_q^2 = 0 \\ \dots\dots\dots\dots\dots\dots\dots\dots \\ \frac{\partial J_{ij}}{\partial a_{j5}^{(i)}} = \sum_{q=1}^{4} F_{1q} \cdot \frac{\partial F_q}{\partial a_{j5}^{(i)}} = \sum_{q=1}^{4} F_q \cdot t_q^5 = 0 \end{cases} \quad (4)$$

where $F_q = \left(y_{ij}^{(q)*} - \sum_{k=0}^{5} a_{jk}^{(i)} t_q^k\right)$.

Solving system (4) alternately, the values of the coefficients $\left\{a_{jk}^{(i)}\right\}$ were defined, the total number of which is 72.

2. The next step is the build procedure for each T_i dependence on $a_k^{(i)} = a_k^{(i)}(P)$. This problem is solved by a method similar to that described in A.1. For this, the coefficients are expressed by the dependence:

$$a_k^i = \sum_{m=0}^{5} \alpha_{km}^{(i)} P^m \quad (5)$$

Using the accuracy functional:

$$J_{ik} = \sum_{j=1}^{3} \left(a_{km}^{(i)*} - \sum_{m=0}^{5} \alpha_{km}^{(i)} P_j^m\right)^2 \to min \text{by coefficients } \alpha_{km}^{(i)}. \quad (6)$$

Denoting the expression in the parentheses of the accuracy functional as:

$$F_j = \left(a_{jk}^{(i)*} - \sum_{m=0}^{5} \alpha_{km}^{(i)} P_j^m\right)^2, \quad (7)$$

For each pair (i, k), differentiating J_{ik} with respect to the variables $\alpha_{km}^{(i)}$ and, equating these partial derivatives to zero, are obtained $i \cdot k = 24$ SLAE, from the solution of which a set of coefficients of the relation (5) (Table 2):

Table 2. Coefficients $\alpha_{km}^{i} = \sum_{n=0}^{5} \beta_{mn}^{(k)} T^n$, defined by the SLAE decision

T_1	a_k^1	$\alpha_{k0}^{(1)}$	$\alpha_{k1}^{(1)}$	$\alpha_{k2}^{(1)}$	$\alpha_{k3}^{(1)}$	$\alpha_{k4}^{(1)}$	$\alpha_{k5}^{(1)}$
T_2	a_k^2	$\alpha_{k0}^{(2)}$	$\alpha_{k1}^{(2)}$	$\alpha_{k2}^{(2)}$	$\alpha_{k3}^{(2)}$	$\alpha_{k4}^{(2)}$	$\alpha_{k5}^{(2)}$
T_3	a_k^3	$\alpha_{k0}^{(3)}$	$\alpha_{k1}^{(3)}$	$\alpha_{k2}^{(3)}$	$\alpha_{k3}^{(3)}$	$\alpha_{k4}^{(3)}$	$\alpha_{k5}^{(3)}$
T_4	a_k^4	$\alpha_{k0}^{(4)}$	$\alpha_{k1}^{(4)}$	$\alpha_{k2}^{(4)}$	$\alpha_{k3}^{(4)}$	$\alpha_{k4}^{(4)}$	$\alpha_{k5}^{(4)}$

SLAE for finding the set of coefficients $\left\{\alpha_{km}^{(i)}\right\}$ have the form:

$$\begin{cases} \frac{\partial J_{ik}}{\partial \alpha_{k0}^{(i)}} = \sum_{j=1}^{3} F_j \cdot \frac{\partial F_j}{\partial a_{k0}^{(i)}} = \sum_{j=1}^{3} F_j = 0 \\ \frac{\partial J_{ik}}{\partial \alpha_{k1}^{(i)}} = \sum_{j=1}^{3} F_j \cdot \frac{\partial F_j}{\partial a_{k1}^{(i)}} = \sum_{j=1}^{3} F_j \cdot P_j = 0 \\ \frac{\partial J_{ik}}{\partial \alpha_{k2}^{(i)}} = \sum_{j=1}^{3} F_j \cdot \frac{\partial F_j}{\partial a_{k2}^{(i)}} = \sum_{j=1}^{3} F_j \cdot P_j^2 = 0 \\ \ldots\ldots\ldots\ldots\ldots\ldots\ldots\ldots\ldots\ldots\ldots\ldots \\ \frac{\partial J_{ik}}{\partial \alpha_{k5}^{(i)}} = \sum_{j=1}^{3} F_j \cdot \frac{\partial F_j}{\partial a_{k5}^{(i)}} = \sum_{j=1}^{3} F_j \cdot P_j^5 = 0 \end{cases} \tag{8}$$

Total number of coefficients $\alpha_{km}^{(i)}$ is – 144.

3. The next step is the build procedure for each $\alpha_{km}^{(i)}$ dependence on $\alpha_{km}^{(i)} = \alpha_{km}^{(i)}(T)$. This problem was solved by a method similar to those described in A.1 and A.2. For this, the coefficients are expressed by:

$$\alpha_{km}^{i} = \sum_{n=0}^{5} \beta_{mn}^{(k)} T^n \tag{9}$$

Introducing the accuracy functional:

$$J_{km} = \sum_{i=1}^{4} \left(\alpha_{km}^{(i)*} - \sum_{n=0}^{5} \beta_{mn}^{(k)} T_i^n\right)^2 \to min \text{ (by coefficients } \beta_{mn}^{(k)}\text{).} \tag{10}$$

Denoting the expression in the parentheses of the accuracy functional as:

$$F_i = \left(\alpha_{km}^{(i)*} - \sum_{n=0}^{5} \beta_{mn}^{(k)} T_i^n\right)^2, \tag{11}$$

For each pair (k, m), with the solution of which many coefficients of the relation differentiating J_{km} on variables $\beta_{mn}^{(k)}$, and equating these partial derivatives to zero, we obtain $(m + 1) \cdot (k + 1) = 36$SLAE with the solution of which is the set of coefficients of the relation (9) (Table 3):

Table 3. Coefficients$a_k^i = \sum_{m=0}^{5} \alpha_{km}^{(i)} P^m$, defined by the SLAE solution

α_k^0	$\beta_{00}^{(k)}$	$\beta_{01}^{(k)}$	$\beta_{02}^{(k)}$	$\beta_{03}^{(k)}$	$\beta_{04}^{(k)}$	$\beta_{05}^{(k)}$
α_k^1	$\beta_{10}^{(k)}$	$\beta_{11}^{(k)}$	$\beta_{12}^{(k)}$	$\beta_{13}^{(k)}$	$\beta_{14}^{(k)}$	$\beta_{15}^{(k)}$
α_k^2	$\beta_{20}^{(k)}$	$\beta_{21}^{(k)}$	$\beta_{22}^{(k)}$	$\beta_{23}^{(k)}$	$\beta_{24}^{(k)}$	$\beta_{25}^{(k)}$
α_k^3	$\beta_{30}^{(k)}$	$\beta_{31}^{(k)}$	$\beta_{32}^{(k)}$	$\beta_{33}^{(k)}$	$\beta_{34}^{(k)}$	$\beta_{35}^{(k)}$
α_k^4	$\beta_{40}^{(k)}$	$\beta_{41}^{(k)}$	$\beta_{42}^{(k)}$	$\beta_{43}^{(k)}$	$\beta_{44}^{(k)}$	$\beta_{45}^{(k)}$
α_k^5	$\beta_{50}^{(k)}$	$\beta_{51}^{(k)}$	$\beta_{52}^{(k)}$	$\beta_{53}^{(k)}$	$\beta_{54}^{(k)}$	$\beta_{55}^{(k)}$

SLAE for finding the set of coefficients $\beta_{mn}^{(k)}$ have the form:

$$\begin{cases} \dfrac{\partial J_{km}}{\partial \beta_{m0}^{(k)}} = \sum_{i=1}^{4} F_i \cdot \dfrac{\partial F_i}{\partial \beta_{m0}^{(k)}} = \sum_{i=1}^{4} F_i = 0 \\ \dfrac{\partial J_{km}}{\partial \beta_{m1}^{(k)}} = \sum_{i=1}^{4} F_i \cdot \dfrac{\partial F_i}{\partial \beta_{m1}^{(k)}} = \sum_{i=1}^{4} F_i \cdot T_i = 0 \\ \dfrac{\partial J_{km}}{\partial \beta_{m2}^{(k)}} = \sum_{i=1}^{4} F_i \cdot \dfrac{\partial F_i}{\partial \beta_{m2}^{(k)}} = \sum_{i=1}^{4} F_i \cdot T_i^2 = 0 \\ \dots\dots\dots\dots\dots\dots\dots\dots \\ \dfrac{\partial J_{km}}{\partial \beta_{m5}^{(k)}} = \sum_{i=1}^{4} F_i \cdot \dfrac{\partial F_i}{\partial \beta_{m5}^{(k)}} = \sum_{i=1}^{4} F_i \cdot T_i^5 = 0 \end{cases} \quad (12)$$

Total number of coefficients $\beta_{mn}^{(k)}$ is – 216.

$$y = \sum_{k=0}^{5}\left(\sum_{m=0}^{5}\left(\sum_{n=0}^{5} \beta_{mn}^{(k)} T_i^n\right) P^m\right) t^k, \quad (13)$$

where $m = \overline{0,5}$. As a result of the research, the PEKTINI software was developed, which allows predicting the yield and quality of pectin.

In Figures 1-4, as an example of program operation, a comparison of the experimental values of the yield of pectin with the calculated models obtained as a result of the work.

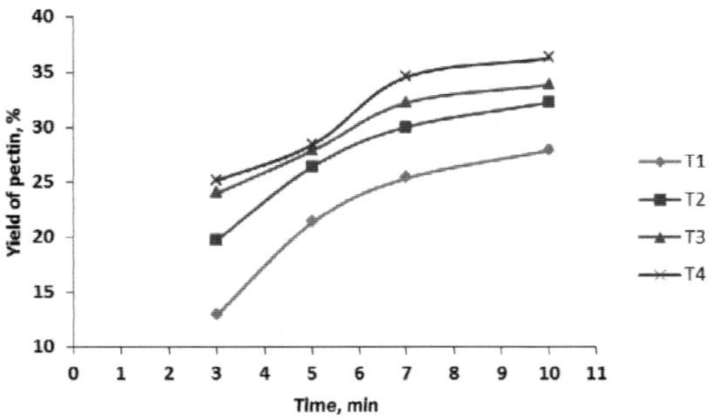

Figure 1. The yield of pectin at P = 2.0 atm. depending on the temperature and duration of the hydrolysis-extraction process (experimental data); T_1 – 100°C, T_2 – 120°C, T_3 – 130°C, T_4 – 140°C.

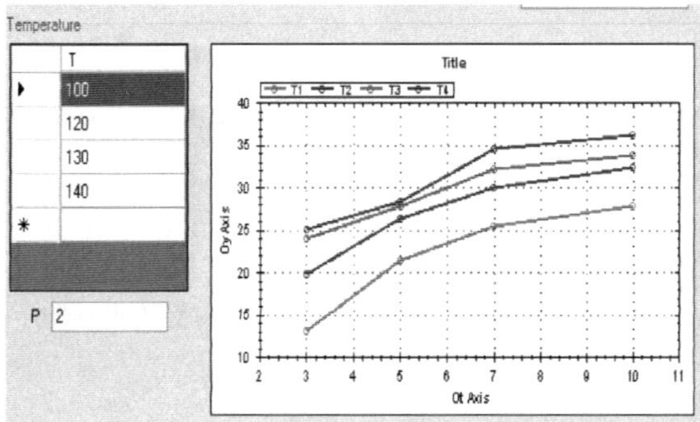

Figure 2. The yield of pectin at P = 2.0 atm. depending on the temperature and duration of the hydrolysis-extraction process (calculated data); T1 – 100°C, T_2 – 120°C, T_3 – 130°C, T_4 – 140°C.

IOP Conf. Series: Journal of Physics: Conf. Series **1045** (2018) 012013 doi:10.1088/1742-6596/1045/1/012013

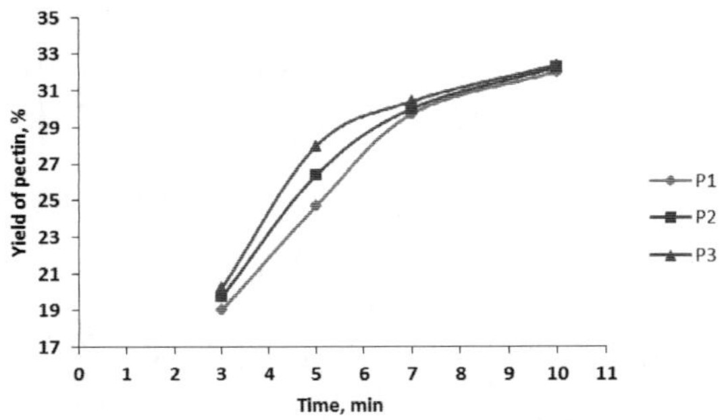

Figure 3. The yield of pectin at 120 °C, depending on the pressure and duration of the hydrolysis-extraction process (experimental data); P_1 – 1,5 atm, P_2 – 2 atm, P_3 – 3 atm

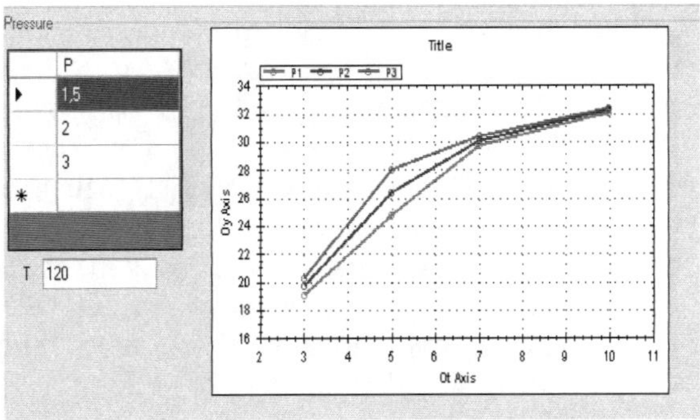

Figure 4. The yield of pectin at 120 °C, depending on the pressure and duration of the hydrolysis-extraction process (calculated data); P_1 – 1,5 atm, P_2 – 2 atm, P_3 – 3 atm

4. Conclusions

Thus, the developed mathematical model of protopectin decomposition under the influence of high temperature and pressure and software allow at the design stage to identify optimal technological conditions, predict the yield and physico-chemical parameters of the target products, and significantly optimize the process of obtaining high-quality pectin.

References

[1] Panchev I N, Kirtchev N A, Kratchanov C1989 *Carbohydr. Polym.* **11** 193
[2] Cho Y J, Hwang J K 2000 *Journal of Food Engineering* **44** 85
[3] Minkov S, Minchev A, Paev K 1996 *Journal of Food Engineering* **29** (1) 107
[4] Masmoudi M, Besbes S, Chaabouni M, Robert C, Paquot M, Blecker C, Attia H 2008 *Carbohydr. Polym.* **74** 185
[5] Andersen N M, Cognet T, Santacoloma P A, Larsen J, Armagan I, Larsen F H, Gernaey K V, Abildskov J, Huusom J K 2017 *Journal of Food Engineering* **192** 61
[6] Duran R, Villa A L, Ribeiro R, Rabi J A 2015 *Chem. Prod. Process Model.* **10** 203
[7] A L Gömze and L N Gömze 2017 *J. Phys.: Conf. Ser.* **790** 012001 https://doi.org/10.1088/1742-6596/790/1/012001
[8] Savchenko N, Sevostyanova I, Sablina T, Gömze L A, Kulkov S 2014 *AIP Conference Proceedings* **1623** 547 http://dx.doi.org/10.1063/1.4899003
[9] Suvorov S A, Musevich V A, Iksanov F R, Slobodov A A 2007 *Refractories and Industrial Ceramics* **48** (5) 362
[10] Slobodov A, Uspenskiy An, Ralys R, Kremnev D 2015 *Epitoanyag – Journal of Silicate Based and Composite Materials* **67** (4) 159 http://dx.doi.org/10.14382/epitoanyag-jsbcm.2015.27
[11] Ralys R V, Uspenskiy A A, Slobodov A A 2016 *J. Non-Equilibrium Thermodynamics* **41** (1) 3
[12] Slobodov A A, Uspenskiy A B, Yavshits S G, Lipin V A, Kritskiy V G 2017 *WIT Transactions on Engineering Sciences* **115** 35
[13] Ageev E I, Andreeva Ya M, Karlagina Yu Yu, Kolobov Yu R, Manokhin S S, Odintsova G V, Slobodov A A, Veiko V P 2017 *Laser Physics* **27** (4) 046001
[14] Gorshkova R M, Nasriddinov A S, Mukhidinov Z K, Khalikov D Kh, Khalikova S, Teshaev Kh I 2010 *News of Russian universities: "Chemistry and chemical technology" series* **53** (6) 87
[15] Gorshkova R M, Mukhidinov Z K, Makhkamov Kh K, Johnmurodov A S, Khalikova S, Khalikov D Kh 2010 *Bulletin of the Tajik Technical University* **3** (11) 55
[16] Gorshkova RM, Mukhidinov ZK, Nasriddinov AS, Valiev MV, Khalikova S, Khalikov DKh, Liu LSh 2010 *News of the Academy of Sciences of the Republic of Tajikistan* **4** (141) 45
[17] Mukhidinov ZK, Gorshkova RM, Valiev MV, Khalikova S, Nasriddinov AS, Khalikov DKh,

Fishman ML 2010 *Reports of the Academy of Sciences of the Republic of Tajikistan* **53** (8) 617

[18] Mukhidinov Z.K, Gorshkova R.M, Valiev M.V, Khalikova S., Nasriddinov A.S, Khalikov D.Kh., Fishman M.L. 2010 *Reports of the Academy of Sciences of the Republic of Tajikistan*. **53** (9) 701

[19] Khalikov DKh, Gorshkova RM, Mukhiddinov ZK, Avloev KhKh, Teshaev Kh I 2002 *Chemistry of natural compounds* 2118

[20] Khalikov DKh, Gorshkova RM, Khalikova S, Avloev KhKh, Mukhiddinov ZK 2007 *Reports of the Academy of Sciences of the Republic of Tajikistan* **50** (4) 340

[21] Gorshkova RM, Yulusova DV, Kholov ShE, Mukhidinov ZK, Khalikov DKh 2015 *State and perspectives of development of organic chemistry in the Republic of Tajikistan: Proceedings of the XII Numanov Readings*. Dushanbe, 130

[22] Yulusova DV, Gorshkova RM, Valiev MV, Khalikov DKh, Khalikova S, Bobodjonova GN 2017 *Proceedings of the VII All-Russian Kargin Conference "Polymers - 2017"*. Moscow. P. 549.

IOP Conf. Series: Journal of Physics: Conf. Series **1045** (2018) 012014 doi:10.1088/1742-6596/1045/1/012014

Physico-chemical and molecular-mass parameters of pectin polysaccharides obtained under high temperatures and pressures

R Gorshkova[1,2], D Khalikov[3], D Slobodova[2], A B Uspensky[2], A Slobodov[2]

[1]Institute of Macromolecular Compounds Russian Academy of Science, St. Petersburg, Russia
[2]ITMO University, St. Petersburg, Russia
[3]V. I. Nikitin Institute of Chemistry, Academy of Sciences of the Republic of Tajikistan, Dushanbe, Tajikistan

Email:gorshkova.raisa@gmail.com

Abstract. Two different methods are used to obtainpectic polysaccharides from food industry waste: traditional and hydrolysis-extraction method under the influence of high temperature and pressure in a short period of time. It is shown that the application of the new method allows to increase the yield of the target products, to optimize the physico-chemical characteristics and the molecular-mass parameters of pectin.

1. Formulation of the problem

Various fruit and vegetable extracts remaining in bulk after production of juices, as well as sunflower heads can become a promising source for the production of pectin polysaccharides, the trend of which in various areas of industry and medicine is increasingly increasing. The preparation of a pectin is a multi-step process involving steps with different functions. The main stages are:

- preparation of raw materials,
- hydrolysis-extraction,
- separation of hydrolyzate solution from cell wall residues,
- purification of the hydrolyzate,
- isolation of water-soluble fraction of pectic polysaccharides,
- drying to the powder state of the finished product,
- in some cases - modification of pectic polysaccharides.

According to most researchers, the main role is played by the hydrolysis-extraction stage. The quality and properties of the target products depend on the way it is carried out, on the choice of parameters at this stage [1-3]. As a consequence, the problem arises of developing modern technologies for the production of pectic polysaccharides with a regulated structure and properties. The available experimental and calculation-theoretical studies about influence of technological process conditions on the physico-chemical and other properties of products [4-9] don't allow to solve this problem, unfortunately. The main reason of it is a complexity of raw materials composition, obtained products, physico-chemical processes taking place during of pectin production, etc.

In connection with this, the purpose of this study was a comparative study of the physico-chemical and molecular-mass parameters of pectins of various origins obtained by the traditional method [10,

IOP Conf. Series: Journal of Physics: Conf. Series **1045** (2018) 012014 doi:10.1088/1742-6596/1045/1/012014

11] and the hydrolysis-extraction method under the influence of high temperature and pressure [12-18].

2. Methodology

Most of the existing methods for the preparation of pectic polysaccharides are based on the principle of hydrolysis-extraction of plant material in the static regime. The process thus proceeds in closed systems, at a temperature of 90-100 °C, a duration of 1-4 hours and continuous mixing. These static conditions provide for the extraction of a significant amount of pectic polysaccharides, but pectin macromolecules diffusing from the plant cell to the solution are exposed to excessive hydrolyzing agent and temperature, which leads to their degradation, deterioration of the physico-chemical parameters, and loss of basic properties. In this paper, the traditional method of hydrolysis-extraction in static mode (Method 1) [10, 11] was used as a reference standard to evaluate the effectiveness of the new developed hydrolysis-extraction method under the influence of high temperature and pressure (Method 2). The raw materials used pomace, pumpkins, stems and leaves of rhubarb, peach, beet pulp, mandarin crusts and sunflower heads.

The traditional method (Method 1): Hydrolysis-extraction of plant raw materials is carried out in closed systems, at a process temperature of 85 ° C for 1-4 hours and continuous stirring. The obtained extract is separated from the fiber residues, neutralized and a water-swelling fraction, conventionally called microgel (MG), is isolated by centrifugation at a speed of 4000-5000 rpm. Subsequent precipitation from the remaining solution with a three-fold volume of ethanol, water-soluble components, conditionally named pectin substances (PS) are isolated and evaporation of the mother liquor yields low molecular weight oligosaccharide (OS) residues [10, 11].

Hydrolysis-extraction method under the influence of high temperature and pressure (Method 2): The plant for the production of pectic polysaccharides is presented in Figure 1. The essence of the method is as follows: the equilibrium swollen plant material, placed in the unit, a hydrolyzing agent with the necessary pH value is added, the lid is sealed and the steam is started. The lid is closed tightly and steam is started. The pressure in the installation is automatically controlled by the steam injector. The time of hydrolysis is read by the stopwatch from the moment the temperature is reached. Hydrolysis in this case can be carried out by varying the temperature from 100 to 140°C, pressure from 1 to 3 atm., Duration from 3 to 10 minutes. In this case, the effect of temperature and pressure ensures a high yield of the desired product without the use of strong acids. It has been experimentally established that the optimal pH of the hydrolysing agent is 2.0 at the same time. Also, the duration of the hydrolysis-extraction process is significantly shortened from 60 to 3-10 minutes [12-18].

The hydrolyzate solution is further divided into MH, MF and OS by the above-described procedure. The content of the elements of galacturonic acid (GA) in pectic substances and the degree of esterification of carboxyl groups of GA (DE) were determined by the methods described in [19, 20], respectively.

The mass-average (Mw), number-average (Mn) molecular weights (MW), z-average MW (Mz) and polydispersity index (Mw/Mn) were determined by high-performance liquid chromatography (HPLC) [21, 22]. Depending on the viscosity of the PV solution, the dried sample was dissolved in 0.05M NaNO3 to a final concentration of 1 or 2 mg/ml, centrifuged at 20,000 g for 20 min, separating the aggregated fraction and filtered through a membrane filter (0.45 μm, Millipore Millex-HN). A 0.1-0.2 ml solution was automatically introduced into the chromatograph. The flow rate of the eluent was 0.8 ml/min. The solvent delivery system consisted of a two-channel vacuum degasser, a high-pressure pump (Waters 1515 Isocratic Pump) and an auto injector (717 Plus Auto Injector, Waters). Two columns containing PL-Aquagel OH40 and PL-Aquagel OH60 were used to separate the PV. MM pectin substances were determined simultaneously using the software ASTRA 5.3.4.13 (Wyatt Technology) and Breez (Waters). Values for Mw were obtained by HPLC using a universal calibration using the molecular standards of Pullulan (Showa Denko K.K., Japan).

IOP Conf. Series: Journal of Physics: Conf. Series **1045** (2018) 012014 doi:10.1088/1742-6596/1045/1/012014

3. Results and Discussion

The yield of pectic polysaccharides depends not only on the source of raw materials, but also on the method of their preparation (Table 1). When using the hydrolysis-extraction method under the influence of high temperature and pressure, the yield of pectin substances and oligosaccharides increases significantly for all the types of raw materials studied.

The main physico-chemical parameters determining the properties of pectic polysaccharides are the content of the elements of galacturonic acid, the degree of esterification and the molecular weight. According to the literature, the effect of high temperature and pressure often leads to a decrease in the content of GA [23]. But the use of the new hydrolysis-extraction method, due to the shortening of the extraction process, does not adversely affect this parameter, on the contrary, the maximum content of galacturonic acid is observed precisely under the influence of high temperature and pressure (Table 1). The degree of esterification, which largely determines the scope of pectic polysaccharides, does not undergo significant changes depending on the hydrolysis-extraction method.

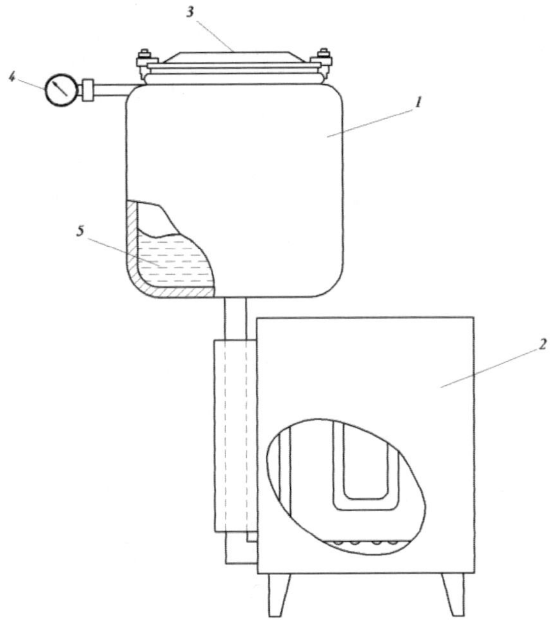

Figure 1. Installation for hydrolysis-extraction under the influence of high temperature and pressure
1 - autoclave; 2 - steam generator;
3 - the hermetic cover; 4 - pressure and temperature sensor; 5 - swollen raw material in hydrolyzate solution.

Table 1. Comparative values of the yield of MG, MF, OS*, degree of esterification and the content of galacturonic acid in pectin substances

The raw material	MG		PS		OS		GA		DE	
	M 1	M 2	M 1	M 2	M 1	M 2	M 1	M 2	M 1	M 2
Apple	3,60	2,60	14,25	23,60	24,65	28,00	70,00	70,00	52,00	52,17
Quince	1,40	4,50	11,40	12,40	25,90	56,50	66,00	67,20	58,24	57,14
Apricot	1,22	3,76	4,86	7,05	37,92	69,19	66,00	67,20	54,65	52,87
Lemon	2,00	2,06	19,00	20,06	23,80	45,84	72,24	74,40	76,74	74,42
Orangegrits	0,80	1,00	20,80	24,00	27,10	32,30	77,76	74,80	74,07	70,60
Rhubarb	1,11	3,18	12,36	17,50	35,03	29,32	81,60	81,60	50,57	50,00
Peach	0,86	4,28	6,49	7,20	37,85	43,36	65,28	66,00	94,94	94,87
Pumpkin	1,19	8,68	2,24	3,77	73,71	76,92	64,80	75,00	85,71	81,69
Orange peel	0,48	1,00	15,19	16,25	18,75	62,40	64,00	62,40	72,73	76,25
Citrus	0,68	1,57	9,23	18,30	24,69	29,97	64,80	70,00	71,59	72,09
Sunflower	1,20	8,20	5,80	24,68	12,80	7,50	56,40	74,00	34,12	26,67
Beet	1,00	1,60	8,20	10,44	48,14	48,96	75,60	78,00	40,23	36,78

*M 1 (Traditional method) - T = 85 °C, hydromodule 1:20, duration 60 minutes, continuous stirring, pH = 1,2; M 2 (Method of hydrolysis-extraction under the influence of high temperature and pressure) - T = 120 °C, P = 1.5 atm., hydromodule 1:20, duration 5 min, pH = 2.0.

There is no clear regularity in the change in the molecular mass parameters of pectin polysaccharides, which requires further study in this field (Table 2, Figure 2). In particular, pectin samples obtained from apples, apricots and pumpkins under the influence of high temperature and pressure are resistant to elevated temperature, have a higher molecular weight compared to similar samples obtained by the traditional method. In these pectins, anomalies associated with the aggregation of macromolecules are observed. As can be seen, the more Mz of pectin, the more its molecules are subject to aggregation, with the exception of lemon pectin. Most likely, this fact is explained by the peculiarities of the structure of the pectin macromolecule of this type of raw material.

Table 2. Molecular Weight Characteristics of PS *

The raw material	$Mw \cdot e^{-3}$,kD		$Mz \cdot e^{-3}$, kD		Mw/Mn		AG, %	
	M 1	M 2	M 1	M 2	M 1	M 2	M 1	M 2
Apple	134,0	309,4	1133	4473	3,80	9,50	23,5	20,0
Apricot	114,5	718,0	755	12500	4,70	10,90	18,5	8,5
Quince	176,0	112,4	7309	613	4,35	2,80	25,0	8,5
Peach	243,8	65,5	2103	145	6,30	3,07	14,5	23,0
Lemon	107,8	82,3	267	387	2,89	9,31	18,5	16,0
Pumpkin	673,9	999,7	2069	4112	1,93	2,74	17,7	40,0
Rhubarb	268,0	114,0	1479	338	4,80	2,30	26,5	49,0
Sunflower	142,4	181,0	1793	1374	3,22	5,00	40,0	33,5

*Mw is the weight average of MW; Mn - number average MW; Mz is the z-average MW; Mw / Mn is the polydispersity index; AG - aggregated fraction

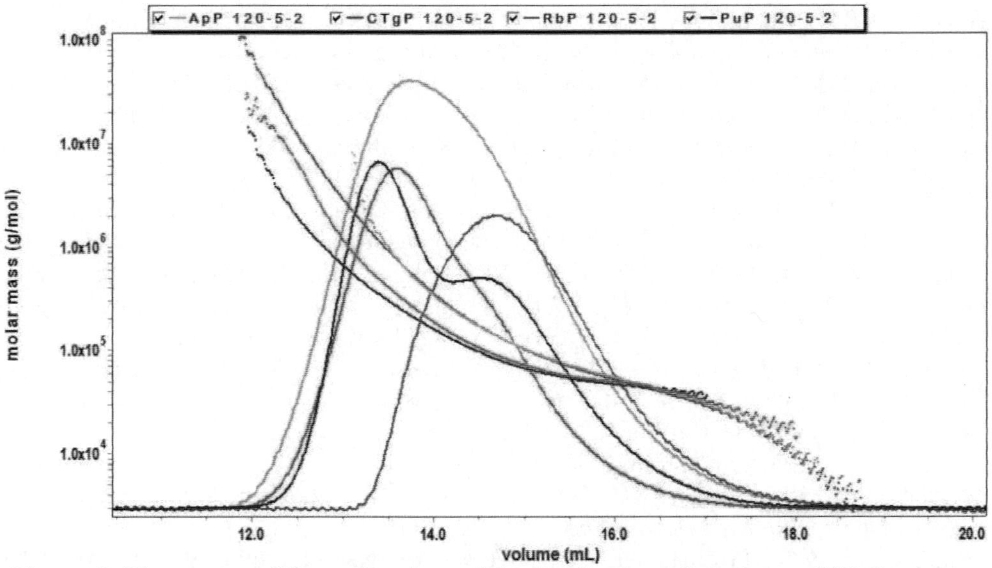

Figure 2. The curves of the molecular weight distribution of pectins on HPLC using a combination of viscosimetric and refractometric detectors

4. Conclusion

Thus, the new developed method of hydrolysis-extraction under the influence of high temperature and pressure allows to increase the yield of pectic polysaccharides, to improve their quality, and to significantly shorten the duration of the production process, to soften the pH of the hydrolyzing agent, which will lead to a reduction in the costs of pectin production and environmental improvement.

References

[1] Karpovich N S, Donchenko L V 1989 *Pectin: production and use* (Kiev: The harvest)
[2] O'Neill MA, York W.S. 2003 The composition and structure of the primary cell walls Rose JKC. editor. - The Plant Cell Wall. - *Blackwell Publishers.* Oxford. - P.1-54.
[3] Sobol I V 2006 *Polytechnical network electronic scientific journal of the Kuban State Agrarian University* **22** 91
[4] L A Gömze and L N Gömze 2013 *IOP Conf. Ser.: Mater. Sci. Eng.* **47** 012033 https://doi.org/10.1088/1757-899X/47/1/012033
[5] Dedova E S, Shutilova E S, Geber R, Gömze L A, Kulkov S N 2016 *IOP Conf. Ser.: Mater. Sci. Eng.* **140** 012007 https://doi.org/10.1088/1757-899X/140/1/012007
[6] Slobodov A, Uspenskiy An, Ralys R, Kremnev D 2015 *Epitoanyag – Journal of Silicate Based and Composite Materials* **67** (4) 159 http://dx.doi.org/10.14382/epitoanyag-jsbcm.2015.27
[7] Slobodov A, Uspenskiy Al, Yavshits S, Mischenko A 2015 *Epitoanyag – Journal of Silicate Based and Composite Materials* **67** (4) 164 http://dx.doi.org/10.14382/epitoanyag-jsbcm.2015.28
[8] Uspensky A B, Ralys R, Kremnev D, Radin M, Slobodov A 2017 *IOP Conf. Ser.: Mater. Sci. Eng.* **175** 012024 https://doi.org/10.1088/1757-899X/175/1/012024
[9] Uspensky A A, Yavshits S, Lipin V, Zhigalov P, Slobodov A 2017 *IOP Conf. Ser.: Mater. Sci. Eng.* **175** 012018 https://doi.org/10.1088/1757-899X/175/1/012018
[10] Khalikov D Kh, Gorshkova R M, Mukhiddinov Z K, Avloev KhKh, Teshaev Kh I 2002 *Chemistry of natural compounds* (2) 118
[11] Khalikov D Kh, Gorshkova R M, Khalikova S, Avloev Kh Kh, Mukhiddinov Z K 2007 *Reports of the Academy of Sciences of the Republic of Tajikistan* **50** (4) 340

IOP Conf. Series: Journal of Physics: Conf. Series **1045** (2018) 012014 doi:10.1088/1742-6596/1045/1/012014

[12] Gorshkova RM, Mukhidinov ZK, Khalikov DKh, Khalikova S, Nasriddinov A S, Johnmurodov A S 2009 *Journal of Public Health in Tajikistan* **3** (303) 14

[13] Gorshkova RM, Nasriddinov AS, Mukhidinov ZK, Khalikov D Kh, Khalikova S, Teshaev KhI 2010 *News of Russian universities: "Chemistry and chemical technology" series* **53** (6) 87

[14] Gorshkova RM, Mukhidinov ZK, Makhkamov KhK, Johnmurodov AS, Khalikova S, Khalikov D Kh 2010 *Bulletin of the Tajik Technical University* **3** (11) 55

[15] Gorshkova RM, Mukhidinov ZK, Nasriddinov AS, Valiev MV, Khalikova S, Khalikov DKh, Liu LSh 2010 *News of the Academy of Sciences of the Republic of Tajikistan* **4** (141) 45

[16] Mukhidinov ZK, Gorshkova RM, Valiev MV, Khalikova S, Nasriddinov AS, Khalikov DKh, Fishman ML 2010 *Reports of the Academy of Sciences of the Republic of Tajikistan* **53** (8) 617

[17] Mukhidinov ZK, Gorshkova RM, Valiev MV, Khalikova S, Nasriddinov AS, Khalikov DKh, Fishman ML 2010 *Reports of the Academy of Sciences of the Republic of Tajikistan* **53** (9) 701

[18] Khalikova S, Gorshkova R M, Valiev MV 2015 *Chemical Journal of Kazakhstan* **2** (50) 202

[19] Filizetti-Cozzi T M C C, Carpite N C 1991 *Journal Anal. Biochem* **197** 157

[20] Kelko, C.P. 2001 Determination of DE. Control metod. March 7 3

[21] Mukhiddinov ZK, Fishman ML, Gorshkova R M, Nasriddinov A S, Khalikov D Kh2008 *Chemical Journal of Kazakhstan: Special Issue* **21** 60

[22] Mukhiddinov Z K, Fishman ML, Gorshkova RM, Nasriddinov AS, Khalikov DKh 2008 *Chemical Journal of Kazakhstan: Special Issue* **21** 83

[23] Ivanova N V, Popova O V, Babkin V A 2003 *Chemistry of plant raw materials* **4** 43

IOP Conf. Series: Journal of Physics: Conf. Series **1045** (2018) 012015 doi:10.1088/1742-6596/1045/1/012015

Influence of the parameters of hydrolysis-extraction high-temperature process on epyyield, physico-chemical and molecular mass characteristics of pectin substances

R Gorshkova[1,2], D Slobodova[2], A A Uspensky[2], A Slobodov[2]

[1]Institute of Macromolecular Compounds Russian Academy of Science, St. Petersburg, Russia
[2]ITMO University, St. Petersburg, Russia

Email: gorshkova.raisa@gmail.com

Abstract. The effect of the duration of high-temperature hydrolysis-extraction on the yield and parameters of pectin from various sources has been studied. It is shown that the variation in the duration of the process leads to a phased conversion of high-molecular aggregated pectin macromolecules into samples enriched with galacturonic acid units with a narrow molecular weight distribution. The results give an additional opportunity to directively control the hydrolysis-extraction process in order to obtain the desired product in high yield and specified parameters.

1. Formulation of the problem

Pectin substances are contained in the cell wall of all higher flowering plants and are one of the most complex and dynamically structured class of biopolymers. Their structure and properties depend on a number of both internal (type, preliminary preparation, term and conditions of storage of plant raw materials, etc.) and external factors (duration of the hydrolysis-extraction process, temperature, pH of the hydrolysing agent, etc.) [1]. Unique gel-forming and sorption properties of pectin substances cause their wide application in various areas of food industry and medicine [2, 3]. This trend is increasing every year, in turn, increasing the need for the production of pectin. Currently, in the market of pectins and pectin-containing products, the needs of consumers significantly exceed the offers of producers. The reason for this is the fact that most of the existing technologies for the production of pectin substances are based on the use of strong acids (HCl, HNO_3, H_3PO_4, H_2S0_4), aluminum chloride, ammonium hydroxide, a mixture of alcohol with acid and alcohols of various concentrations. The production process takes place in aggressive media with variations in extraction and hydrolysis time from 3 to 6 hours and a general process cycle of up to 12 hours or more, which causes significant energy costs. Isolation of water-soluble pectins is carried out with the help of huge amounts of ethyl alcohol [4]. These technological factors significantly increase the cost of the targeted products, making them virtually inaccessible to a wide range of consumers.

Also it should be noted that the available experimental, computational, model methods for research of multicomponent physicochemical systems [5-10] don't cover the whole variety of the proceeding processes. In this connection, the task arises of developing modern energy and resource-saving technologies that allow regulate the process of obtaining pectin substances with the specified structure and properties.

2. Methodology

Pomace, sunflower heads, pumpkin fruits, stems and leaves of rhubarb were used as a raw material for pectin substances. Hydrolysis-extraction was carried out by the method developed by us [11-17]. The process proceeded at 120 °C, a pressure of 1.5 atm., the hydraulic module 1:20. As a hydrolysing agent, a solution of HCl with pH = 2.0 was experimentally determined. The duration of the hydrolysis-extraction process varied from 3 to 10 minutes. Water-soluble pectin substances were isolated according to the method described in [18, 19]. For the samples obtained, physicochemical characteristics were determined: the content of galacturonic acid, the degree of esterification by the methods described in the works, respectively [20, 21]. The mass-average (Mw), number-average (Mn) molecular weights (MW), z-average MW (Mz) and polydispersity index (Mw / Mn) were determined by high-performance liquid chromatography (HPLC) [22, 23].

3. Results and Discussion

Earlier, we showed that the hydrolysis-extraction method under the influence of high temperature and pressure leads to a significant increase in the yield of pectin substances, positively affecting their quality [11, 14]. But, the parameters of the process for obtaining pectin, in particular the duration of extraction, are selected depending on the characteristics of the processed raw materials. In this connection, the effect of the duration of the hydrolysis-extraction process under the influence of high temperature and pressure on the yield and parameters of pectin substances was studied.

For all the types of raw materials studied, with increasing duration of the process, the yield of pectin substances naturally increases, and somewhat stabilizes in the region of 7-10 minutes (Table 1).

Table 1. The yield of pectin substances of different raw materials (% with respect to the weight of the feedstock), the content of galacturonic acid and the degree of esterification of the samples (% relative to the mass of the PV)

Time,	Apple			Sunflower			Rhubard			Pumpkin		
min	Yeld	GA	DE	Yeld	GA	DE	Yeld	GA	DE	Yeld	GA	DE
3	21,50	68,40	54,35	19,00	70,00	38,88	15,00	78,00	49,44	2,24	72,00	84,51
5	23,60	70,20	52,17	24,68	74,00	26,37	17,50	81,60	50,00	1,77	75,00	81,69
7	26,30	72,00	50,55	29,71	76,00	23,91	18,00	84,00	51,14	3,77	70,80	82,35
10	26,80	74,40	50,55	32,00	76,20	21,98	18,98	85,20	52,27	3,92	61,20	84,29

The content of galacturonic acid in pectin substances varies according to a similar pattern, with the exception of pumpkin pectin. With the increase in the duration of the hydrolysis-extraction process, the change in the content of GA units in pumpkin pectin is extreme. In this case, the maximum is for a 5-minute extraction. Further, the content of HA decreases, reaching 61.20% at the end of the process, which gives grounds to suggest that the process of decomposition of the pectin macromolecule of this type of raw material is uneven. For other sources of raw materials, as the duration of the process grows, the pectin substances are continuously enriched with GA units. It should be noted that the PS rhubarb is characterized by a high content of galacturonic acid units, which indicates the predominance of linear homogalacturonan chains in pectin macromolecules of this type of raw material.

The value of the degree of esterification remains practically unchanged. Exceptions are the pectin substances of the sunflower heads, the value of which with the increase in the duration of the process is somewhat reduced. The reason for this can be the de-esterification of the methyl groups of the galacturonic acid units and the breakdown of the macromolecule of the polysaccharide matrix, especially the esterified GA residues (Table 1).

Calculation of molecular and hydrodynamic parameters of pectin substances was carried out using the software ASTRA 5.3.4.13 and Breez (Waters). The change in the molecular mass parameters of various pectins is of a dissimilar character and indicates the influence of the raw material source on the properties of the target products.

In particular, for pectin substances of apples and sunflower, the molecular weight and the polydispersity index decrease with increasing duration of the hydrolysis-extraction process under the influence of high temperature and pressure. This fact, as well as low values of yield of PS in three-minute hydrolysis, indicate extraction at the beginning of the process of native, yet not degraded pectin macromolecules. Further increase in the duration of the process leads to the destruction of aggregated pectin molecules with a simultaneous decrease in the polydispersity of the samples, which indicates the enrichment of the pectin substances in the hydrolysis-extraction process by linear homogalacturonan chains.

Table 2. Molecular-mass parameters of pectin substances of apples, sunflower heads and rhubarb.

Time, min	Apple				Sunflower				Rhubard		
	$Mw\ e^{-3}$, kD	Mw/Mn	$Mz\ e^{-3}$, kD	AG*, %	$Mw\ e^{-3}$, kD	Mw/Mn	$Mz\ e^{-3}$, kD	AG*, %	$Mw\ e^{-3}$, kD	Mw/Mn	AG*, %
3	560	11,5	7736	31,5	5195	70	86660	36,5	98	2,4	9,9
5	370	8	7310	7,5	175	5	1120	33,5	114	2,3	9,8
10	190	6,7	3806	15	98	1,5	351	19	127	7,5	6,34
60**	134	3,8	1133	23,15	142	3,22	1692	40	268	4,8	5,3

*aggregated fraction;

** samples were obtained at atmospheric pressure, temperature 85 °C, hydromodule 1:20, duration 60 minutes, continuous stirring, pH = 1.2

From the differential molecular weight distribution (MWD) curves shown in Figures 1 and 2, it can be seen that at the beginning of the process the pectin substances of apples and sunflower have a broad MMP. But, with increasing duration of hydrolysis, high-molecular aggregates gradually turn into samples with a narrow molecular-mass distribution. Further increase in the duration of the process to 10 min leads to the formation of molecules with a bimodal distribution. If the low temperature and long duration of the process lead to the production of low molecular weight pectin substances with a narrow molecular weight distribution, the effect of high temperature and pressure even in a short period of time exerts a stronger effect on the degradation of the polysaccharide matrix of these raw materials.

The rhubarb pectin substances obtained at atmospheric pressure, on the other hand, have a higher molecular mass and are more polydisperse than those obtained under the influence of high temperature and pressure. With increasing duration of the MMP process, the rhubarb pectin becomes broad, and the molecular weight and polydispersity values increase. Taking into account the high value of Mw of pectin obtained at 60 min. and low temperature, it can be concluded that the high molecular weight fractions of pectins in the rhubarb are localized in the secondary cell wall. In addition, PS of rhubarb are different from samples obtained from other sources of raw materials in that they are more thermolabile and are degraded at 120 °C, even in a short period of time (Table 2, Figure 3).

For pectin substances of pumpkin, regardless of the method of production and the duration of the process, a bimodal molecular weight distribution is characteristic (Table 3, Figure 4). The appearance of a second peak in the differential MWD curves indicates the presence of a second low-molecular fraction, which indicates the heterogeneity of the structure of the PV of the given feedstock.

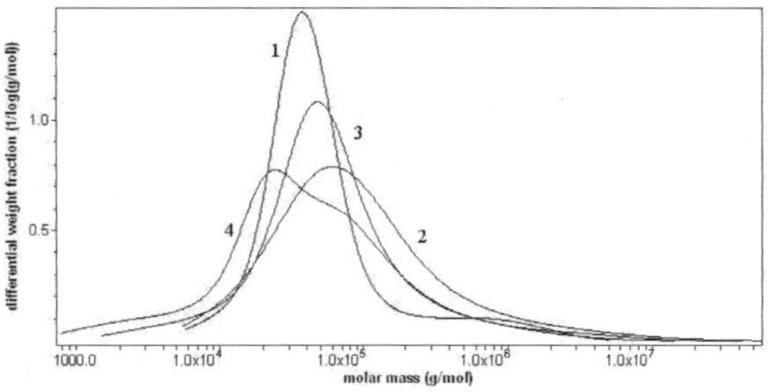

Figure 1. Differential curves of the molecular weight distribution of the pectin substances of apples obtained at T = 85 °C (1) and under the influence of high temperature and pressure (2 - t = 3 min, 3 - t = 5 min, 4 - t = 10 min)

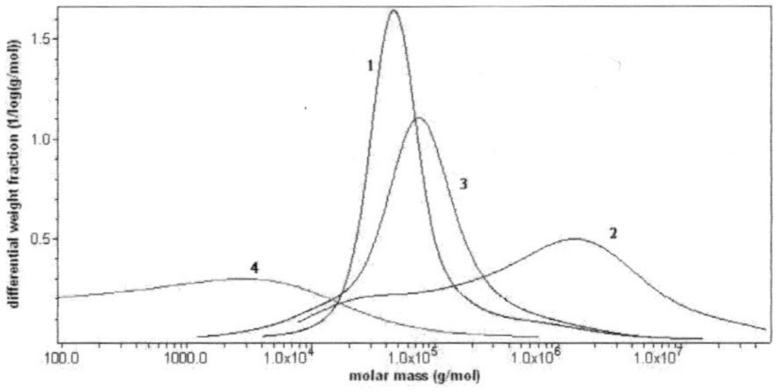

Figure 2. Differential curves of the molecular weight distribution of pectin substances of sunflower obtained at T = 85 °C (1) and under the influence of high temperature and pressure (2 - t = 3 min, 3 - t = 5 min, 4 - t = 10 min)

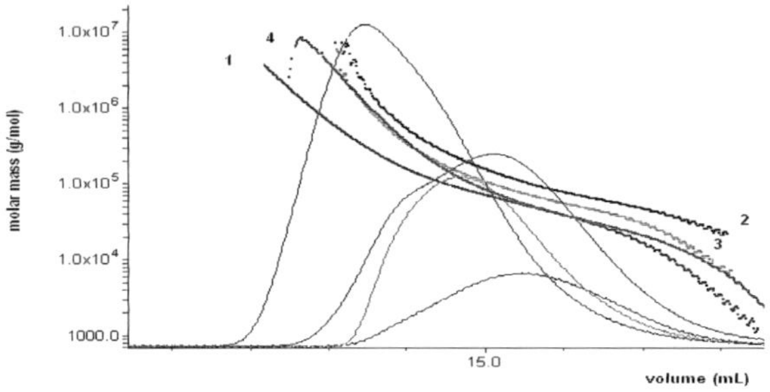

Figure 3. The curves of the molecular weight distribution of rhubarb pectin substances obtained at T = 85 °C (1) and under the influence of high temperature and pressure (2 - t = 3 min, 3 - t = 5 min, 4 - t = 10 min)

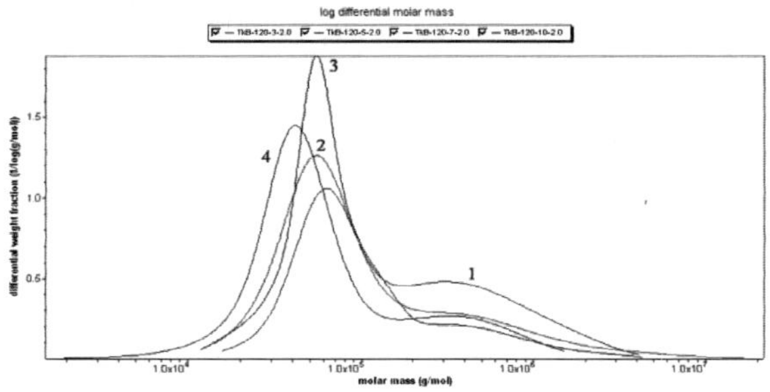

Figure 4. Differential curves of the molecular weight distribution of pectin substances of pumpkin, obtained under the influence of high temperature and pressure (1 - t = 3 min, 2 - t = 5 min, 3 - t = 7 min, 4 - t = 10 min)

Table 3. Molecular-mass and hydrodynamic parameters of pectin substances of pumpkin

Time, min.	№	Mw e^{-3}, kD	M_w/M_n	Mw e^{-3}, kD	$[\eta]^{*}$, ml/g	$R_{h\,(w)}^{**}$, nm	b^{***}	AG, %
60	1	673,90	1,93	2069,00	747,0	39,7	0,57	17,70
	2	72,02	1,20	82,99	251,0	13,2	-	-
3	1	661,00	1,80	1267,00	689,4	39,0	0,61	37,50
	2	65,71	1,15	74,29	162,4	11,2	-	-
5	1	769,50	3,40	3969,00	375,1	33,1	0,47	40,00
	2	58,58	1,20	69,01	125,8	10,0	-	-
7	1	622,10	1,86	1317,00	471,8	33,4	0,56	21,60
	2	62,61	1,28	77,84	153,3	10,9	-	-
10	1	403,50	1,56	619,40	492,4	33,1	0,50	34,95
	2	41,72	1,31	48,64	180,0	10,2	-	-

*viscosity

** hydrodynamicradius;

*** coefficient, found from the curve of the dependence of the hydrodynamic radius and the molecular mass, which determines the conformation of the macromolecule in the solution.

The main fraction of pumpkin pectin substances obtained in the low temperature process has a high molecular weight of 673900 kD and Mz (2069000 kD), a low polydispersity index (Mw / Mn = 1.93), and a low aggregation capacity. At the beginning of the hydrolysis-extraction process under the influence of high temperature and pressure, the pumpkin pectin substances have molecular weight and polydispersity values close to those of samples obtained at low temperature. But, at the same time, there is an almost twofold decrease in the Mz index and an increase in the aggregation capacity of the main fraction of pectin substances. With an increase in the duration of the hydrolysis-extraction process, the change in the molecular weight of both Pumpkin PT fractions is extreme. The maximum is 5 minutes. Similarly, the Mz index and the polydispersity of the samples change.

The value of the coefficient b also has a high value, regardless of the method for producing the pumpkin pectin, with the exception of samples obtained by 5-minute extraction under the influence of high temperature and pressure. The average values of the coefficient b for the mixture of extracted pectins are found to indicate the conformation of random coils with an elongated shape. The PS macromolecules obtained with a 5-minute extraction have a compact conformation, because of their high aggregation, as indicated by the high values of AG and Mz.

An analysis of the data obtained shows that the appearance of maxima in the change in the content of the elements of galacturonic acid and the molecular mass parameters in the region of 5 minutes indicates that up to this time value the cell wall decays and the extraction of native pectin macromolecules. With the further increase in the duration of the process, along with the continuing disintegration of elements of the cell wall, the decomposition of already extracted high molecular molecules begins.

4. Conclusion

The studied effect of the duration of the hydrolysis-extraction process under the influence of high temperature and pressure gives additional information on the structure of the pectin macromolecule of various plant materials. The revealed regularities can serve as a basis for development of technology for obtaining target products with specified properties.

References

[1] Voragen A G J, Pilnik W, Thibault J-F, Axelos M A V, Renard M G C, Stephen A M (Ed) 1995 *Food polysaccharides and their applications*, Marcel Dekker, pp. 287-339

[2] Ovodov Yu S 2006 *North: science and prospects for innovative development*. Syktyvkar: Publishing house of the Komi Scientific Center of the UrB RAS, pp. 236-255

[3] Endress H. U. 1991 *The Chemistry and Technology of Pectin*. Ed. Walter B.H. San Diego: Acad. Press, pp. 251-268

[4] Karpovich NS, Donchenko L.V. 1989 *Pectin: production and use*. Kiev: The harvest. 88 p.

[5] Savchenko N, Sevostyanova I, Sablina T, Gömze L, Kulkov S 2014 *AIP Conference Proceedings* **1623** 547 http://dx.doi.org/10.1063/1.4899003

[6] Gömze L A, Gömze L N 2013 Advanced Hetero-Modulus and Hetero-Viscous Complex Materials *Materials Science Forum* **729** 43

[7] Slobodov A, Uspenskiy An, Ralys R, Kremnev D 2015 *Epitoanyag – Journal of Silicate Based and Composite Materials* **67** (4) 159 http://dx.doi.org/10.14382/epitoanyag-jsbcm.2015.27

[8] Ralys R V, Uspenskiy A A, Slobodov A A 2016 *J. Non-Equilibrium Thermodynamics* **41** (1) 3

[9] Uspensky A B, Ralys R, Kremnev D, Radin M, Slobodov A 2017 *IOP Conf. Ser.: Mater. Sci. Eng.* **175** 012024 https://doi.org/10.1088/1757-899X/175/1/012024

[10] Ageev E I, Andreeva Ya M, Karlagina Yu Yu, Kolobov Yu R, Manokhin S S, Odintsova G V, Slobodov A A, Veiko V P 2017 *Laser Physics* **27** (4) 046001

[11] Gorshkova R M, Mukhidinov Z K, Khalikov D Kh., Khalikova S, Nasriddinov A S, Johnmurodov A S 2009 *Journal of Public Health in Tajikistan* **3** (303)14

[12] Gorshkova R.M, Nasriddinov A.S., Mukhidinov Z.K, Khalikov D.Kh., Khalikova S., Teshaev

Kh.I. 2010 *News of Russian: "Chemistry and chemical technology" series* **53** (6) 87

[13] Gorshkova RM, Mukhidinov ZK, Makhkamov KhK, Johnmurodov AS, Khalikova S, Khalikov D Kh 2010 *Bulletin of the Tajik Technical University* **3** (119) 55

[14] Gorshkova RM, Mukhidinov ZK, Nasriddinov AS, Valiev MV, Khalikova S, Khalikov DKh, Liu LSh 2010 *News of the Academy of Sciences of the Republic of Tajikistan* **4** (141) 45

[15] Mukhidinov ZK, Gorshkova RM, Valiev MV, Khalikova S, Nasriddinov AS, Khalikov DKh, Fishman ML 2010 *Reports of the Academy of Sciences of the Republic of Tajikistan* **53** (8) 617

[16] Mukhidinov Z.K, Gorshkova R.M, Valiev M.V, Khalikova S., Nasriddinov A.S, Khalikov D.Kh., Fishman M.L. 2010 *Reports of the Academy of Sciences of the Republic of Tajikistan.* **53** (9) 701

[17] Khalikova S., Gorshkova R. M., Valiev M.V. (2015) *Chemical Journal of Kazakhstan* **2** (50) 202

[18] Khalikov D Kh, Gorshkova R M, Mukhiddinov Z K, Avloev Kh Kh, Teshaev Kh I 2002 *Chemistry of natural compounds* **2** 118

[19] Khalikov D Kh , Gorshkova R M, Khalikova S, Avloev Kh Kh, Mukhiddinov Z K 2007 *Reports of the Academy of Sciences of the Republic of Tajikistan* **50** (4) 340

[20] Filizetti-Cozzi T M C C, Carpite N C. 1991 *Journal Anal. Biochem.* **197** 157

[21] Kelko C P 2001 Determination of DE. *Control metod.* March 7 3

[22] Mukhiddinov Z K, Fishman M L, Gorshkova R M, Nasriddinov A S, Khalikov D Kh 2008 *Chemical Journal of Kazakhstan: Special Issue.* **21** 60

[23] Mukhiddinov ZK, Fishman ML, Gorshkova RM, Nasriddinov A S, Khalikov DKh 2008 *Chemical Journal of Kazakhstan: Special Issue.* **21** 83

IOP Conf. Series: Journal of Physics: Conf. Series **1045** (2018) 012016 doi:10.1088/1742-6596/1045/1/012016

Galvanic Corrosion Occurs Heat Experiments by Thermographic Camera

F Haraszti and T Kovacs

Óbuda University, Bánki Donát Faculty of Mechanical and Safety Engineering
Material Science Department, 1081 Budapest, Hungary

Email: kovacs.tunde@bgk.uni-obuda.hu

Abstract. A part of the electric connections has a corrosion aptitude. We can find galvanic corrosion process hazard in case of the contact between different electron potential metals. This process impairs mechanical properties and the connection increasing resistance. The conduction parameters decrease in some under power parts because of the heat rise. This heat effect can be unperceived. By thermographic camera experiments, we can discover this irregularity on time. In this work, the authors present this investigation and the emerging problems.

1. Introduction

The electric connection constructions have different metal elements with different mechanical properties. We know different corrosion forms for different materials. Between the different metals, a corrosion reaction can occur if we have different electrode potentials. This form of the corrosion is knows like galvanic corrosion. This corrosion is a chemical corrosion that occurs between two different metals in the presence of a liquid electrolyte. The name of this corrosion type came from the name of the chemical process, because two electrochemically different metals and the electrolyte create a galvanic cell. The low electrode potential metal dissolves in the electrolyte and the surface damaged. This reaction effect impairs the electric current transfer because it increases the momentary resistance [1-3].

2. Theory basis of the investigation

2.1. Electrical contact testing possibilities

In our modern age, we cannot live without electric current, to assure the continuous current supply is a critical infrastructure task [4]. The metal connection bounding lifetime calculation is very difficult because of the corrosion phenomenon and is not calculable easily. The electrical contact problem can cause big problems in the industrial and personal life too. Prediction of the electric device-bounding lifetime is a challenge for our century. The simplest tests are the visual inspection and the temperature measurement. The bounding impairment causes temperature increase. The temperature measurement is not easy by the traditional ways and measuring methods, since for this process we switch off the current. The power stop can cause impairment in industries and homes. In grace of the modern technology innovations, we can find new devices for safe and exact temperature detection.

2.2. Thermographic camera usability

Temperature measuring the infrared radiation by the thermographic camera depend on the bodies heat effect capabilities. The heat radiation process influences the material heat movement that produce electromagnetic waves. The heat energy emission can be in form of conduction, convection and radiation. The bigger part of the heat emission for other bodies can be reflection or transmission. A completely black body absorbs all radiation. The sample that reflects all radiation is a completely reflective body. The one, which lets through all radiation, is the transparent body. The heat radiation like electromagnetic wave transfer does not need any media. The heat radiation fundamental law is the Stefan-Boltzmann law (1) [5]. Where temperature T [K], the radiated energy P_{Rad} [J], e [-] is the emissivity of the object, A_s [m^2] is the surface area, and σ=5.67 10^{-8} [J/m^2K^4] is the Stefan-Boltzmann constant. The emissivity constant depends entirely on the material of the object and is capped at 1 for an ideal black body [5].

$$P_{Rad} = e\sigma A_s T^4 \,[\mathrm{J}] \tag{1}$$

Even that e, A_s and σ are constant, the equation will be simple (2):

$$P_{Rad} = \sigma \cdot T^4 \,[\mathrm{J}] \tag{2}$$

By this law, the radiation of the black body on T temperature for unit area and during unit time is proportionate with T^4. In an ideal event, the black body is able to absorb and emit all heat energy. It is only a model. Different materials can emit different energy. In case of measuring, the black body like material gives exact results. In case of the different materials, we need to correct the result.

3. Experiments

3.1. Used samples and preparation

The electrical bounding is very common in the copper and steel contacts. To investigate the galvanic corrosion phenomenon, we made a bounding for our laboratory simulation test. The measured resistance before the corrosion is from 0.01Ω. Before the test we treated this bounding by surface preparation and corrosion load effect in 3% salt spray during 168 hours on 40 °C temperature. Figure 1 shows the used samples.

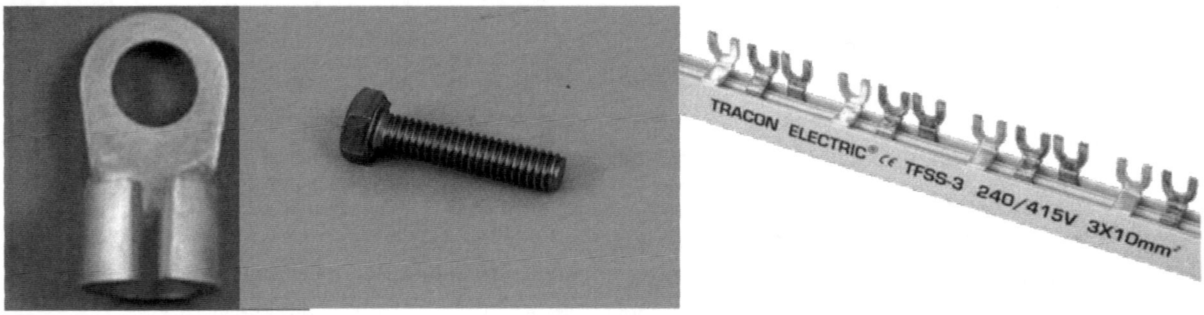

Figure 1. Copper sample, steel screw, single-phase comb busbar max. 63 A

3.2. Experiment process

Fig. 2 shows the experimental setup with integrated thermographic camera (TESTO 880). In case of the heating device's different position, the used power is different. The electric current is transferring on the prepared bounding, that causes a measurable heat because its resistance changes during the treating preparation (5600 Ω). The thermographic camera supports a safe and exact heat measuring process [6, 7]. In this experiment, we measured the temperature of the bounding as a function of the electric current change.

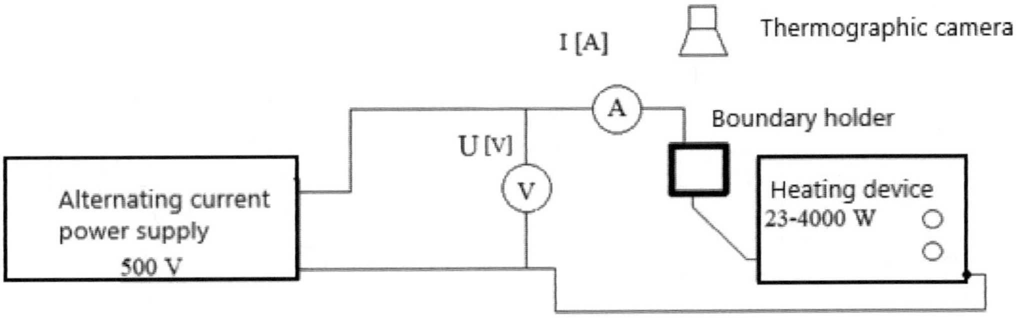

Figure 2. Experimental setup

The thermographic camera registered result shows on Fig. 3.

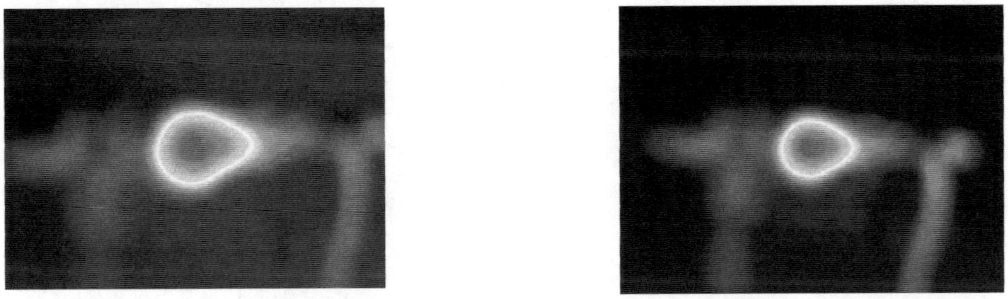

Figure 3. Image of the tested bounding

3.3. Experiment results

Tab.1. shows the experiment results, which shows a relationship between the change in temperature and the current. High-density electric power causes heat. We supposed based on of the equation (3) a relationship between the temperature and the power. The equation (3) where P [W] is the electric power, I [A] is the electric current, U [V] is the voltage:

$$P = U \cdot I \ \text{[W]} \tag{3}$$

Table 1. Experimental results

Number	Current [A]	Temperature [°C]
1	0	27,2
2	0,08	27,4
3	0,15	28,4
4	3,9	44,3
5	7,66	75
6	11,58	125

Use the fundamental equation, where R is the ohmic resistance (4):

$$U = R \cdot I \ \text{[V]} \tag{4}$$

The equation (3) we modified according the equation (4) to write the next (5):

$$P = R \cdot I^2 \ \text{[W]} \tag{5}$$

The result equation (5) shows the relationship between the power (P) and the current (I). From this equation, we can see that the electric power in our experiment and the heat energy has square relationship with current (Fig. 4).

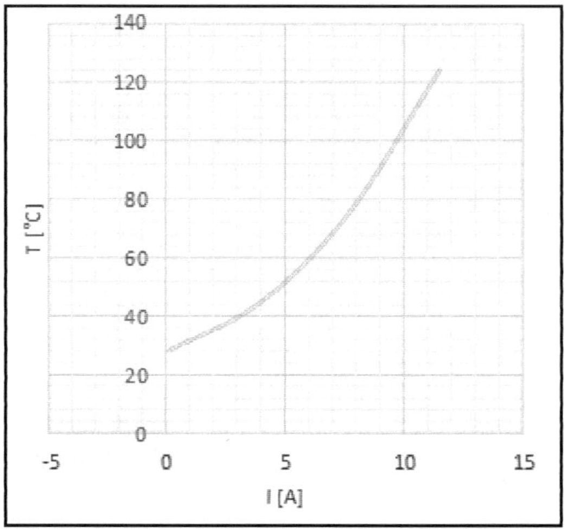

Figure 4. The relationship between current and temperature

4. Conclusions

The experiment results supported our hypothesis about the relationship between the change in temperature and the current. The reason of this relationship is that the electrical energy transforms to heat energy, according to Newton's energy conservation law.

I. The temperature of the used bounding changed during the experiment. Based on this result we can conclude that in real application we can detect the same phenomenon.
II. The galvanic corrosion causes the change of the resistance. The galvanic corrosion results an extremely high resistance in the bounding.
III. The changing temperature is measurable and it can be detected a by thermographic camera device. The thermographic camera is very useful because it can measure the temperature without any contact meaning that the measuring process is safe.

References

[1] Haraszti F 2016 The bases of corrosion's investigation, Papers on Technical Science EME kiadó, Kolozsvár pp. 185-188
[2] Haraszti F 2016 Corrosion investigation of steel smaples , Papers on Technical Science EME kiadó, Kolozsvár pp. 189-192
[3] Kovács-C T, Völgyi B, Sikari-Nágl I 2013. Papers on Technical Science EME kiadó, Kolozsvár pp. 201-204
[4] Nyikes Z, Rajnai Z 2015 IEEE 13th International Symposium on Intelligent System and Informatics Proceedings pp. 217-218.
[5] Wellons M.: The Stefan-Boltzmann Law, Physics Department, The College of Wooster, Wooster, 2007. Ohio 44691, USA
[6] Kittel C, Kroemer H.: Thermal Physics (W. H. Freeman and Company, 1980) ASTM Standard Practice in A 262 for Detecting Susceptibility to Intergranular Corrosion in Austenitic Stainless Steels
[7] Lei Fu, et al: Effect of 1,2,4-triazole on galvanic corrosion between cobalt and copper in CMP based alkaline slurry, Journal of Semiconductors, Volume 39, Number 4. 2018.

IOP Conf. Series: Journal of Physics: Conf. Series **1045** (2018) 012017 doi:10.1088/1742-6596/1045/1/012017

Stress Distribution of a New Type of Perforated Implants in Femur Neck Fracture: Finite Element Model

Afgan Jafarov[1], Chingiz Ali-Zadeh[2], Zafer Özer[3], Amirullah M. Mamedov[4,5], Ekmel Özbay[4]

[1]Modern Hospital, Baku, Azerbaijan
[2]Azerbaijan Scientific-Research Institute of Traumatology and Orthopedics, Baku, Azerbaijan
[3]Mersin University, Mersin Vocational High School Electronic and Automation Department, Mersin, 33335, Turkey
[4]Bilkent University, Nanotechnology Research Center (NANOTAM), Ankara, 06800, Turkey
[5]International Scientific Center, Baku State University

E-mail: mamedov46@gmail.com

Abstract. Osteosynthesis of femoral neck fractures (FNF) is accompanied by a large percentage of various complications. The aim of this research was to study biomechanical implications arising from the stress and strain distributions by use of new device –perforated H-beam implant. A reference 3D femur model has been developed using digital femur geometry. The stress and deformity characteristics after the application of force were determined by ANSYS, which entails the finite element analysis (FEA) method. This method was used for both the perforated and non-perforated implant models and compared the application of constant strength. Three types of FNF were realized according to the Pauwell classification system.

We concluded that, the perforations in the implants make the distribution of strength and pressure homogenous along the whole implant functioning as wave breaker to reduce the pressure on the bone. This contributes to implant stability and the minimization of bone destruction.

1. Introduction

Scientists and clinicians have long been interested in the assessment of the mechanical behavior of human whole bones. The general impetus has been twofold, namely to understand the various geometric and material factors that affect gross mechanical behavior under various physiological loading conditions and to determine which orthopedic repair techniques and implants can best preserve or improve intact whole bone biomechanical strength and stability. To these ends, both experimental methodological and theoretical models have been reported in the literature [1-3]. For example, in the assessment of the performance of orthopedic implants to identify areas of high stress, and to facilitate implant design, biomechanical testing has often been augmented by the use of finite element models (FEM) of bone-implant systems.

Finite element analysis of a whole bone was first performed in [1-2] to gain a perspective on the stresses encountered by the human femur (2D geometry), although later models were 3D in nature [3].

Whole bone FEM studies of these types ignore the structure of bone, instead treating it as continuum. Nevertheless, this approach is useful for implant design purposes, where comparative, rather than absolute, values of stress are required. More fundamental properties, such as the microstructure, loading mode, and elastic properties of tissue have been obtained using FEM. On the other hand, the most important problem for elderly patients is the problem of the implant's internal fixation method and its geometry.

The aim of the present investigation was to gain insight into the biomechanical implification arising from the stress and strain distributions induced by the implantation of specific fixation devices: perforated and non-perforated H-beam implants.

The FEM that has been accepted as a standard tool for the analysis of a wide range of biomechanical problems since its introduction into this research field during the early 1990s was used to investigate the stress in the implants and the surrounding bone material under typical loading conditions. These stress patterns were compared with those found in a physiological uninjured femur, exposed to the same loading environment (Fig. 1)

1.1 Model Geometry, Implants, and Fracture zone

A major simplification of the essentially 3D nature of the problem nature of the problem to be studied can be based on the observation that the external forces corresponding to appropriate representative load systems (such as the single leg stance phase of normal gait) are acting mainly in the mid frontal plane. Accordingly, the relevant mechanical responses of the bone-implant configurations are dominated by in plane deformations and stresses. Thus, provided that the tensile, compressive, shear and bending stiffnesses of the bone-implant system (Fig. 3) are adequately represented, two-dimensional plane stress FE-models are sufficient for investigating this problem. Specific layers of elements for each constituent of the femur- implant system (i.e. steel implant, cortical bone and trabecular bone) were introduced. The thickness of each individual finite element being selected according to the real geometry of the implants and a representative femur, taking into account the medullary cavity of the femoral shaft, and the reduction of bone material thickness caused by implantation as well as the percentage of cancellous and cortical bone in the femoral head. By using these overlay techniques, the numerical requirements were significantly reduced without seriously compromising the realism of the models.

Therefore, the objectives of this study are as follows:

1. To examine the perforated and non-perforated H-beam implants (see, Fig. 3)
2. Investigate the physico-mechanical behavior of the H-beam implants using the finite element method.
3. To examine various stress, strain, and changes in the geometry of the different implant groups.

Three osteotomies through the trochanteric region, inclined at an angle of 30°, 50°, and 70° respectively, formed the quasi-real model of an unstable trochanteric fracture. Three variations of this idealized wedge shape fracture were used to model clinically relevant configurations. This configuration was approximated by reducing the fracture area to a zigzag line, along which frictionless contact between the surfaces was assumed, resulting in a non-linear contact problem. In a second scenario, a callous bridging was assumed, filling the wedge-shaped fracture zone in the cancellous as well as in the cortical layer with callus of relatively low stiffness. The third configuration addressed the fully healed state of the femur with completed bony bridging.

2. The model and calculation method

The calculations of the stress and shape change condition in the implant were carried out by the finite element method. The newly developed hip prosthesis, shown in Figure 3, was investigated for mechanical behavior in the modeled femur bone, the Pauwel's classification system (PCS) type1 in Figure 1, type 2 and type 3 fractures.

The key points used in the reference model of the femur bone were obtained from the two-dimensional digital picture in the CAD environment. When creating a 3D model, the bottom-up

IOP Conf. Series: Journal of Physics: Conf. Series **1045** (2018) 012017 doi:10.1088/1742-6596/1045/1/012017

modeling approach was combined with key points splines, fields using splines, and volumes were created using fields.

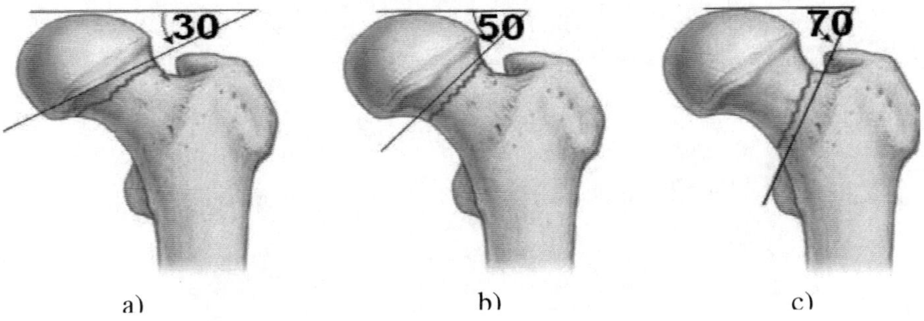

Figure 1. Pauwel's classification system (PCS-femur bone)
a) Type 1, b) Type 2 c) Type 3

The results of the surgical titanium prosthesis were applied to the femur and the result was evaluated by the finite element method, with a vertical force of up to 5 times the weight of a normal weight person during walking. The anatomical femur was investigated using a three-dimensional finite element model (3D-FEA). The measurements of the 2-dimensional femur bone for modeling were transferred to CAD software.

The 4000 N force, which occurred during the walking movement of a person of normal weight and up to 5 times its weight, was applied to 4 different points in 1000N. This is based on one leg of the press during the walk, and the load transferred from the hip joint to the femur head sphere was applied in accordance with the mechanical axis. The total number of elements and nodes in the implanted models are as shown in Table 1. The material properties used in the analysis were taken from the literature [17]. Figure 2 shows the applied force and boundary conditions on the femur. For FEA, the Solid 92 element in a 3-D 10-node tetrahedral structure (triangular finite elements with six nodes and semi-circular side were used) suitable for dividing irregular structures such as the hipbone into meshes was used.

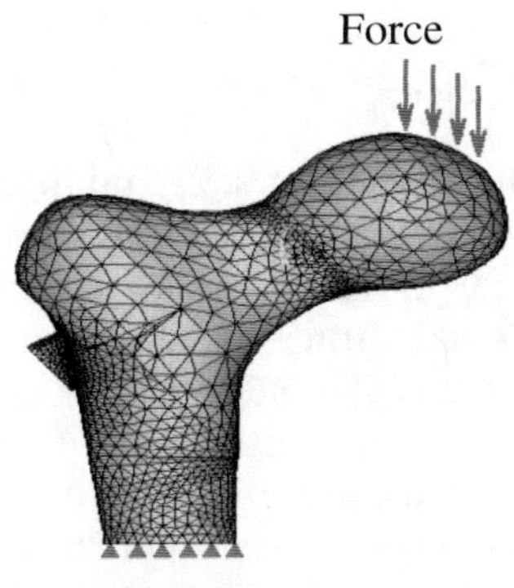

Figure 2. Directions of applied forces and boundary conditions

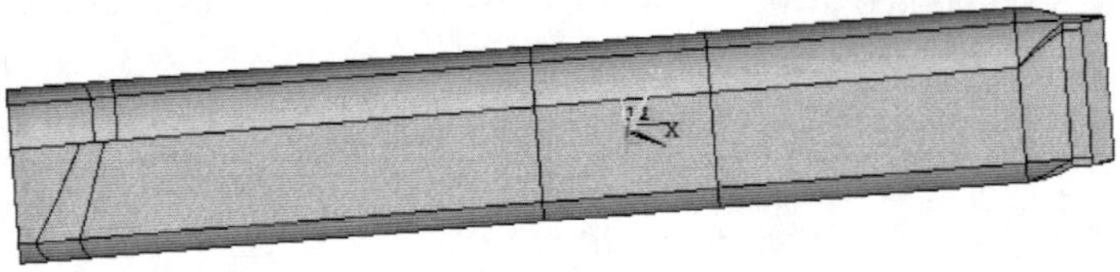

a.

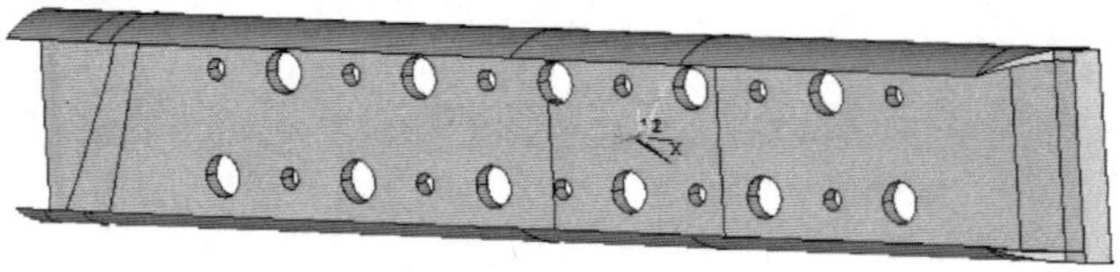

b.

Figure3. ImplantModels: a) non-prtforated-model 1 and b) perforated-model 2

Table1. Input data for FEA

	Numbers of elements	Numbers of nodes
Type 1 (Model 1)	67141	94381
Type 2 (Model 1)	71592	97715
Type 3 (Model 1)	100954	137006
Type 1 (Model 2)	136230	186452
Type 2 (Model 2)	74900	109912
Type 3 (Model 2)	162737	230258

3. Results and Discussions

According to the results obtained, the amount of Von Mises stress consisting of both models (hole-free and perforated) in parallel to the increase of the fracture type from 1 to 3 as increased. Maximal stress is increasing in the case of cortical and trabecular bone models with the increase of the non-hanging angle of the model. Only in Model 2/Type 2 class Q types, changing is calls in the maximal stress of cortical bone (Tables 2-3).

In all fracture types of Model 1, the amount of stress falling on the cortical bone of the maximal Von Mises that fall into the trabecular bone region was observed at an average of 1/3 percent. Model 2/Type 1 fracture type also continues to 1/3 ratio. However, in the Model 2/Type 2 and type 3 fracture types, the maximal ratio of trabecular and cortical bone weaving was observed in 1/2. These results indicate that there are holes in the implants and the bathtubs that fall into the trabecular bone models even increase. In Model 1 and 2 samples, in comparison with type 3 fractures, the most significant increase in stress figures on the implant and trabecular bone tissue was calculated (521.324/285.339 on the implant); 58,111/32,056) on the trabecular tissue. In both cases, there was a 1.8 fold increase. In the fracture area of Pauwel's Type 3 fractures, the implant and trabecular bone area were observed to decrease the forces. Tables 2-3 summarize the numerical results for two types of implants. A comparative analysis of the results obtained by using two types of implants showed a good convergence deviations ranging in reasonable limits. The plane model is suitable for carrying out a more detailed study of the state of stresses, using finite element analysis, behavior under mechanical load for proximal area. As a result of the action of mechanical strain forces in the femoral head and the greater trochanter in the elements that compose the structure is generated axial efforts (tensile and compression). Therefore, by way of finite element analysis, the state of stresses identifying the

IOP Conf. Series: Journal of Physics: Conf. Series **1045** (2018) 012017 doi:10.1088/1742-6596/1045/1/012017

maximum loaded area can be shown. To characterize this state of stresses, Von-Mises equivalent stress distributions are summarized in table 2. Von-Mises equivalent stress values and the maximum and minimum principal stresses depending on application force F.

Trabecular structure is shown taking over the mechanical load and is distributed to the cortical structure and especially toward diaphysis where the maximum stress is registered, the critical area – femoral neck – thereby subjected to a lower load level. It is noted as an important role that it plays in the trabecular structure and integrity of this structure depending on the integrity of the structure of the human femoral bone (proximal area).

Table 2. Von Mises stress distribution on Model 1 implant.

	Maximum Von Mises Stress [MPa]		
	Type1	Type2	Type3
Prosthesis	210.673	278.53	285.339
Trabecular Bone	23.949	31.281	32.056
Cortical Bone	70.630	90.094	95.377

Tablo 3. Von Mises stress distribution on Model 2 implant.

	Maximum Von Mises Stress [MPa]		
	Type1	Type2	Type3
Prosthesis	209.821	300.996	521.324
Trabecular Bone	23.915	33.705	58.111
Cortical Bone	70.392	67.117	116.012

In the Model 1 and 2 samples, in comparison with type 3 fractures, the most significant increase in stress figures on the implant and trabecular bone tissue was calculated (521.324/285.339 on the implant); 58,111/32,056) on the trabecular tissue. In both cases there is a relatively 1.8 fold increase. After all, in the fracture area of Pauwel's Type 3 fractures, the implant and trabecular bone area was observed to decrease the forces.

4. Conclusion

The finite element method is an easy and safe method to describe the weight of the hips of the hip play, the geometry of the ink in the different phases of the surface, and the load distribution on the implants used in different fracture types [18-19].

In the computer environment, it is easier to perform stress analysis with FEA and predict errors for the prostheses and implants designed. FEA was initially developed for the aircraft industry, and it was also used in the analysis of implants developed in medicine over time. 3D-FEA allows us to understand the biometric properties of implants better, thereby helping the designer in the design and improvement of the material used in the implant biomechanical functions.

In our study, a finite element method was used to investigate the load distribution around bone tissue and implant with holes on the new implant that were planned for osteosinthesis of femur neck fractures. The findings of the perforated and hollow implant models were compared and different and interesting results were obtained. After loading in perforated implant models, more strain was observed on the implant. This strain also affects the trabecular tissue regions.

References

[1] Ström O, Borgstrom F, Kanis J A, Compston J, Cooper C, McCloskey E V, and Jonsson B 2011. *Arch. Osteoporos* **6** 59

[2] Raaymakers E L 2006 *Acta Chir. Orthop. Traumatol. Cech.***73** 45
[3] Borqvist L, Lindelöw G, and Thorngren K G 1991.*Acta orthop. Scand* **62** 39
[4] Lagerby M, Asplund S, and Ringqvist I 1998 .*Acta Orthop. Scand.* **69** **(4)** 387
[5] Khoo CCH 2014 *Malays Orthop. J.* **8** **(2)** 14 PMCID: PMC4181088
[6] Osarumwense D,Tissingh E, Wartenberg K, Aggarwal S, Ismail F, Orakwe S, and Khan F 2015 *Clin. Orthop. Surg.***7** (1) 22
[7] Boraiah S, Paul O, Gardner MJ, Parker RJ, Barker JU, Helfet D, and Lorich D 2010 *Arch. Orthop. Trauma Surg.* **130** (12) 1523
[8] Shields E and Kates S L 2014 *Arch. Orthop. Trauma Surg.* **134** 1667
[9] Gjertsen J E, Vinje T, Engesaeter L B, Lie S A, Havelin L I, Furnes O, and Fevang J M 2010 *J. Bone Joint Surg. Am.* **92** **(3)** 619
[10] Wild M, Jungbluth P, Thelen S, Laffrée Q, Gehrmann S, Betsch M, Windolf J, Hakimi M 2010 *Orthopedics.* **11** 33
[11] Xu M, Zhang L, Mao Z, Wang H, Chen H, Guo Y, Tao S, Zhang Q, Liang X, Tang P 2010 *Zhongguo Xiu Fu Chong Jian Wai Ke Za Zhi.* **24** (12) 1419
[12] Bhandari M, Devereaux P J, Tornetta P, Swiontkowski M F, Berry D J, Haidukewych G, Schemitsch E H, Hanson B P, Koval K, Dirschl D, Leece P, Keel M, Petrisor B, Heetveld M, and Guyatt G H 2005 *J. Bone Joint Surg Am.* **87** (9) 2122
[13] Strömqvist B, Hansson LI, Nilsson L T,and Thorngren K G 1987. *Clin. Orthop. Relat. Res.* **218** 58
[14] Rehnberg L and Olerud C 1989 *Acta Orthop Scand* **60** (5) 579
[15] Lu-Yao G L, Keller R B, Littenberg B and Wennberg W E 1994 *J Bone Joint Surg* **76-A**15–25
[16] Mattsson P Larsson S 2003 *Scandinavian Journal of Surgery* **92** 215
[17] Atik F, Özkan A and Uygur İ 2012 *Sakarya Üniversitesi Fen Bilimleri Enstitüsü Dergisi* **16** (3) 249
[18] Voo L, Armand M, and Kleinberger M. 2004 *Johns Hopkins APL Technical Digest* **25** 223
[19] Bougherara H, Bureau M N, and Yahia L 2010 *J. Biomed. Mat. Res. Part A* **92** 164

IOP Conf. Series: Journal of Physics: Conf. Series **1045** (2018) 012018 doi:10.1088/1742-6596/1045/1/012018

Numerical simulation of mechanical behaviour of concrete under shock wave loading

V V Karakulov[1], I Yu Smolin[1,2] and S N Kulkov[1,2]

[1]National Research Tomsk State University, 634050 Tomsk, Russia
[2]Institute of Strength Physics and Materials Science SB RAS, 634055 Tomsk, Russia

E-mail: valery@ftf.tsu.ru

Abstract. In this paper results of numerical simulation of the mechanical behaviour of heavy concrete under preliminary compression and shock wave loading with intensity up to 1.2 GPa are submitted. The model specimen of concrete consisted of cement paste and granite filler with volumetric concentration above 65%. Mechanical behaviour of the representative volume of concrete was described by the model of the heterogeneous medium with stochastic distribution of filler particles. It was supposed that cement matrix can be considered as elastic-plastic material and granite of inclusions is a brittle material. The yield limit of cement paste was accepted to be 0.08 GPa, and the strength limit of granite equals 0.1 GPa. The absence of macroscopic and mesoscopic voids and cracks in the material was assumed. Ideal adhesion of filler material and matrix was supposed at the initial condition. Results of numerical simulation of the mechanical behaviour of concrete under shock loading have shown the formation of non-stationary and essentially non-uniform fields of deformations. The generation of a dissipative structure on the meso-scale level of the concrete under reshock wave loading was revealed in the simulation.

1. Introduction

The estimation of reliability of large scale concrete structures under natural and man-caused dynamic loadings remains to be one of urgent scientific and technical problems. The basic problem of the estimation of reliability of massive concrete constructions (dams, bearing elements of multi-storey buildings, concrete lining of tunnel and excavation) at intensive impacts is closely connected with prediction of the mechanical behaviour of concrete under combination of static and dynamic loads. Experimental data on properties of various grades of concretes in similar loading conditions practically are absent.

In the simulation of deformation of concrete constructions, concrete is often taken as homogeneous or quasi-homogeneous material. However, this material presents a complex of components (cement paste and filler) with different physical and mechanical properties. The components form the structure of concrete. The structure and its evolution during deformation can have a significant influence on the mechanical behaviour and the effective properties of concrete. However, the degree and nature of this influence is still incompletely understood.

In this connection, the present paper aims at the numerical simulation of the mechanical behaviour of concrete at macro-scale and meso-scale levels under preliminary compression and shock wave loading with intensity up to 1.2 GPa and study of effective mechanical properties of concrete with consideration for structural evolution during deformation of the material.

IOP Conf. Series: Journal of Physics: Conf. Series **1045** (2018) 012018 doi:10.1088/1742-6596/1045/1/012018

2. Simulation of the mechanical behaviour of concrete

The paper employs a plate of concrete consisting of a matrix (cement paste) and reinforcing inclusions (particles of granite filler) with volumetric concentration 65% to simulate loading by a plane shock wave. The simulation was performed on a rectangular fragment of the plane section of the plate along the direction of the shock wave front.

The concrete plate was loaded by a shock wave and an additional loading wave (reshock wave). The amplitude of particle velocity in shock wave equals 33 km/s, and the amplitude of particle velocity in additional loading wave equals 99 km/s. The additional wave was generated 23 micro seconds later than the loading by the first shock wave.

The mechanical behaviour of the concrete under the considered loading conditions was described using the physical and mathematical model of the two-phase condensed heterogeneous medium with an explicit description of its structure [1–3]. The used model assumes the heterogeneous medium as a set of interrelated structural elements, namely, a matrix and inclusions. Inclusions are arbitrary in shape and randomly distributed in the matrix. Mechanical interaction between structural elements occurs along internal contact surfaces, i.e., matrix - inclusion interfaces. Within interfaces of each structural element the medium is taken as homogeneous and isotropic while in transition through the interface physical and mechanical properties of the medium change abruptly.

It was supposed that cement matrix can be considered as an elastic-plastic material and granite of inclusions is a brittle material. The yield limit of cement paste was accepted to be 0.08 GPa and the strength limit of granite is equal to 0.1 GPa. The absence of macroscopic and mesoscopic voids and cracks in the material was assumed. Ideal adhesion of filler material and matrix was supposed at an initial condition.

Geometry of the simulated area and the number of structural elements was so chosen as to determine effective values of parameters of the mechanical state of the medium by averaging of calculated local values. The simulated area of the two-phase heterogeneous medium with the model structure is shown in Figure 1.

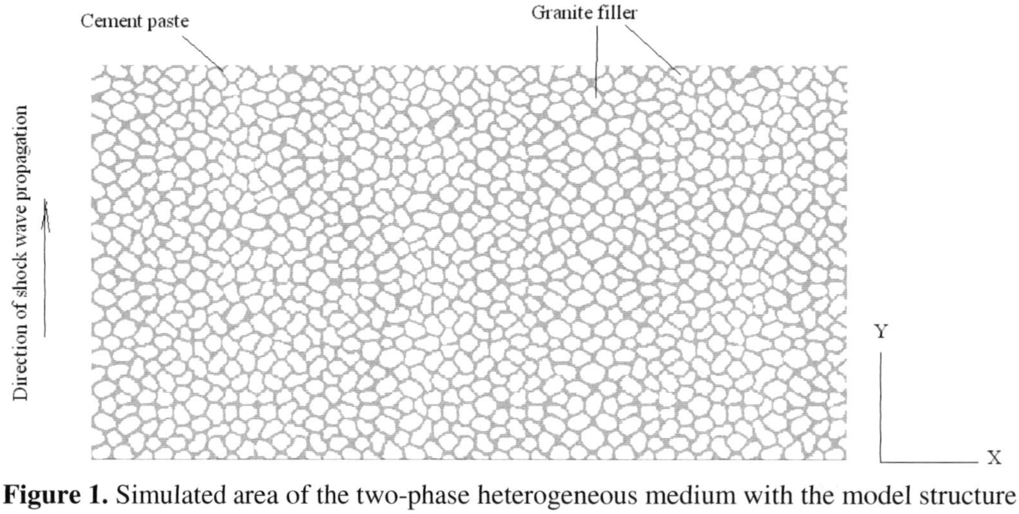

Figure 1. Simulated area of the two-phase heterogeneous medium with the model structure of the concrete composed of the matrix (dark region) and arbitrary-shaped inclusions (light regions). The volume concentration of inclusions is 65%

Effective parameters of the mechanical state of the heterogeneous medium loaded by a plane shock waves was determined by volume averaging of local values of the state parameters in thin flat layers perpendicular to the shock front direction. A method for determining effective parameters of the mechanical state was described in [4, 5].

IOP Conf. Series: Journal of Physics: Conf. Series **1045** (2018) 012018 doi:10.1088/1742-6596/1045/1/012018

3. Investigation of the mechanical behaviour of concrete

Results of numerical simulation of the mechanical behaviour of concrete under shock loading have shown a formation of non-stationary and essentially non-uniform fields of stress and deformation. Inelastic deformation of concrete under dynamic loading is a result of cement paste deformation and the movement of filler particles at the meso-scale level. Results are shown in Figure 2 and 3.

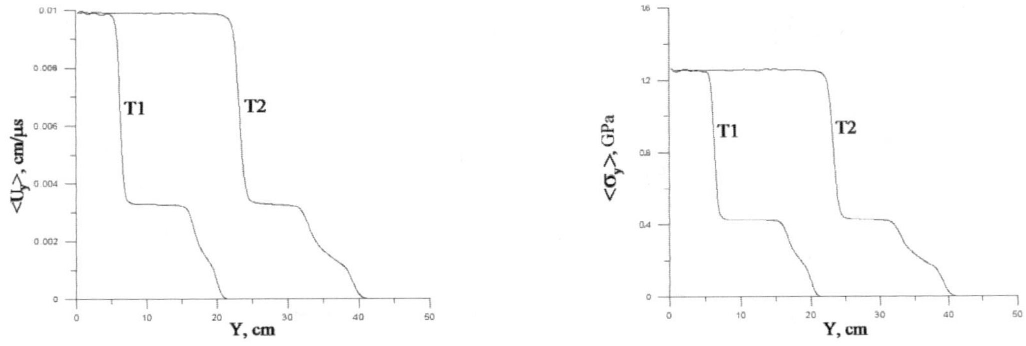

Figure 2. Calculated values of effective particle velocity $<U_y>$ (a) and effective stress $<\sigma_y>$ (b) under reshock wave loading of concrete

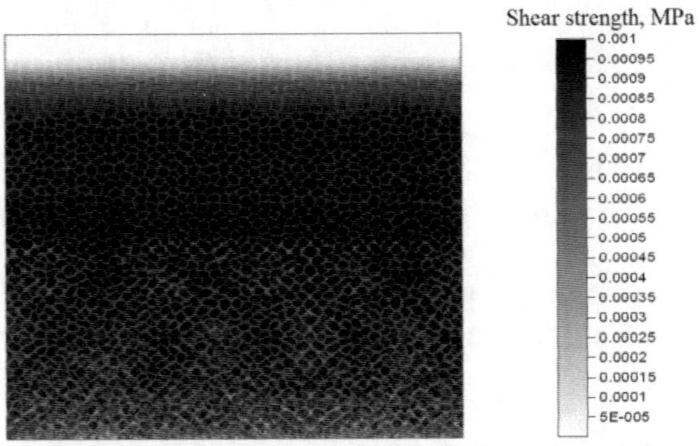

Figure 3. Calculated field of shear strength in the concrete plate under shock compression

The simulation results have shown the formation of a dissipative structure at meso-scale level of the concrete under loading by the reshock wave with amplitude of 1.2 GPa, as one can see in Figure 4.

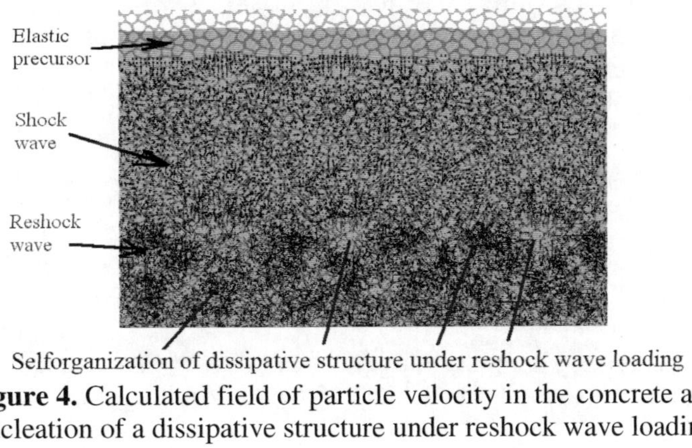

Figure 4. Calculated field of particle velocity in the concrete and nucleation of a dissipative structure under reshock wave loading

Nucleation of a dissipative structure leads to the change of distribution function of particle velocities at the meso-scale level in front of reshock wave. Results of numerical simulation have revealed the formation of bimodal distribution of velocities of material particles under loading of concrete by the reshock wave with amplitude of 1.2 GPa as seen in Figure 5.

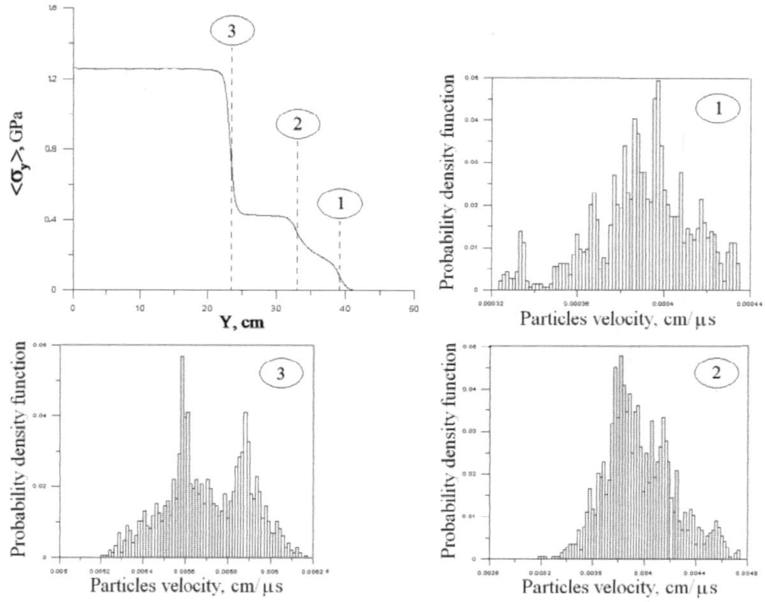

Figure 5. Distribution of particle velocity in shock wave.
Formation of bimodal distribution in front of reshock wave

The simulation results have shown that distribution function of particle velocities behind the front of reshock wave is similar to logarithmically normal distribution function.

4. Conclusion

The effective Hugoniot elastic limit of concrete depends on the shear strength of a cement paste. Inelastic deformation of concrete under dynamic loading is a result of cement paste deformation and movement of filler particles at the meso-scale level.

Formation of the dissipative structure leads to the change of distribution function of particle velocities at the meso-scale level. The logarithmically normal distribution function takes place at shock waves with amplitudes less than 0.4 GPa.

Formation of a dissipative structure under loading of shock wave with amplitude of 1.2 GPa is accompanied by the formation of the bimodal distribution of velocities of material particles in the shock wave. Distribution function of particle velocities behind the front of reshock wave is similar to logarithmically normal distribution function.

References

[1] Platova T M, Skripnyak V A and Karakulov V V 1995 *Vychislitel'niye Tekhnologii* **4** (1) 200
[2] Skripnyak V A and Karakulov V V 2004 *Fiz. Mezomekh.* **7** (Spec.Iss., Part I) 329
[3] Skripnyak V A and Karakulov V V 2005 *Vestnik Tomsk. Gos. Univer.* **50** 7
[4] Karakulov V V, Smolin I Yu and Skripnyak V A 2013 *Vestnik Tomsk. Gos. Univer. Mat. Mekh.* **4** 70
[5] Karakulov V V, Smolin I Yu and Skripnyak V A 2014 *AIP Conference Proceeding* **1623**, ed V E Panin *et al* (Melville, NY: AIP Publishing) 237

IOP Conf. Series: Journal of Physics: Conf. Series **1045** (2018) 012019 doi:10.1088/1742-6596/1045/1/012019

Surface layer evolution under the external heating

Anna Knyazeva

Tomsk Polytechnic University, 634050, Tomsk, Russia
Institute of Strength Physics and Material Science, 634055, Tomsk, Russia

Email: anna-knyazeva@mail.ru

Abstract. Generalized equation system is written to describe the surface layer evolution under electron beam action. Chemical reactions and porosity evolution are taken into account. The model was used to describe the electron beam treatment of thin layer with varied thickness. The review is made for some previous particular models of electron beam surface treatment with chemical reactions and shrinkage. It was pointed that the model permits the existence of self-sustaining regimes and quasi-stationary technology regimes.

1. Introduction

Electron-beam and laser technologies are interested for the development of new materials production and for the treatment and modification of the surfaces of metals and alloys. The last technologies connect immediately with additive manufacturing [1]. Thermo-physical phenomena determining the details properties in additive manufacturing based on electron and laser beam employment are typical for high temperature technologies [2]. However, in powder additive manufacturing applications, it is necessary to know and predict the possible shrinkage of a powder layer during Selective Laser or Electron Beam Melting. This phenomenon in different technologies can be linear or nonlinear in connection with nonlinear thermal conductivity, nonlinear absorptance or phase transitions, porosity evolution [3, 4]. There are many physical factors that affect the densification of the products during sintering of metallic powders [5, 6]. It is not possible practically to take into account all phenomena in generalized model because different physical phenomena are characterized by different spatial and temporal scales [7]. Known simplified thermal models of surface treatment and selective laser melting describe some effects [8, 9]; there are laser welding models [10], where reliable changes of melting pool due to nonlinear surface tension have been modeled. Thermal-kinetic models of surface layer evolution during EB-treatment were suggested in [11-13]. Porosity evolution in the heated zone is taken into account explicitly in [14], and new approach is suggested in [15] to describe the evolution of thin surface layer under traveling energy source action.

In this paper, the interrelation between chemical reactions and mechanical processes are taken into account to describe the surface layer formation on substrate under moving electron beam action. The model generalizes the previous investigations [11-16].

2. Key equations

Similarly to [15, 16], we assume that the evolution of the mixture consisting of powders, melt, and new solid and liquid phases and containing the pores can be described by Maxwell model of viscoelastic body [17]. In this model, components of deviators of stress s_{ij} and strain e_{ij} tensors are connected by relations

$$e_{ij} = \frac{1}{2\mu}\frac{ds_{ij}}{dt} + \frac{1}{2\eta_v} s_{ij} \tag{1}$$

and the usual relation between first invariants of stress and strain tensors

$$\sigma_{kk} = 3K(\varepsilon_{kk} - \omega) \tag{2}$$

takes a place. Here

$$e_{ij} = \varepsilon_{ij} - \frac{1}{3}\varepsilon_{kk}\,;\; s_{ij} = \sigma_{ij} - \frac{1}{3}\varepsilon_{kk}\,; \tag{3}$$

μ is the shear module, K is the bulk module, η_V is the viscosity; the function ω depends on the temperature and composition

$$\omega = 3\left[\alpha_T(T - T_0) + \sum_{k=1}^{n} \alpha_k (C_k - C_{k0})\right], \tag{4}$$

where α_T is the thermal expansion coefficient, C_k – mass concentrations of species (reagents and products), α_k are concentration expansion coefficients; index "0" relates to the initial state.

Substituting (2) into (1) we obtain

$$\frac{d\sigma_{ij}}{dt} + \frac{\mu}{\eta_v}\sigma_{ij} = 2\mu\frac{d\varepsilon_{ij}}{dt} + \delta_{ij}\left[\lambda\frac{d\varepsilon_{kk}}{dt} - K\frac{d\omega}{dt} + \frac{K\mu}{\eta_v}(\varepsilon_{kk} - \omega)\right], \tag{5}$$

where λ, μ – Lame coefficients; $K = \lambda + \frac{2}{3}\mu$ – bulk module; η_V – viscosity.

If $\eta_V \to \infty$, we come to the model of solid body. If $\eta_V \to 0$, we obtain ideal liquid. If we take

$$\eta_V = \eta_V(T,\theta),\; \mu = \mu(T,\theta),\; \lambda = \lambda(T,\theta), \tag{6}$$

where θ – is the porosity changing during shrinkage, then rheology of the reacting mixture will depend on temperature and composition evolution.

Motion and continuity equations are usual for hydrodynamics:

$$\rho\frac{d\mathbf{v}}{dt} = \mathbf{F} + \nabla\cdot\boldsymbol{\sigma}, \tag{7}$$

$$\frac{d\rho}{dt} + \rho\nabla\cdot\mathbf{v} = 0. \tag{8}$$

Energy equation in the form of thermal conductivity equation will contain the terms connecting with viscous dissipation, coupling effect and energy sources due to chemical reactions happen during surface treatment

$$\boldsymbol{\sigma}\cdot\frac{d\boldsymbol{\varepsilon}}{dt} + c_\varepsilon\rho\frac{dT}{dt} = \nabla\cdot(\lambda_T\nabla T) + W - 3KT\alpha_T\frac{d\varepsilon_{kk}}{dt} \tag{9}$$

Here c_ε is the heat capacity for constant strains; λ_T is effective thermal conductivity coefficient; ρ is the density; $\mathbf{v}$ is the velocity vector. Thermal properties depend on porosity also, and some kinetic law should be formulated for porosity evolution

$$\frac{d\theta}{dt} = \varphi(\theta, T, \sigma_{kk}) \tag{10}$$

The term W includes chemical heat release and the heat from external heating, if energy absorption takes a place in the volume. Chemical source is the sum

$$W_{ch} = \sum_{i=1}^{r} Q_i \varphi_i \tag{11}$$

where Q_i and φ_i - heats and rates of the reactions, $i=1,2,\ldots,r$.

Molar concentrations of species (connecting linearly with mass concentrations) follow from the equation system of chemical kinetics corresponding to the stages of chemical reaction under study. Boundary conditions correspond to experimental situation.

3. Reactions in thin layer

If reaction occurs in thin layer placed on some substrate and external source is developed in line and moves in the direction $0x$ of Cartesian coordinate system, we can assume $\varepsilon_{yy} = 0$; $\sigma_{zz} = 0$ and other components of strain and stress tensor depend on coordinate x only. In his case we obtain four equations from (5) and equilibrium equation (only one equilibrium equation remains here if inertia forces are negligible small) to find the values $\varepsilon_{xx}, \varepsilon_{zz}$ and σ_{xx}, σ_{yy} depending on temperature and composition. Then integrating the thermal conductivity equation along the thickness we obtain

$$U\frac{\partial T}{\partial t} + c_B\rho_B h_B u'\frac{\partial T}{\partial x} = \frac{\partial}{\partial x}\left(W\frac{\partial T}{\partial x}\right) + W_{ext}{}' + h_B W(T) - H(x,t), \tag{12}$$

where the term $H(x, y, t)$ depends on the solution of mechanical part of the problem

$$H(x,t) = \sigma_{xx}\frac{d\varepsilon_{xx}}{dt}(h_B + h_A) + 3BT\frac{d\varepsilon_{kk}}{dt} \tag{13}$$

and designations

$$U = c_A\rho_A h_A + c_B\rho_B h_B;\ W = \lambda_A h_A + \lambda_B h_B,$$

$$u' = u + \frac{\lambda_B}{c_B\rho_B}\frac{1}{h_B}\frac{\partial h_B}{\partial x},\ B = K_A\alpha_{TA}h_A + K_B\alpha_{TB}h_B$$

are introduced. Index "A" relates to the substrate, "B" relates to the treated layer; where h_A and h_B are their thicknesses.

For porous substance, we can assume [15]

$$\varepsilon_{kk} = ln\frac{\rho_0}{\rho} = ln\frac{1-\theta_0}{1-\theta} \tag{14}$$

where $\rho = (\rho_A h_A + \rho_B h_B)/(h_A + h_B)$ – is the averaging density. The density $\rho_B = \rho_S(1-\theta)$ depends on the composition changing during reaction.

Assuming that mass does not change during the treatment with the reactions in condensed phases we can write simple relation for the thickness of layer B:

$$h_B = h_{B0}\frac{1-\theta_0}{1-\theta}. \tag{15}$$

Particular variants of one-dimensional problems describing surface layer evolution were investigated in [11-16, 18 and 19]. So, the model of synthesis process in the layer containing $TiAl_3$ and Ti, placed on substrate made of Fe and heated by moving electron beam was suggested in [16]. Shrinkage process did not analyze. The paper [18] was devoted to the study of coupling problem, where mechanical stresses accompanying the chemical reactions in the system $TiNi_3 + Ti$ were taken into account. Shrinkage of powder layer from Ti-Nb alloy in two-dimensional formulation was analyzed in [14], where the melting process was modeled in term of two-phase zone. The intermixing of mixture from *Ni* and *Al* with substrate alloyed by electron-beam was analyzed in [19].

The question appears: "Will possible self-sustaining regime in reacting mixture initiated by electron-beam?" It was shown in [16] for simple formulation that self-sustaining combustion regime can appear, however it was not detected there numerically due to high heat losses to substrate.

When electron beam moves along the surface of large size, quasi-stationary regime establishes. Quasi-stationary models were analyzed in laser and electron-beam welding, for example, it is described in [20]. Here going to the coordinate system connecting with moving electron beam we come to nonlinear thermal conductivity equation with coefficients depending on mechanical properties and porosity. This problem demands a special investigation.

4. Conclusion

So, generalized equation system is written to describe the surface layer evolution under electron beam action. Chemical reactions and porosity evolution are taken into account. The equations were simplified for thin layer of variable thickness. It was shown that previous particular models of electron beam surface treatment with chemical reactions and shrinkage are contained in general model. It was pointed that the model permits the existence of self-sustaining regimes and quasi-stationary technology regimes that should be investigated specially.

Acknowledgments

The study was supported by the Russian Foundation of Fundamental Investigation (grant No. 16-58-00116).

References

[1] Beaman J J (Chair), Atwood C, Bergman T L, Bourell D, Hollister S, Rosen D 2004 WTEC Panel Report on Additive/subtractive manufacturing research and development in Europe, (World Techn. Ev. Center, Inc. 2809 Boston Street, Suite 441. Baltimore, Maryland 21224) 136
[2] Nicholas P L et al 2014 *Sustainable Design and Manufacturing* **7** 668
[3] Sercombe T B and Hopkinson N 2006 *Advanced engineering materials* **8** 260
[4] Ranogajec J et al 2000 *Ceramics – Silikáty* **44** 71
[5] Herderick E 2011 *Materials Science and Technology* (MS&T) ASM International 176252
[6] Lawrence E et al 2012 *Mater. Sci. Technol.* **28** 1
[7] Rishi Ganeriwala, Tarek I Zohdi 2014 *Procedia CIRP* **14** 299
[8] Tiebing Chen,Yuwen Zhang 2007 *Appl. Phys.* **86** 213
[9] John Romano et al 2015 *Procedia Manufacturing* **1** 238
[10] Isaev V I et al 2017 *AIP Conference Proceedings* **1893** 030108
[11] Knyazeva A G et al 2007 *Physical mesomechanics* **10** 207
[12] Knyazeva A G, Pobol I L, Demidov V N 2016 *Additive Technologies IOP Conf. Series: Materials Science and Engineering* **140** 012021
[13] Knyazeva A G, Bookrina N V 2003 *Materials of 12th International conference on radiation physics and chemistry of inorganic materials* (23–27 September, 2003, Tomsk: TPU). 167
[14] Knyzeva A G and Sharkeev Yu P 2017 *AIP Conference Proceedings* **1893** 030105
[15] KnyazevaA G 2016 *J. Phys.: Conf. Ser.* **754** 042009 https://doi.org/10.1088/1742-6596/754/4/042009
[16] Sorokova S N and Knyazeva A G 2008 *Theor. Foundations of Chemical Engineering* **42** 443
[17] Boley B A and Weiner J H 1960 Theory of Thermal Stresses (Wiley, New York) 540
[18] Sorokova S N and Knyazeva A G 2010 *Theor.Foundations of Chemical Engineering* **44** 172
[19] Knyazeva A G and Kryukova O N 2017 *J. Phys.: Conf. Ser.* **899** 072001 https://doi.org/10.1088/1742-6596/899/7/072001

IOP Conf. Series: Journal of Physics: Conf. Series **1045** (2018) 012020 doi:10.1088/1742-6596/1045/1/012020

Model of layers bonding during LOM-process

Anna Knyazeva[1], Nahum Travitzky[2]

[1] Tomsk Polytechnic University, 634050 Tomsk, Russia
Institute of Strength Physics and Material Science, 634055, Tomsk, Russia
[2] Department of Materials Science, Institute of Glass and Ceramics, University of Erlangen-Nuremberg, 91058 Erlangen, Germany

Email: anna-knyazeva@mail.ru

Abstract. The model of LOM-process with chemical reaction in intermediate layer is suggested in this paper. Maximal stress in contact zone and half-thickness of contact area are given from contact problem of mechanics. It was shown that this problem is similar to the problem of different material conjugation using synthesis in solid phase. Dimensionless variables and dimensionless criteria are identical for these processes. The qualitative examples of temperature field evolution for self-sustaining mode are demonstrated.

1. Introduction

Today additive manufacturing is a well-known technology for making real three dimensional objects from different materials, which may be subjected to various applications. Several AM techniques such as Stereo lithography (SLA) and its derivatives, Fused Deposition Modeling (FDM), Selective Laser Sintering or Melting (SLS/SLM), Laminated Object Manufacturing (LOM and similar sheet stacking processes), 3D-Printing (3DP) were applied for the manufacturing of structures with complex geometry [1-3]. Laminated object manufacturing (LOM) is one of categories of the sheet lamination processes in which material sheets are either cut by using laser or combined by using ultrasound. LOM-technique can be further classified based on the bonding mechanism between the sheets such as adhesive bonding, thermal bonding, and clamping [2] and includes several stages. For example, the LOM process of LZSA (LiO_2–ZrO_2–SiO_2–Al_2O_3) laminates included [4] three basic stages: tape casting of LZSA slurries, lamination by LOM machine accompanied by intermediate layers formation, and sintering accompanied by shrinkage.

LOM-technique is characterized by some advantages in comparison with other methods [2-3]: the variety of used organic and inorganic materials including multi-materials and multi-colors; high fabrication speed; low warping and internal stresses; no support structures are needed. Using this method it is possible to create the multilayered structures with different composition and properties of layers. To improve the layer joining, special materials can be used between layers. In this case the process can be accompanied by chemical reactions between components. It has been used to develop composite synthesis with the help of combustion during lamination or post-processing [5]. Today this possibility has the limitations connecting with bad controllability [6]. Note the chemical reactions can proceed during laser cut when detail is formed based on CAD-model.

There are numerous theoretical papers devoted to the description of mechanical and thermal properties of LOM-composites, for example [7-9]. However the physical and chemical processes accompanying the intermediate layers formation during LOM stages are modeled scarce. Appearing here problems need a special investigations. In this paper the mathematical model of different

IOP Conf. Series: Journal of Physics: Conf. Series **1045** (2018) 012020 doi:10.1088/1742-6596/1045/1/012020

materials conjugation (with different thermal physical properties) is studied numerically. It is assumed that chemical reaction starts in intermediate layer. Substance flow is not taken into account in a first approximation.

2. Mathematical problem formulation

Schematically the LOM-process is presented in fig.1a,b and post-processing accompanied by the chemical reactions in Fig.1c. The roll with temperature T_s moves along the surface of above layer A. the contact area between roll and A-layer depends on roll-diameter 2R and applied pressure P.

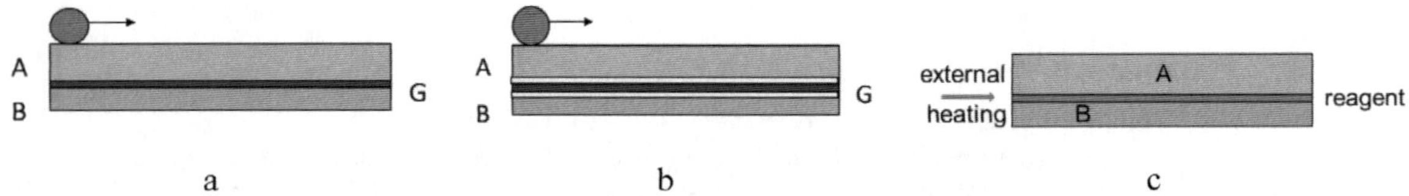

Fig.1 schematic presentation of some stages of LOM-process: a) process without intermediate layers formation; b) process with intermediate layers formation; c) post-processing with external heating

Maximal stress in contact zone can be evaluated by formula

$$\sigma_{max} = \left[\frac{P}{2R(\alpha_1 + \alpha_2)}\right]^{1/2} \tag{1}$$

and half-thickness of contact area – by formula

$$s = 0.798[2PR(\alpha_1 + \alpha_2)]^{1/2} \tag{2}$$

where

$$\alpha_k = \frac{1 - \nu_k^{\;2}}{E_k}, \tag{3}$$

ν_k, E_k are the Poisson coefficients and elastic modulus for contacting materials, R – is roll radius. Elastic modulus of layered composite is

$$\frac{H}{E} = \frac{H_A}{E_A} + \frac{H_B}{E_B} + \frac{H_G}{E_G}. \tag{4}$$

Linear velocity of roll V is given.

Temperature field follows from conjugate thermal conductivity problem solution that includes equations

$$c_k \rho_k \frac{\partial T_k}{\partial t} = \lambda_k \left[\frac{\partial^2 T_k}{\partial x^2} + \frac{\partial^2 T_k}{\partial y^2}\right] + W_k, \tag{5}$$

where $k=A, G, B$, c_k, ρ_k, λ_k are the heat capacities, densities and thermal conductivity coefficients of layers A, B, G; t is the time; x, y are the spatial coordinates; $W_A = W_B = 0$;

$$W_G = Qz_0 \varphi(a) exp\left(-\frac{E}{RT_G}\right), \tag{6}$$

E, z_0, Q are the activation energy, pre-exponential factor, heat release for total chemical reaction in layer G; R is universal gas constant; a is conversion level following from the equation

$$\frac{\partial a}{\partial t} = z_0 \varphi(a) exp\left(-\frac{E}{RT_G}\right), \tag{7}$$

$\varphi(a)$ is some kinetic function representing the reaction mechanism in the micro scale level.

The conditions of ideal thermal contact are correct in the interfaces, where temperatures of layers and heat fluxes are equal; the adiabaticity condition takes a place in low surface; heat exchange is given in lateral surfaces or adiabaticity condition also. For $y = H = H_A + H_B + H_G$ we have

$$T = T_S, \text{ if } Vt + x_0 - s \le x \le Vt + x_0 + s, \tag{8}$$

and $\partial T / \partial y = 0$ outside this interval.

At the initial time moment, we write:

$$T = T_0, \quad a = 0.$$

In this problem it is necessary to calculate the temperature field and study the possible conversion modes in layer G varying the properties, thicknesses of layers and heating conditions.

The post-processing (Fig 1. c) is characterized by similar (the moving heat source instead roll) or other heating condition. The mathematical problem formulation is the same with other boundary conditions including constant temperature T_S on the left surface during given time interval.

3. Dimensionless variables

Toinvestigatethemodel qualitatively it is convenient to go to dimensionless variables:

$$\theta_k = \frac{T_S - T_k}{RT_S^2} E, \xi = \frac{x}{H_G}, \eta = \frac{y}{H_G}, \tau = \frac{t}{t_a} \tag{9}$$

where t_a isadiabatictimeortimeduringwhichthechemicalreactionwillcompletedinadiabaticalconditions with given temperature [10],

$$t_a = \frac{c_G \rho_G R T_S^2}{z_0 EQ} exp\left(\frac{E}{RT_S}\right), \tag{10}$$

In new variables, the problem takes the form:

$$K_{CB} \frac{\partial \theta_B}{\partial \tau} = \frac{K_{\lambda B}}{\delta}\left[\frac{\partial^2 \theta_B}{\partial \xi^2} + \frac{\partial^2 \theta_B}{\partial \eta^2}\right], \; 0 < \eta < \Delta_B; \tag{11}$$

$$\frac{\partial \theta}{\partial \tau} = \frac{1}{\delta}\left[\frac{\partial^2 \theta}{\partial \xi^2} + \frac{\partial^2 \theta}{\partial \eta^2}\right] + \varphi(a) F(\theta), \; \Delta_B < \eta < 1 + \Delta_B; \tag{12}$$

$$\frac{\partial a}{\partial \tau} = \gamma \varphi(a) F(\theta) = \gamma \varphi(a) exp\left(\frac{\theta}{1 + \beta\theta}\right), \tag{13}$$

$$K_{CA} \frac{\partial \theta_A}{\partial \tau} = K_{\lambda A}\left[\frac{\partial^2 \theta_A}{\partial \xi^2} + \frac{\partial^2 \theta_A}{\partial \eta^2}\right], \; 1 + \Delta_B < \eta < 1 + \Delta_B + \Delta_A; \tag{14}$$

$$\eta = 0: \partial \theta_B / \partial \eta = 0; \tag{15}$$

$$\eta = \Delta_B : K_{\lambda B} \frac{\partial \theta_B}{\partial \eta} = \frac{\partial \theta}{\partial \eta}; \; \theta_B = \theta; \tag{16}$$

$$\eta = \boldsymbol{\Delta}_B + 1 : \frac{\partial \theta}{\partial \eta} = K_{\lambda A} \frac{\partial \theta_A}{\partial \eta};\ \theta = \theta_A; \tag{17}$$

$$\eta = 1 + \boldsymbol{\Delta}_B + \boldsymbol{\Delta}_A :$$

$$\xi_0 + \omega\tau - \xi_A < \xi < \xi_0 + \omega\tau + \xi_A :\ \theta = 0; \tag{18}$$

$$\xi \le \xi_0 + \omega\tau - \xi_A \text{ and } \xi \ge \xi_0 + \omega\tau + \xi_A :\ \partial\theta_A / \partial\eta = 0;$$

$$\xi = 0, L : \partial\theta_k / \partial\xi = 0; \tag{19}$$

$$\tau = 0 : \theta_B = -\theta_0; \theta_A = -\theta_0; \theta = -\theta_0; a = 0. \tag{20}$$

This problem includes dimensionless parameters

$$\beta = \frac{RT_S}{E};\ \gamma = \frac{c_G \rho_G R T_S^2}{EQ};\ \theta_0 = \frac{T_S - T_0}{RT_S^2} E;\ \delta = \frac{H_G^2 c_G \rho_G}{\lambda_G t_a},$$

typical for combustion theory [10], geometrical parameters

$$\boldsymbol{\Delta}_A = \frac{H_A}{H_G};\ \boldsymbol{\Delta}_B = \frac{H_B}{H_G}$$

and parameters characterizing the heating rate

$$\xi_0 = \frac{x_0}{H_G};\ \omega = \frac{Vt_a}{H_G};\ \xi_A = \frac{s}{H_G}.$$

Until the substances do not heated to temperature T_S, the temperature θ_k<0. If $\omega < 1$, roll runs during the time t_a the distance less then layer G thickness; if $\delta > 1$, the layer with thickness H_G does not have a time for the heating; and etc.

For situation presented in the Fig.1 c, the conditions change

$$\eta = 1 + \Delta_B + \Delta_A : \partial\theta_A / \partial\eta = 0;$$

$$\xi = 0 : \theta = 0,\ \tau < \tau_0 \text{ and } \partial\theta_k / \partial\xi = 0 \text{ for } \tau \ge \tau_0.$$

The both problems are solved numerically.

4. Example of calculation

The calculations show that various regimes of reaction propagation between materials A and B are possible depending on the heating conditions. When the heat in the heated zone is enough for self-sustaining mode, the reaction front propagates along the layer G. When materials are given, to stabilize the reaction front, one can vary the sizes of layers. On the Fig. 2,a the sizes of A and B layers are identical. The no symmetrical reaction front appears because the thermal conductivity of layer B is high. In the Fig 2,b, the upper layer has the same thickness as the reagent. In this case the ignition point is earlier that connects with the fact that the top layer stores more heat and thereby it facilitates to the reaction acceleration. The reaction front stays not symmetrical

Note that similar problems were investigated in [11, 12].

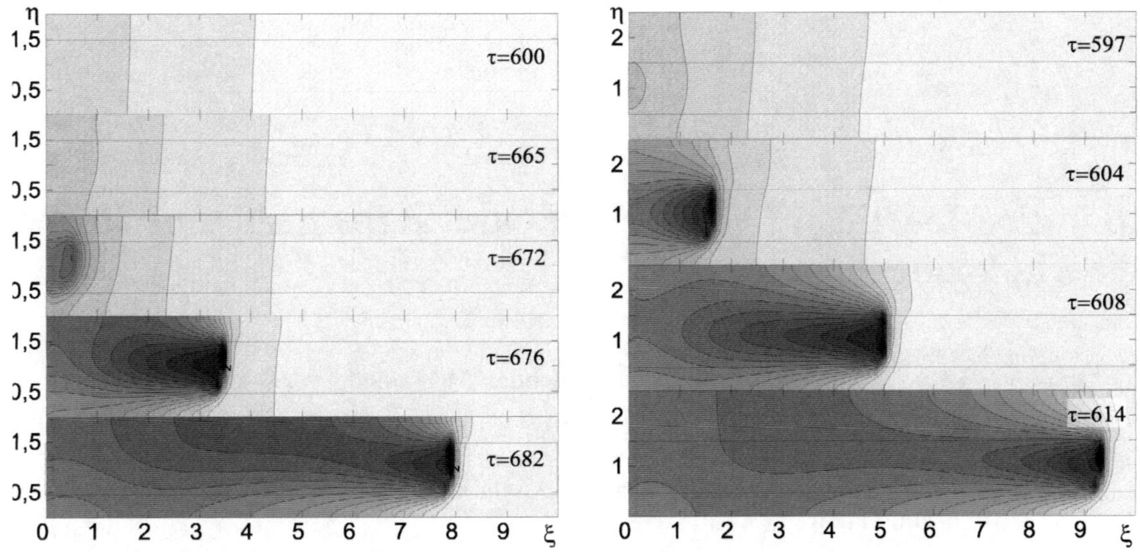

a б

Fig. 2 Temperature field evolution with time for various sizes of the layers.
Parameters: $\theta_0 = 10$; $\beta = 0.03$; $\gamma = 0.03$; $Fr = 20$; $K_{\lambda A} = 4$; $K_{\lambda B} = K_{cA} = K_{cB} = 1$.

5. Conclusion

So, the problem on reaction propagation between inert layers during LOM-process was formulated. It was shown that this problem is typical combustion problem appearing during materials conjugation using chemical reacting connecting materials. The reaction front is non-symmetrical due to the different properties of conjugated layers and due to different thicknesses of the same layers. Reaction front can be stable because stored heat in conjugated materials can facilitate the reaction propagation.

6. Acknowledgments

The work was supported by Ministry of Science and Education of Russia, project number is 10.9944.2017/5.2

References

[1] Kruth J-P, Levy G, Klocke F, Childs T H C 2007 *Annals of the CIRP* **56** 2730
[2] Jian-Yuan Lee, Jia An, Chee Kai 2017 *Applied Materials Today* **7** 120
[3] Wei Gao et al 2015*Computer-Aided Design* **69** 65
[4] Cynthia M Gomesa et al. 2008 *Journal of materials processing technology* **206** 194
[5] Zhan Y et al. 2001 *Mater Sci Eng A* **299** 218
[6] Kumar S, Kruth J-P 2010 *Materials and Design* **31** 850
[7] Samaratunga D, Jha R, Gopalakrishnan S 2014 *Finite Elements in Analysis and Design* **89** 19
[8] Xiaoping Shu, Rongliang Wang 2017 *Composite Structures* **174** 366
[9] Mauricio F Caliri Jr, Antonio J M Ferreira , Volnei Tita 2016 *Composite Structures* **156** 63
[10] RogachevA S, Mukasyan A S, Combustion for materials synthesis. Introduction to structure macro kinetics 2013 (in Russian, Moscow, Fizmatlit,) 400
[11] Chachshina A A, Knyazeva A G 2005 *Chemistry for sustainable development* **13** 339
[12] Aligozhina K A and Knyazeva A G 2017 *Combustion, Explosion, and Shock Waves* **53** 411

IOP Conf. Series: Journal of Physics: Conf. Series **1045** (2018) 012021 doi:10.1088/1742-6596/1045/1/012021

Ab initio Modeling of Elastic and Optical Properties of Sb and Bi Sesquioxides

H Koc[1], Chingiz G Akhundov[2], Amirullah M Mamedov[2, 3], Ekmel Ozbay[3]

[1]Department of Physics, Siirt University, 56100 Siirt, Turkey
[2]International Scientific Center, Baku State University, Baku, Azerbaijan
[3]Nanotechnology Research Center (NANOTAM), Bilkent University, 06800 Bilkent, Ankara, Turkey

E-mail: mamedov46@gmail.com

Abstract. First-principle calculations performed the structural, mechanical, electronic, and optical properties of Sb_2O_3 and Bi_2O_3 compounds in monoclinic (claudetite and α-Bi_2O_3) and orthorhombic (valentinite) structures. Local density approximation has been used for modeling exchange-correlation effects. The lattice parameters, bulk modulus, and the first derivate of bulk modulus (to fit to the Murnaghan's equation of state) of considered compounds have been calculated. The second-order elastic constants have been calculated, and the other related quantities have also been estimated in the present work. The electronic bands structures and the partial densities of states corresponding to the band structures are presented and analyzed. The real and imaginary parts of dielectric functions and energy-loss function are calculated. Our structural estimation and some other results are in agreement with the available experimental and theoretical data.

1. Introduction

Sb_2O_3 and Bi_2O_3, the members of compounds with the general formula $A_2^V B_3^{VI}$ (A=Bi, Sb, As and B=S, Se, Te), are important semiconductors with wide band gaps intensified recent years [1], due to their attractive physical properties. Sb_2O_3 includes three crystalline structures: cubic α-phase (senarmonitite), orthorhombic β-phase (valentinite), and a very recently found new phase (γ-phase). As_2O_3 also include two crystalline structures: cubic arsenolite and monoclinic claudetite. Bi_2O_3 include six crystalline structures: monoclinic α-phase, tetragonal β-phase, cubic γ-phase, cubic δ-phase, or orthorhombic ε-phase, and triclinic ω-phase [2-5]. Sb_2O_3 is used extensively in industry as a flame retardant in polymer, coatings, and textiles while Bi_2O_3 is used in the field of gas sensors, fuel cells, optical coatings or ceramic glass manufacturing [1, 5].

The positions corresponding to these compounds have been obtained from experimental and theoretical data [2, 6-8]. In the past, some detailed works [1, 2, 5, 9-13] have been carried out on the theoretical or experimental works of Sb_2O_3 and Bi_2O_3 compounds. Matsumoto et al. [2] systematically investigated the relationships between the electronic structures, energetic and atomic arrangements of three sesquioxides (As_2O_3, Sb_2O_3, and Bi_2O_3) using first-principles lattice-dynamics calculations. Pereira et al. [5] investigated experimentally and theoretically under high pressure the Sb_2O_3 in cubic phase (senarmontite) by means of X-ray diffraction (XRD) and the density-functional theory (DFT). Condurache-Bota et al. [1] studied the structural characteristics of Sb_2O_3 and Bi_2O_3 by means of X-ray diffraction. Fruth et al. [9] analyzed the structural changes of α- Bi_2O_3 by means of powder X-ray diffraction, SEM/EDX analysis and infrared spectroscopy. Chouinard et al. [10] observed the structural transition of α-Bi_2O_3 under hydrostatic pressure using different experimental techniques. Pereira et al. [11, 12] investigated the structural and vibrational properties using the density functional theory of α- Bi_2O_3 under pressure. Li et al. [13] investigated the atomic-scale interfacial structure and the electronic-scale interface properties between α- Bi_2O_3 and β- Bi_2O_3 homo-junction using the density functional theory within the generalized gradient approximation.

In this work, we have especially focused on the electronic, mechanical and optical properties of Sb_2O_3 and Bi_2O_3 compounds by using ab initio total energy calculations. To our knowledge, the elastic constants, Young's modulus, shear modulus, Poisson's ratio, sound velocities, Debye temperature (outside the Debye temperature of α-Bi_2O_3), and optical properties have not been reported in detail for Sb_2O_3 and Bi_2O_3 so far.

2. Method of calculation

Our calculations have been performed using the density functional formalism and local density approximation (LDA) [14] through the Ceperley and Alder functional [15] as parameterized by Perdew and Zunger [16] for the exchange-correlation energy in the SIESTA code [17, 18]. This code calculates the total energies and atomic forces using a linear combination of atomic orbitals as the basis set. The basis set is based on the finite range pseudoatomic orbitals (PAOs) of the Sankey Niklewsky type [19], generalized to include multiple-zeta decays.

The interactions between electrons and core ions are simulated with separable Troullier-Martins [20] norm-conserving pseudopotentials. We have generated atomic pseudopotentials separately for atoms, Sb, Bi and O by using the $5s^2 5p^3$, $6s^2 6p^3$ and $2s^2 2p^4$ configurations, respectively. The cut-off radii for present atomic pseudopotentials are taken as s: 3.82 au, p: 2.71 au, 2.92 au for the d and f channels of Bi, 2.35 for the s, p, d and f channels of Sb, 1.47 au for the s, p, d and f channels of O.

Siesta calculates the self-consistent potential on a grid in real space. The fineness of this grid is determined in terms of an energy cut-off E_c in analogy to the energy cut-off when the basis set involves plane waves. Here by using a double-zeta plus polarization (DZP) orbitals basis and the cut-off energies between 100 and 500 Ry with various basis sets, we found an optimal value of around 350 Ry for Sb_2O_3 and Bi_2O_3 in claudetite, α-Bi_2O_3 and valentinite structures. For the final computations, 100 k-points for Sb_2O_3 and Bi_2O_3 were enough to obtain the converged total energies ΔE to about 1meV/atoms.

3. Results and discussion

3.1 Structural properties

For Sb_2O_3 and Bi_2O_3, structures that are monoclinic (claudetite and α-Bi_2O_3) and orthorhombic (valentinite) are considered. The equilibrium lattice parameters, the bulk modulus, and its pressure derivative were calculated by means of Murnaghan's equation of states (eos) [21], and the results are shown in Table 1 along with the experimental and theoretical values. The lattice constants are found to be a= 4.90 Å, b= 12.61 Å, c=5.39 Å in valentinite structure and a= 10.13 Å, b= 5.07 Å, c=15.11 in claudetite structure for Sb_2O_3 and a= 11.71 Å, b= 5.70 Å, c=5.65 Å in valentinite structure and a= 5.88 Å, b= 8.22 Å, c=7.48 Å in α-Bi_2O_3 structure for Bi_2O_3. The lattice parameters obtained are in good agreement with the experimental and theoretical values [2, 4, 7, 11, 13 and 22]. In all our calculations, we have used the computed lattice constants. In the present case, the calculated bulk moduli are 140.93 GPa (valentinite) and 123.56 GPa (claudetite) for Sb_2O_3 and 133.54 GPa (valentinite) and 113.74 GPa (α-Bi_2O_3) for Bi_2O_3. The bulk modulus of the solid as fundamental physical properties provides valuable information, including the average bond strengths of the atoms for the given crystals [23]. In the present case, the largest value of bulk modulus is obtained for Sb_2O_3 in valentinite structure, and it implies that this structure is the least compressible one among the others. The obtained bulk modulus for α-$Bi_2O_3$is in good agreement with the experimental value, but is larger (%20.78) than the theoretical value [11].

Table 1. The calculated equilibrium lattice parameters (a, b, and c), bulk modulus (B), and the pressure derivative of bulk modulus (B') together with the theoretical and experimental values for Sb_2O_3 and Bi_2O_3 compounds in monoclinic (claudetite and α-Bi_2O_3) and orthorhombic (valentinite) structures

Compound	Reference	Structure	a (Å)	b (Å)	c (Å)	$B(GPa)$	B'
Sb_2O_3	Present	valentinite	4.90	12.61	5.39	140.93	4.72
	Experimental[4]		4.91	12.46	5.41		
	Theory (GGA-VASP) [2]		5.20	12.56	5.52		
Sb_2O_3	Present	claudetite	10.13	5.07	15.11	123.56	4.57
	Theory (GGA-VASP)[2]		9.73	4.87	14.51		
Bi_2O_3	Present	valentinite	11.71	5.70	5.65	133.54	4.54
	Theory (GGA-VASP)[7]		11.81	5.74	5.69		
Bi_2O_3	Present	α-phase	5.88	8.22	7.48	113.74	4.55
	Experimental[11]		5.85	8.17	7.51	107.07,	1.6,
	Experimental[22]		5.84	8.16	7.50	90.1^{GGA}	4.8^{GGA}
	Theory (GGA-VASP)[7]		5.93	8.28	7.54		
	Theory (GGA-CASTEP)[13]		5.84	8.02	7.34		

3.2. Elastic Properties

The mechanical behavior of materials reflects the resistance or the deformation of the material against the applied load or force. The stiffness, strength, hardness and ductility comprise the basic parameters used in the design. The elastic constants of the material establish a relationship between the mechanical and dynamic behavior of solids, and gives information about the stability and mechanical hardness of the materials. The magnitude of the elastic constants are a measure of resistance (so, of bond strength between atoms) against separation from each other of the neighboring atoms [24-26].

SIESTA for elastic constants makes calculations using the "volume-conserving" technique [27]. The elastic constants for Sb_2O_3 and Bi_2O_3 in the valentinite, claudetite, and α-Bi_2O_3 structures are given in Table 2. Unfortunately, there are no theoretical results for comparing the present work.

Table 2. The calculated elastic constants (in GPa) for Sb_2O_3 and Bi_2O_3compounds

Compound	Structure	C_{11}	C_{12}	C_{13}	C_{15}	C_{22}	C_{23}	C_{25}	C_{33}	C_{35}	C_{44}	C_{46}	C_{55}	C_{66}
Sb_2O_3	claudetite	202.1	56.2	94.1	10.6	236.3	111.6	29.6	240.1	15.7	49.5	18.7	72.3	52.9
	valentinite	234.2	100.2	103.2	-	228.8	98.7	-	153.0	-	69.5	-	55.1	77.9
Bi_2O_3	α-phase	222.1	80.6	83.2	-2.1	198.2	55.2	29.1	249.3	31.9	59.2	16.7	78.1	91.1
	valentinite	215.3	99.6	88.2	-	203.5	73.2	-	251.9	-	77.7	-	97.9	83.9

The elastic constants C_{11}, C_{22}, and C_{33} measure the *a*-, *b*-, and *c*-direction resistance to linear compression, respectively. The C_{33} for valentinite structureis lower than the C_{11} and C_{22} while the C_{11} for claudetite structure of Sb_2O_3 is lower than the C_{22} and C_{33}. The calculated C_{22} for both structure of Bi_2O_3 is lower than the C_{11} and C_{33}. Thus, Sb_2O_3 compound is more compressible along c-axis for valentinite structure and a-axis for claudetite structure while Bi_2Se_3 compound is more compressible along the b-axis for both structures. The large C_{44} shows a strong resistance against the monoclinic shear distortion in (100) plane. The C_{66} relates to the shear resistance in the direction <110>. In the Sb_2O_3 and Bi_2O_3 compounds, C_{44} and C_{66} in valentinite and α-Bi_2O_3 structures of Bi_2O_3 compound have the highest value.

Table 3. The calculated isotropic bulk modulus (B, in GPa), shear modulus (G, in GPa), Young's modulus (E, in GPa), Poisson's ratio, sound velocities (v_l, v_t, v_m, in m/s) and Debye temperature for Sb_2O_3 and Bi_2O_3compounds.

Compound	Reference	Structure	B_H	G_H	E	υ	v_l	v_t	v_m	θ_D (K)
Sb_2O_3	Present	claudetite	129.3	59.1	153.8	0.302	5970	3181	3554	414.9
	Present	valentinite	132.7	59.3	154.8	0.305	5882	3113	3479	412.7
Bi_2O_3	Present	α-phase	118.9	72.2	180.1	0.248	4679	2709	3007	357.2
	Exp. [42]									316
	Present	valentinite	132.2	77.7	194.9	0.254	4983	2860	3177	373.2

The polycrystalline elastic moduli are calculated from Voigt and Reuss approximation methods [28-34] using elastic constants. We use the Hill average [34] to calculate Young's modulus (E) and Poisson's ratio (v) using the refs. [35, 36]. The calculated bulk modulus, isotropic shear modulus, Young's modulus and Poisson's ratio are given in Table 3. The bulk modulus is a measure of resistance against the volume change under an applied pressure. The size of bulk modulus shows the hardness of solid. The calculated bulk moduli of Sb_2O_3(Bi_2O_3)-valentinite and Sb_2O_3-claudetite (α- Bi_2O_3) structures are 132.7 (132.2) GPa and 129.3 (118.9) GPa. In general, the calculated bulk modulus for either phase is Sb_2O_3> Bi_2O_3. Therefore, Sb_2O_3 for either phase is harder than Bi_2O_3. The shear modulus is a measure of resistance to reversible deformations upon shear stress [37]. The calculated shear modulus for Bi_2O_3 is higher than Sb_2O_3 compound (see Table 3). Young's modulus is a measure of the ratio of stress and strain. The material is stiffer if the value of Young's modulus is high. The Young's modulus (194.9 GPa for valentinite and 180.1 for α-Bi_2O_3) of Bi_2O_3 compound is relatively stiffer than Sb_2O_3 (154.8 GPa for valentinite and 153.8 GPa for claudetite). The value of the Poisson's ratio is small (υ =0.1) for covalent materials, whereas for ionic materials a typical value of υ is 0.25 [38]. The calculated Poisson's ratios of Sb_2O_3(Bi_2O_3)-valentiniteand Sb_2O_3-claudetite (α-Bi_2O_3) structures are approx. 0.305 (0.254) and 0.302 (0.248). Therefore, the ionic contribution to inter atomic bonding for these compounds is dominant.

The Debye temperature and sound velocity that separates [39-41] the low and high temperature region of solids are calculated for Sb_2O_3 and Bi_2O_3 compounds by use of the elastic constants. The calculated values of the longitudinal, transverse, and average sound velocities are shown in Table 3 along with the Debye temperature. For materials, usually, the higher Debye temperature, the larger microhardness. The calculated Debye temperature for Sb_2O_3 in both structures is higher than Bi_2O_3. The Debye temperature obtained is greater than experimental value (see Table 3).

3.3. Electronic properties

The electronic band structures of Sb_2O_3 and Bi_2O_3single crystals in monoclinic (claudetite and α-Bi_2O_3) and orthorhombic (valentinite) structures have been calculated along high symmetry directions in the first Brillouin zone (BZ).

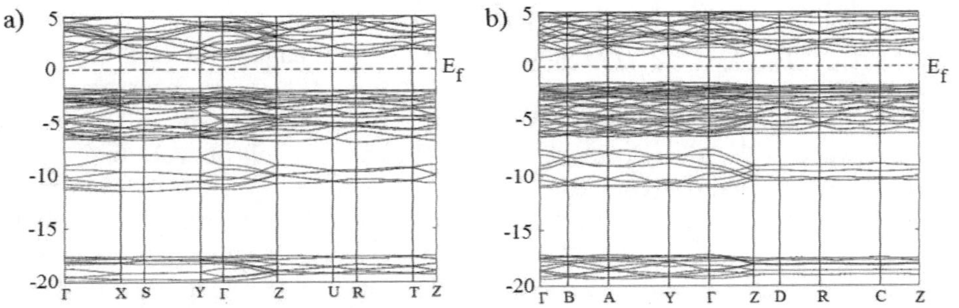

Figure 1. Energy band structures for Sb_2O_3 in a) valentinite and b) claudetite structures

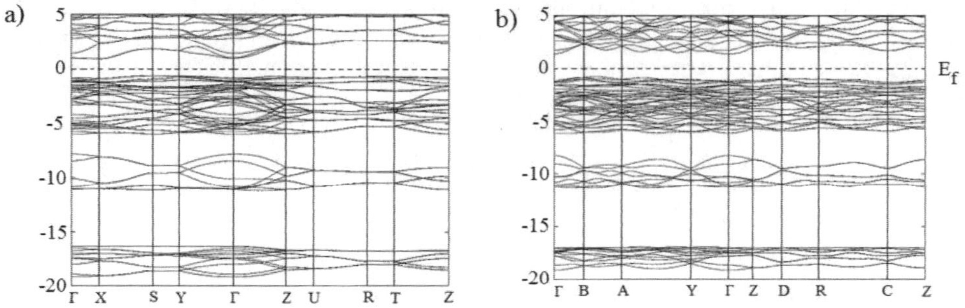

Figure 2. Energy band structures for Bi_2O_3 in a) valentinite and b) α-Bi_2O_3 structures

The energy band structures calculated using LDA for Sb_2O_3 and Bi_2O_3 are shown in Fig. 1 and Fig. 2. As can be seen in Fig. 1a, the Sb_2O_3 compound in valentinite structure has a direct band gap (at the Γ point) semiconductor with the value 1.97 eV. The band gap is 0.23 eV smaller than that obtained by theory [2]. The band gap of Sb_2O_3 compound in claudetite structure (see Fig. 1b) has the different character with that of valentinite structure, that is, it is an indirect band gap semiconductor. The top of the valance band positioned at the Y point of BZ, the bottom of the conduction band is located of BZ. The indirect band gap values of claudetite structure are 2.24eV. The band gap is 0.27 eV smaller than that obtained by theory [2].The calculated band structures of Bi_2O_3 are given in Fig. 2. As can be seen from the figure, valentinite and α-Bi_2O_3structures have the same character, that is, it have an indirect band gap semiconductor with the value 1.54 eV and 2.14 eV, respectively. For valentinite structure, the top of the valance band and the bottom of the conduction band are located at the S point and X point of BZ, respectively. The band gap is 0.2 eV smaller than that obtained by theory [2]. Similarly, the top of the valance band for α-Bi_2O_3structures positioned at the nearly B point between B and A point of BZ, the bottom of the conduction band is located at the Γ point of BZ. The band gap is 0.15 eV, 0.66 eV and 0.57 eV smaller than that obtained by theory [2, 13] and experiments, respectively [43]. The total and partial densities of states of Sb_2O_3 and Bi_2O_3 compounds were calculated too. The lowest valence bands (approx. -20 and -17 eV) are dominated by O 2s states. While the highest occupied valance bands are essentially dominated by O 2p states. The middle valance bands (approx. -12 and -7eV) are dominated by Sb 5s and Bi 6s states. The lowest unoccupied conduction bands just above Fermi energy level is dominated by Sb 5p and Bi 6p. The band structures of Sb_2O_3 and Bi_2O_3 single crystals are compared, band structures of these crystals are highly resemble one another. Thus, on formation of the band structures of Sb_2O_3 and Bi_2O_3 the 2s 2p orbitals of O atoms are more dominant than 5s5p and 6s6p orbitals of Sb and Bi atoms.

3.4. Optical properties

The linear optical properties obtained from the complex dielectric function [44, 45, 24] of Sb_2O_3 and Bi_2O_3 are investigated. The imaginary parts of the frequency-dependent linear dielectric function were calculated. The real part of the linear dielectric function can be obtained by the Kramers-Kronigrelations [45] using the imaginary component. We first calculated the real and imaginary parts of the x- and z-components of the frequency-dependent linear dielectric function. The ε_1^x behaves mainly as a classical oscillator. It vanishes (from positive to negative) at approx. 7.34 (6.77) eV and 11.75 (11.35) eV, whereas the other function ε_1^z is equal to zero at approx. 6.20 (6.23) eV and 12.58 (11.92) eV for Sb_2O_3 in valentinite (claudetite) structure. For Bi_2O_3, the ε_1^x is equal to zero at approx. 6.66 eV, 11.37 eV, 13.11 eV, 13.33 eV, 16.89 eV, and 18.04 eV in valentinite structure and at approx. 6.47 eV, 10.74 eV, 14.53 eV, and 18.66 eV in α-Bi_2O_3 structure, whereas the other function ε_1^z is equal to zero at approx. 6.93 eV, 11.76 eV, 16.87 eV, 16.98 eV, 17.57 eV, and 18.58 eV in valentinite structure and at approx. 6.83 eV, 10.80 eV, 14.36 eV, and 19.08 eV in α-Bi_2O_3 structure. The peaks of the ε_2^x and ε_2^z correspond to the optical transitions from the valence band to the conduction band and are in agreement with the previous results. The maximum peak values of ε_2^x and ε_2^z for Sb_2O_3 in valentinite (claudetite) structures are around 5.03 (3.84) eV and 3.95 (4.27) eV, respectively, whereas the maximum values of ε_2^x and ε_2^z for Bi_2O_3 in valentinite (α-Bi_2O_3) structures are around 4.52 (4.16) eV and 4.68 (4.41) eV, respectively. The corresponding energy-loss functions, $L(\omega)$, were also calculated. The function $L(\omega)$ describes the energy loss of fast electrons traversing the material. The sharp maxima in the energy-loss function are associated with the existence of plasma oscillations [46]. The curves $L(\omega)$have a maximum near 11.78 (11.43) eV and 12.74 (12.03) eV for Sb_2O_3 in valentinite (claudetite) structures, respectively and 20.41 (19.32) eV and 20.90 (19.57) eV for Bi_2O_3in valentinite (α-Bi_2O_3) structures, respectively.

4. Conclusion

In this study, the structural, mechanical, electronic, and optical properties of Sb_2O_3 and Bi_2O_3 compounds in valentinite, claudetite, and α-Bi_2O_3 structures are investigated by using the local density approximation. The calculated lattice parameters are in agreement with experimental and theoretical values. The elastic constants obtained show that Sb_2O_3compound is more compressible along the c-axis for the valentinite structure and the a-axis for the claudetite structure while Bi_2Se_3 compound is more compressible along the b-axis for both structures. From the calculated bulk modulus using elastic constants, it can be said that Sb_2O_3 for either phase is harder than Bi_2O_3. The calculated Poisson's ratio of Sb_2O_3(Bi_2O_3)-valentiniteand Sb_2O_3-claudetite (α- Bi_2O_3) structures is approx. 0.305 (0.254) and 0.302 (0.248). Therefore, the ionic contribution to inter atomic bonding for these compounds is dominant. The Debye temperature and sound velocities have been calculated, and the calculated Debye temperature for Sb_2O_3 in both structures is higher than Bi_2O_3. The obtained electronic band structures show that $Sb_2O_3$and Bi_2O_3 compounds are semiconductor in nature. Lastly, we have examined photon-energy dependent dielectric functions and the energy-loss function along the x-and z-axes.

5. References

[1] Condurache-Bota S, Rusu G I, Tigau N, Nica V, and Drasovean R 2009 *Journal of Optoelectronics and Advanced Materials* **11** 2159
[2] Matsumoto A, Koyama Y, Togo A, Choi M, and Tanaka I 2011 *Physical Review B* **83** 214110.1-10
[3] Rubbens A, Drache M, Roussel P, and Wignacourt J P 2007 *Materials Research Bulletin* **42** 1683
[4] Svensson C 1975 *Acta Cryst. B* **31** 2016
[5] Pereira L J, Gracia L, Santamaría-Pérez D, Vilaplana R, Manjón F J, Errandonea D, Nalin M, and Beltrán A 2012 *Physical Review B* **85** 174108.1-14
[6] Svensson C 1974 *Acta Crystallogr. Sect. B* **30** 458
[7] Matsumoto A, Koyama Y, and Tanaka I 2010 *Physical Review* B **81** 094117.1-11
[8] Ivanov S A, Tellgren R, Rundlof H, and Orlov V G 2001 *Powder Diffraction* **16** 227
[9] Fruth V, Popa M, Berger D, Ionica C M, and Jitianu M 2004 *Journal of the European Ceramic Society* **24** 1295
[10] Chouinard C and Desgreniers S 2000 *Solid State Communications* **113** 125
[11] Pereira A L J, Errandonea D, Beltán A, Gracia L, Gomis O, Sans J A, Garcia-Domene B, Miquel-Veyrat A, Manjón F J, Muñoz A, and Popescu C 2013 *J. Phys. Condens. Matter.* **27** 475402.1-12
[12] Pereira A L J, Gomis O, Sans J A, Porres J P, Manjón F J, Beltán A, Hernándes P R, and Muñoz A 2014 *J. Phys. Condens. Matter.* **26** 225401.1-15
[13] Li Q Y and Zhao Z Y 2015 *Phys. Lett.* A. **379** 2766-2771
[14] Kohn J W and Sham L J 1965 *Phys. Rev.* **140** A1133 -A1138
[15] Ceperley D M and Adler M J 1980 *Phys. Rev. Lett.* **45** 566-569
[16] Perdew P and Zunger A 1981 *Phys. Rev. B.* **23** 5048-5079
[17] Ordejón P, Artacho E, and Soler J M 1996 *Phys. Rev. B.* **53** R10441-R10444

IOP Conf. Series: Journal of Physics: Conf. Series **1045** (2018) 012021 doi:10.1088/1742-6596/1045/1/012021

[18] Soler J M, Artacho E, Gale J D, García A, Junquera J, Ordejón P, and Sánchez-Portal D 2002 *J. Phys. Condens. Matt.* **14** 2745-2779

[19] Sankey O F and Niklewski D J 1989 *Phys. Rev.B.* **40** 3979

[20] Troullier N and Martins J L 1991 *Phys. Rev. B.* **43** 1993

[21] Murnaghan F D 1944 *Proc. Natl. Acad. Sci U.S.A.* **50** 244

[22] Ivanov S A, Tellgren R, Rundlöf H, and Orlov V G 2001 *Powder Diffr.* **16** 227

[23] Maradudin A A, Montroll E W, Weiss G H, and Ipatova I P 1971 Theory of Lattice Dynamics in the Harmonic Approximation 2nd edition., New York, London: *Academic Press*

[24] Koc H, Deligöz E, andMamedov A M 2011 *Philosophical Magazine* **91** 3093

[25] Koc H**,** Yildirim A, Tetik E, and Deligoz E 2012 *Computational Materials Science* **62** 235

[26] Deligoz E, Ozisik H, Colakoglu K, Surucu G, and Ciftci Y O 2011 *J. Alloy. Comp.* **509** 1711

[27] Wallace D C 1972 Thermodynamics of Crystals, Chap. 1, where finite Lagrangian strains ηij are discussed. In the case of infinitesimal strains these reduce to our εij of classical elasticity theory, *Wiley*, New York

[28] Watt J P 1979 *J. Appl. Phys.* **50** 6290

[29] Watt J P 1980 *J. Appl. Phys.* **51** 1520

[30] Nye J F 1985 Physical properties of crystals, Oxford, *Oxford University Press*

[31] Wu Z, Zhao E, Xiang H, Hao X, Lui X, and Meng J 2007 *Phys. Rev. B.* **76** 054115.1-15

[32] Voight W 1928 Lehrbook der kristallphysik Leipsig: *Teubner* 962

[33] Reuss A and Angew Z 1929 *Math. Mech.* **9** 49

[34] Hill R 1952 *Proc. Phys. Soc. London Sect A* **65** 349

[35] Panda K B andChandran K S R 2006 *Acta Mater.* **54** 1641

[36] Ravindran P, Fast L, Korzhavyi P A, Johansson B, Wills J, and Eriksson O 1998 *J. Appl. Phys.* **84** 4891

[37] Shein I R and Ivanovskii A L 2008 *J. Phys. Considens Matter* **20** 415218.1-9

[38] Bannikov V V, Shein I R, and Ivanovskii A L 2007 *Phys. Stat. Sol.(RRL)* **3** 89

[39] Johnston I, Keeler G, Rollins R, and Spicklemire S 1996 Solids state physics simulations, the consortium for upperlevel physics software. *Wiley*, New York

[40] Anderson O L 1963 *J. Phys. Chem. Solids.* **24** 909

[41] Schreiber E, Anderson O L, and Soga N 1973 Elastic constants and their measurements. McGraw-Hill, New York

[42] Fredenburg D A and Thadhani N N 2011 *J. Appl. Phys.* **110** 063510.1-5

[43] Hou J, Yang C, Wang Z, Zhou W, Jiao S, and Zhu H 2013 *Appl. Catal. B: Environ.***504** 142

[44] Levine Z H and Allan D C 1989 *Phys. Rev. Lett.* **63** 1719

[45] Philipp H R and Ehrenreich H 1963 *Phys. Rev.* **129** 1550

[46] Marton L 1956 *Rev. Mod. Phys.* **28** 172

IOP Conf. Series: Journal of Physics: Conf. Series **1045** (2018) 012022 doi:10.1088/1742-6596/1045/1/012022

Modeling of Bone Tissue Structure and Porous Ceramics

T V Kolmakova, S P Buyakova and S N Kulkov

Institute of Strength Physics and Materials Science SB RAS, Tomsk, Russia
Tomsk State University, Tomsk, Russia

E-mail: kolmakova@ftf.tsu.ru

Abstract. The results of computer modeling of the structure and mechanical behavior of porous ceramics, microvolumes of compact bone tissue and mesovolumes of bone containing compact and spongy bone tissue are presented. The structure and mineral content of bone tissue fragments are determined in which the most uniform distribution of deformations is realized under axial tension and compression. It was found that the geometry of the porous space of ceramic samples significantly influences the character of the distribution of the regions in which microfractures are possible.

1. Introduction

The long-term experience of the synthetic materials' use to repair the bone elements by means of endoprosthesis replacement points out that the biochemical compatibility of the implant with the body is an essential, although insufficient, condition for functioning of the implant - bone system as a whole. Their biomechanical compatibility is an important condition. It is known that a difference in the mechanical behaviour of the bone and implant can result in the loss of the latter due to the subsequent resorption of the bone tissue, which is in contact with the implant.

As regards the biochemical compatibility with the body, the preferred materials for the endoprosthesis replacement of the bone tissue are those referred to the ceramics class. Ceramics are identical to the inorganic bone matrix in the type of chemical bond. The strong interest in the use of ceramics for biomedical engineering applications has been developed from the end of the 20th century. New ceramics [1-4], with very improved properties, contribute to increasing the possibilities of using ceramics as osteoimplants.

Using the computer modeling methods creates the preconditions for better understanding of the laws of the human bone functioning and promotes to the development and selection of an appropriate prosthesis for a specific individual.

The results of computer modeling of the structure and mechanical behavior of porous ceramics, microvolumes of compact bone tissue and mesovolumes of bone containing compact and spongy bone tissue are presented.

2. Models of bone fragments and ceramic samples

The computer simulation of the mechanics of compact bone microvolumes, bone mesovolumes and ceramic samples was performed in the software package ANSYS.

The geometrical model of the compact bone tissue microvolume (Figure 1) was based on a real image of the natural bone microstructure.

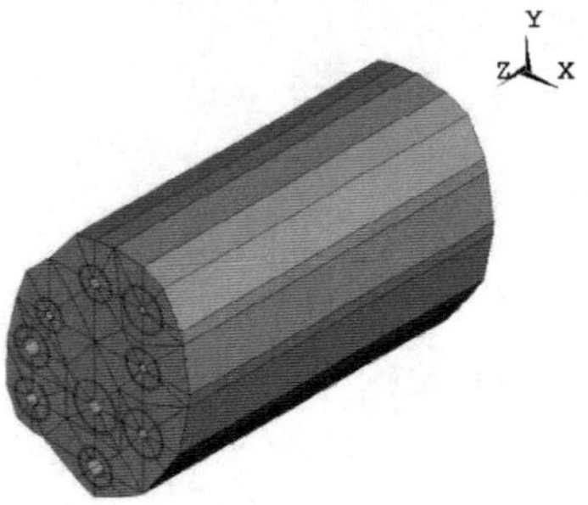

Figure 1. Geometrical model of compact bone tissue

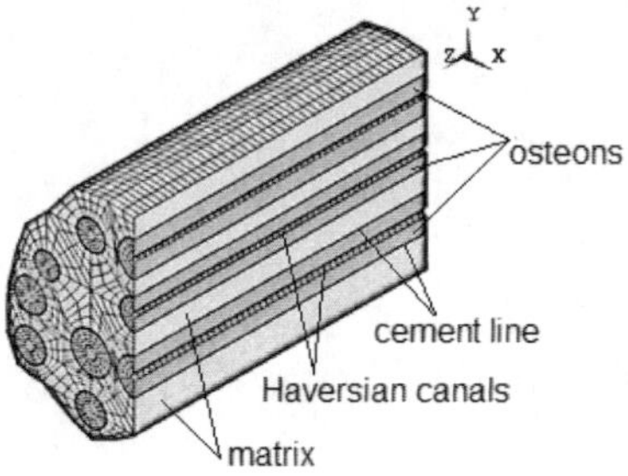

Figure 2. Finite element model of compact bone tissue microvolume

The microvolume was simulated as a reinforced composite material comprising the reinforcing structural elements, i.e. the osteons with the Haversian canals inside, the matrix and the cement streak separating the osteons and the matrix (Fig. 2). The total porosity of the simulative fragment of the compact bone P was derived by Haversian canals being explicitly simulated (Fig. 2), and Volkmann's canals being implicitly considered in the calculation of the effective mechanical characteristics of the structural elements of the simulative microvolume.

The material of the structural elements of the simulative microvolume of the compact bone was considered transversely isotropic. The effective mechanical properties of the matrix material and the reinforcing elements, osteons, were calculated subject to the volume ratio of the collagen and mineral components, the porosity due to the presence of Volkmann's canals, the direction of the collagen-mineral fibers [5] (I type - transverse fibers, I±45° type - fibers are arranged at an angle of ±45° in the neighboring osteon lamellas, II type – the fibers change their direction from perpendicular on parallel to a bone axis in the osteon lamellas, III type - longitudinal fibers).

Bone fragments (mesovolumes) were simulated as layered composite materials. Structural layers were associated with compact and spongy bone tissues and intermediate layer interfacing those tissues. The ratio of compact (c) and spongy (s) bone tissues volumes Vc / Vs varied from 0.25 to 4.0 (Fig. 3).

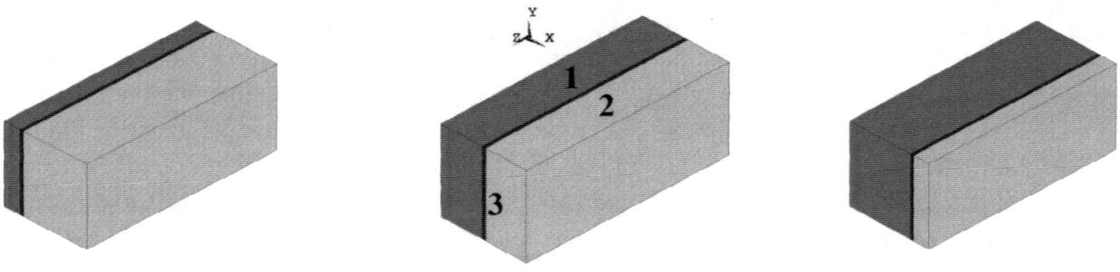

Figure 3. Geometrical models of bone mesovolumes (bone fragments)
(1– layer of compact bone tissue, 2 – layer of spongy bone tissue, 3 – intermediate layer)

Modulus of elasticity of compact (Ec) and spongy (Es) components of the modeled sample was set up depending on their density ρ (g/cm^3) and mass fraction of minerals α in accordance with Hernandez model [6]. The density of compact bone ρ_c varied from 1.6 to 1.9 g/cm^3, and the density of spongy bone ρ_s varied from 0.3 g/cm^3 to 1.0 g/cm^3 [7]. The mineral content in the compact bone tissue varied from 0.3 to 0.6, and for the spongy bone tissue it was from 0.25 to 0.53 [8]. Poisson's ratios of the compact and spongy layers were considered to be 0.32 and 0.2 respectively. Modulus of elasticity and Poisson's ratio of the intermediate layer were defined as arithmetical mean between corresponding values of the compact and spongy layers.

IOP Conf. Series: Journal of Physics: Conf. Series **1045** (2018) 012022 doi:10.1088/1742-6596/1045/1/012022

The geometric models of zirconium ceramic samples with porosity 7% were constructed based on the histogram of pore size distribution of ceramics obtained experimentally. Figure 4 shows the geometric models of ceramic samples with different geometry of pore space.

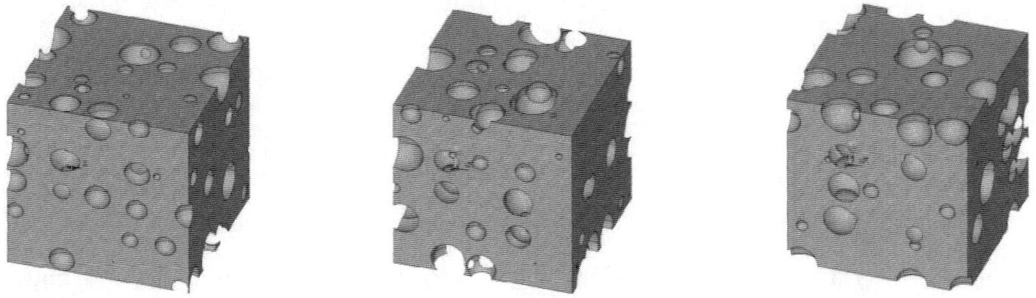

Figure 4. The geometric models of zirconium ceramic samples with different geometry of pore space

The material was considered homogeneous and isotropic. The modulus of elasticity of the material was assumed equal to 200 GPa, Poisson's ratio 0.3. The accepted characteristics correspond to the non-porous ceramic ZrO_2 (Y_2O_3).

The stress-strain state of the model samples of the bone fragments and ceramics at a compressive or tensile stress was calculated using the method of the finite elements. The tasks were solved within the context of the linear elasticity theory.

3. Results of calculation

The structure and mineral content mesovolumes and microvolumes of compact bone tissue, at which the most uniform distribution of strain components is realized in the case of axial compression or stretching, was determined. The ratio of the maximum value of the strain components to the minimum value, max ε_z / min ε_z, is considered as a quantity that determines the uniformity of the distribution of deformation components. The maximum value of the strain components in bone meso volumes belongs to the compact layer, and the minimal belongs to the spongy layer. For microvolumes of compact bone tissue, the maximum value of the strain components belongs to the matrix, the minimum belongs to osteons.

It was found that the most uniform distribution of deformation components in axial compression is realized in the meso-volume of bone with a ratio of the volumes of compact and spongy layer 0.5 at a ratio of the densities of these layers of 2.75 and the ratio of their mass fractions of minerals 1.2. The most uniform distribution of strain components at axial compression is realized in a microvolume of compact bone tissue with a porosity of 3%, a mass fraction of 35% of minerals and the direction of the collagen-mineral fibers perpendicular to the axis of the osteons (bone axis) (Fig. 5, a). The most uniform distribution of strain components at axial tension is realized in a microvolume of compact bone tissue with a porosity of 15%, a mineral mass fraction of 60%, and the parallel to the osteon axis direction of the collagen-mineral fibers (Fig. 5, b).

The stress-strain state of model samples of porous zirconium ceramics is calculated at a compressive stress of 1300 MPa. Figure 6 presents the calculation results of the regions of model samples of zirconium ceramics (light gray color) in which the compressive normal stresses σ_y exceed the ultimate compression strength of non-porous ceramic 2000 MPa. It is believed that the micro fractures of the material are possible in these regions.

Regions in which micro-fractures of a material are possible can be concentrated both near individual pores and also connect several pores with each other.

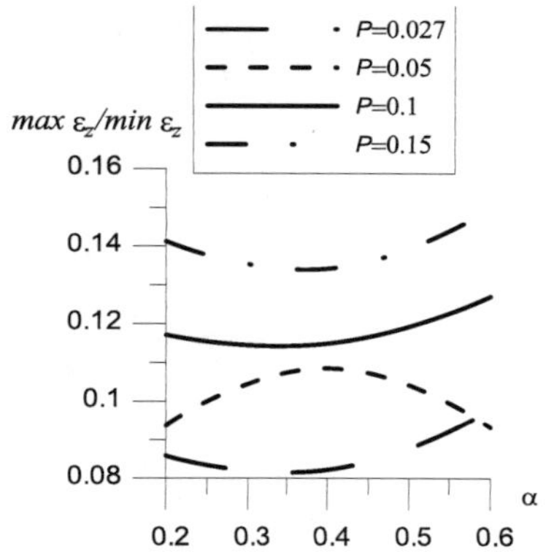

(a) samples of I type under axial compression

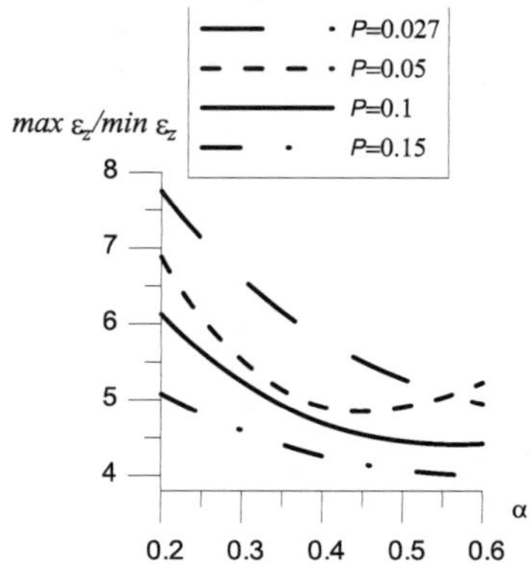

(b) samples of III type under axial tension

Figure 5. Dependences of the ratio of maximum and minimum deformation of compact bone microvolumes of different porosity from the mass fraction of minerals

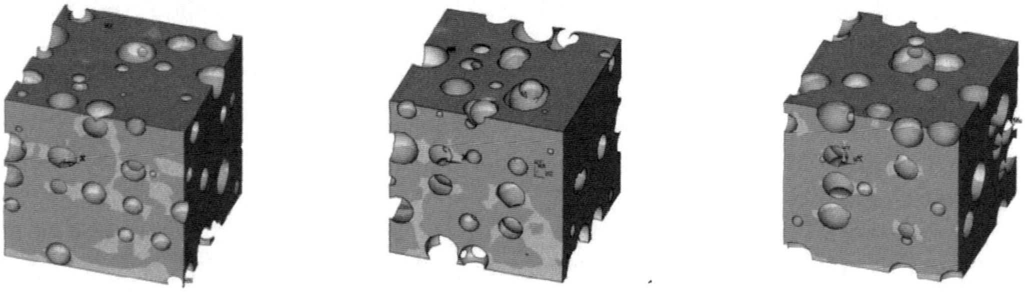

Figure 6. Regions of model samples of zirconium ceramics in which the compressive normal stresses σ_y exceed the ultimate compression strength of 2000 MPa

4. Conclusion

The structure and mineral content of bone tissue fragments are determined in which the most uniform distribution of deformations is realized under axial tension and compression. It was found that the geometry of the porous space of ceramic samples significantly influences the character of the distribution of the regions in which micro fractures are possible. Regions in which micro-fractures of a material are possible can be concentrated both near individual pores and also connect several pores with each other.

References

[1] Gömze L A 2016 *Applied materials science* I. *Compilation of Selected Scientific Papers* (IGREX Ltd, Hungary) p 189 ISBN 978-963-12-6600-9
[2] Gömze L A and Gömze L N 2017 *IOP Conf. Ser.: Mater. Sci. Eng* **175** 012001 https://doi.org/10.1088/1757-899X/175/1/012001
[3] Kurovics E and Gömze L A 2017 *MultiScience - XXXI. microCAD Int. Multidiscip.Sci.Conf.* (University of Miskolc, Hungary)
[4] Buyakov A S, Kulkov S N 2017 *IOP Conf. Ser.: Mater. Sci. Eng.* **175** 012025 https://doi.org/10.1088/1757-899X/175/1/012025
[5] Bromage T *et al* 2003 *Anat. Rec.B: New Anat.* B **274** 157
[6] Hernandez C J, Beaupre G S, Keller T S and Carter D R 2001 *Bone* **29** 74
[7] Liebschner M A K 2004 *Biomaterials* **25** 1697
[8] Keller T S 1994 *J. Biomech.* **27** 1159

IOP Conf. Series: Journal of Physics: Conf. Series **1045** (2018) 012023 doi:10.1088/1742-6596/1045/1/012023

Modeling three-dimensional flow of the polymer melts in a converging channel

A Kuznetsov[1], K Koshelev[2], N Cherpakova[3] and G Pyshnograi[1]

[1]Altai State University, avenue Lenin, 61, 656049, Barnaul, Russia
[2]Institute for Water and Environmental Problems SB RAS, Molodejnaya, 2, 656049, Barnaul, Russia
[3]Altai State Technical University, avenue Lenin, 46, 656049, Barnaul, Russia

E-mail: pyshnograi@mail.ru

Abstract. Temperature and volume rate impact of the polymer melt on the sizes of the vortex area has been considered. The vortex area appears during the flow at the inlet of the slotted channel. A modified Vinogradov and Pokrovskii rheological model is used for the mathematical modeling of three-dimensional flow of the melt in the convergent plane-parallel channel. This model was extended to take account of the nature of non-monotonic gradient dependence of elongation viscosity. On the solid wall the sticking conditions were used for velocity. The temperature dependence of the initial shear viscosity of the polymer melt has Arrhenius form. The initial relaxation time was rated by comparison with experimental data for agradient based stationary viscosity in uniaxial tension, and on the basis of the molecular-kinetic approach. Discrete analogues of a system of equations of the dynamics for the polymer liquid were obtained via the method of controlling the volume with the division due to physical processes. The numerical algorithm was implemented so that the technology of parallel computing based on GPU was conveniently used. Calculations of hydrodynamic characteristics of polymer melt flows at different temperatures in a convergent channel with a rectangular cross section have shown that the sudden narrowing of the channel leads to the appearance of eddy flows. The sizes of these flows pass through a maximum with increasing melt temperature which is detected in the experiments. It also confirms that the rheological model can be used for the flows of polymer melts in areas with complex geometry. The results are proving the effectiveness of CUDA technology parallel computing for unsteady calculations of three-dimensional flows of nonlinear viscoelastic fluids with rheological law of behavior in differential form. The above calculations show that some convergent flow of polymermelts may show substantial three-dimensional picture, which manifests itself in a velocity component in a neutral flow direction. This should be taken into account when arranging experiments, as there are methods that cannot measure all velocity components. Also note that the velocity profile in the slit of the channel is set at a considerable distance from the entrance to the channel.

1. Introduction

The paper considers properties of hydrodynamic structure for 3D flows of thermoplastic melts in a plane-parallel channel with sudden convergence. Thermoplastics can be heated to melts. They solidify upon cooling in the shape of the acquired object without changing its structure. Therefore, they can undergo melting several times. Such polymers are: polyethylene, polypropylene, polystyrene, polyvinylchloride etc. The mathematical study of two and three-dimensional flows of nonlinear viscoelastic medium is of great importance in polymer melts and solutions rheology. It is known that such flows may be observed in different types of polymer processing (e.g., extrusion, casting, blowing, injection molding and others). Mathematical modeling of such flows requires refinements of a rheological model. The accuracy of rheological models was tested by calculating two and three-dimensional flows especially in domains with complex setting.

In this paper the Vinogradov and Pokrovskii rheological model is employed to solve the problem of mathematical description for three-dimensional nonlinear viscoelastic fluid flows in a plane-equivalent channel with a sudden constriction. The control volume method including the division due to physical operations is used to obtain discrete analogs of initial particular differential equations. The numerical application is carried out via the GPU-setup equivalent calculating technology CUDA.

2. The mathematical model

It is well-defined that melts of polymers are non-linear viscoelastic media. A perfect foundation of the relaxation processes theory for polymer melts can be presented by the microscopic approach. The approach originates from researching the motion of the entangled macromolecules system. All the consequences of this approach cannot be easily calculated due to the complicated mathematics and many extra assumptions must be introduced in the calculation. The microscopic method gives justification to the phenomenological results.

To use the mesoscopic (intermediate) approach successfully in relation to polymer melts and solutions single macromolecule approximation is employed. It means that the single macromolecule is moving surrounded by other macromolecules and the solvent. It is considered to be an efficient field. Some phenomenological parameters should be introduced into the mesoscopic theory in order to characterize this effective field [1].

$$\sigma_{\mathrm{ik}} = -p\delta_{\mathrm{ik}} + 2\eta_1\gamma_{\mathrm{ik}} + 3\frac{\eta_0}{\tau_0}a_{\mathrm{ik}}, \tag{1}$$

$$\frac{d}{\mathrm{dt}}a_{\mathrm{ik}} - \nu_{\mathrm{ij}}a_{\mathrm{jk}} - \nu_{\mathrm{kj}}a_{\mathrm{ji}} + \frac{1+(\kappa-\beta)I}{\tau_0}a_{\mathrm{ik}} = \frac{2}{3}\gamma_{\mathrm{ik}} - \frac{3\beta}{\tau_0}a_{\mathrm{ij}}a_{\mathrm{jk}}. \tag{2}$$

Where σ_{ik} – tenser of stress for the polymer system; a_{ik} – dimensionless tensor of extra stress; p – hydrostatic pressure; η_0 – shear viscosity initial value, τ_0 – time of initial relaxation; β and κ – coefficients of scalar anisotropy which consider shape and size of macromolecular coil; $\eta_1$viscosity of residual shear; v_{ik} – tensor of velocity gradient; γ_{ik} – gradient tensor of symmetrized velocity; I – trace of tensor a_{ik}.

It can be expected that this model tested on simple flows will be suitable for the numerical study of polymer behavior in difficult conditions of deformation, which are typical for polymer processing. These are steady and unsteady flows in circular ducts, flows in channels with a sharply varying cross-sectional area and flows with a free surface. An essential feature of such flows is their two- and three-dimensional character. Earlier attempts of modeling such flows on the basis of the model have been made [1-3].

Work [3] considers the two- and three-dimensional flows under the constant pressure gradient in channels with a rectangular cross-section. For calculating real flows on the basis of this model the equations of momentum and massconservation should be added to the rheological model.

3. Results and Discussion

Temperature is one of the most important parameters affecting the quality of products made of polymers. The influence of polymer melts temperature on the size of the vortex area, appearing in the flow at the slit channel entrance was studied.

Model parameters ($\beta = 0.1$ and $p_0 = 0.005$) were obtained by investigating similarity between numerical data and theoretical dependences for shear and elongation viscosity. The comparative results for these curves are shown in Figure 1. Recalculation of values η_0 and τ_0 for pure shear is made by the following equations:

$$\eta_0(T) = \eta_0(T_0)\cdot 10^{\frac{E_a}{2{,}3R}\left(\frac{1}{T}-\frac{1}{T_0}\right)}, \tau_0(T) = \frac{\eta_0(T)\cdot M}{\rho \mathrm{RT}}, \tag{3}$$

Here R – the ideal gas constant, $E_a = 58$ KJ/mol – activation energy, $T_0 = 150^{\circ}$C.

IOP Conf. Series: Journal of Physics: Conf. Series **1045** (2018) 012023 doi:10.1088/1742-6596/1045/1/012023

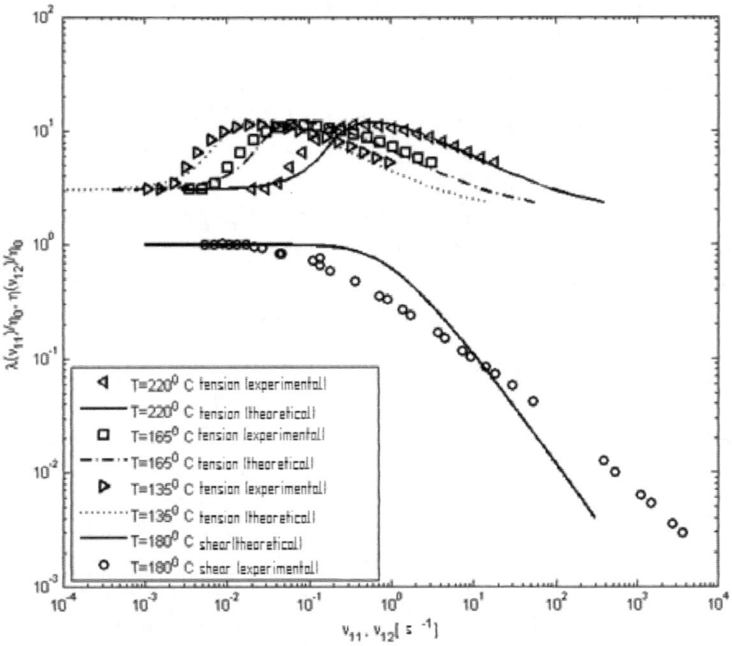

Figure 1. Comparing theoretical and experimental dependences on elongation speed of stationary shear viscosity and stationary tension viscosity at different temperatures.

Values of τ_0 have been chosen so that the initial section of elongation viscosity growing has been described properly, while values of p_0 have been selected for the decreasing region of elongation viscosity.

Let us notice that the influence of p_0 parameter on the view of stationary and non-stationary viscometric functions was studied in [2]. They have shown that this parameter slightly affects the stationary shear viscosity function. It has been added for the description of non-monotonic type of the stationary viscosity function under axial tension.

The value of density was given in [4] and it is 918 kg/m^3 for the low density polyethylene. Note, that this parameter has an insignificant effect on the final solution, since the Reynolds number for the described flows is rather small (about 10^{-6}). As for residual viscosity η_1, its values were obtained according to $\eta_1 = 0.005 \cdot \eta_0$ and its influence is not visible in Figure 1. This parameter was added to the model for preventing a rheological model from transition into a perfect fluid model under large velocity gradients. It can be observed when velocity field and stress field undergo significant transformations. For residual viscosity detection in polymer melts at viscometric experiments, a range of shear velocities needs to be extended significantly at the same time as this parameter for concentrated solutions can be obtained easily and is not only a viscosity solvent. It also depends on polymer concentration. The effect on final results of η_1 is negligible but it provides a much bigger time step at the first computation period when the available approximation needs significant realignment.

We refer to paper [4] for comparison between experimental data and computation results. Firstly, it should be noted that stationary values could have been attained neither at calculations nor at experiments because calculated velocity, stress and pressure profiles fluctuate near stationary values. The magnitude of these fluctuations which does not exceed 10 percent for velocity vector components allows making comparison with an experiment. This comparison is shown in Figure 1 (**T**=180 °C). The vortex area before the slit channel entrance has been detected both in the experiment and in the calculation. Because of the irregular mesh type in the numerical research, the small scale is applied for velocity vector display at the right image. In reality velocity vector modules are close, which is granted by a good match between volume rates.

So the size or the vortex area is an essential characteristic of the studied flows. Paper [4] offers to calculate the vortex area via the following technique.

Firstly values of $f(x)$ are obtained, for which$\int_{-7}^{f(x)} u(y)\mathrm{d}y = 0$, and then the area under $f(x)$ curve. Let's notice that good results can be obtained by this approach only if the velocity vector component at the direction neutral to the flow is much lower than at two other directions.

Now let us calculate vortex area areas in$z = \text{const}$section plane. Comparing experimental and calculated dependences of the vortex size on the distance to the channel axis as shown in Figure 2, it can be concluded that the intensity of vortex flows increases while removing from the channel axis. This fact may be explained by the Weissenberg effect. From the above comparison it may be stated that the experimental results and the numerical data agree qualitatively. It is noted that the calculations given for Newton's law of behavior ($\tau_0 = 0; \beta_0 = p_0 = 0$) show the absence of vortex areas where as the calculations for Oldroyd-B viscoelastic fluid ($\beta_0 = p_0 = 0$) demonstrate understated values for vortex area and absence of the increase in vortex intensity while the section plane removes from the axis of the channel [3].

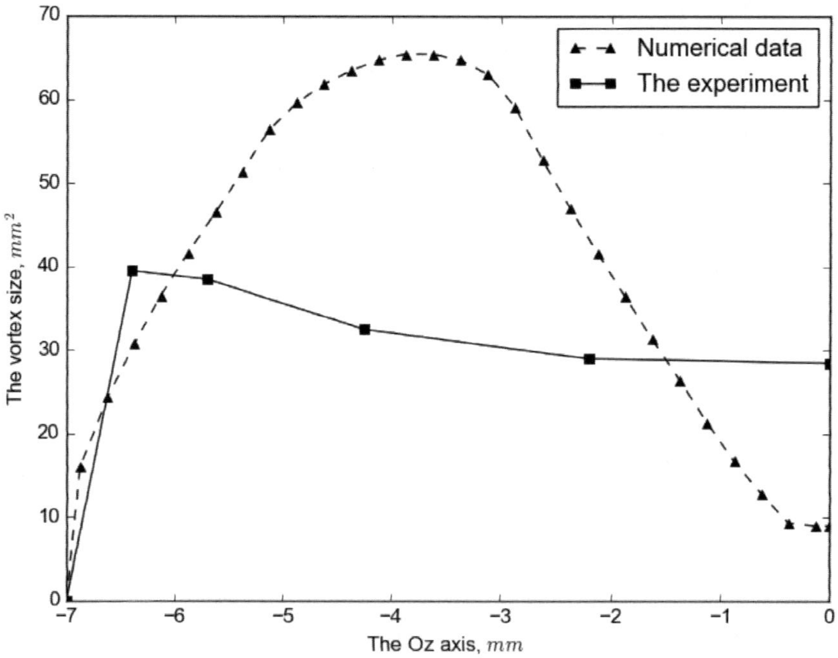

Figure 2. Vortex size at different section planes at T=180°C.

Let us consider the influence of the polymer melt temperature on the vortex area size. The computed values of the vortex area in the axial section plane at different temperatures are shown in Figure 3. Since the temperature affects both initial shear viscosity η_0 and initial relaxation time τ_0, it leads to the reduction of these parameters during the temperature increase.Thus, the temperature influence is quite complicated. Initially the vortex size grows alongside with the temperature increase but then, after going through the maximum it starts to decrease. Such a non-monotonic view of dependence is also observed at experiments, the results shown in Figure 3 too.

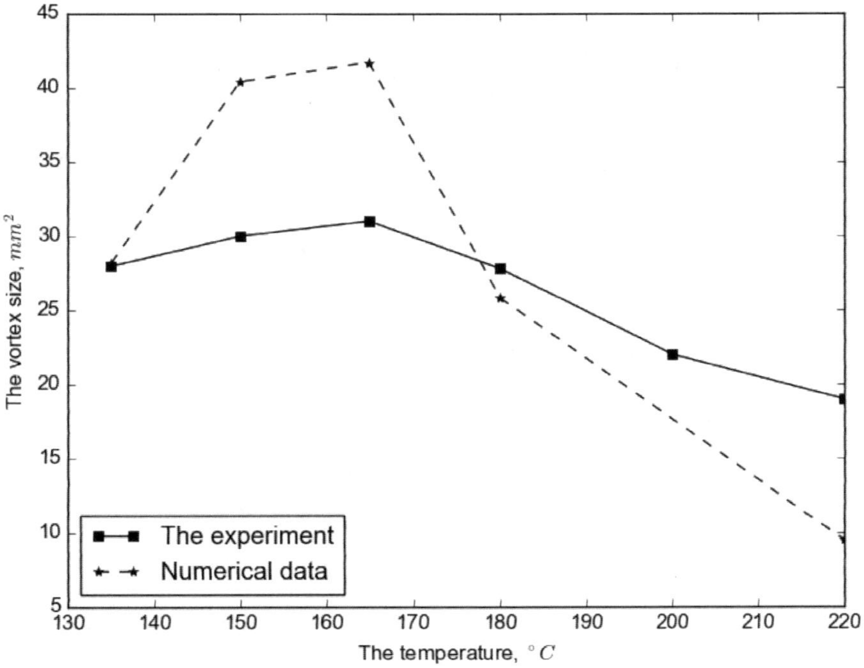

Figure 3. The dependence of the melt temperature on the vortex size.

The analysis of the results shown in Figures 2 and 3 reveals the discrepancy between theoretical and experimental dependences (which can differ by a factor of two). Can this be considered acceptable? Or does such accuracy suggest the inadequacy of rheological model (1)?

Let us notice that in modeling viscometric functions calculated on the basis of model (1) as showed in [1, 2]. These discrepancies can reach 5-10 times for simple shear stress and when axial elongation stress is plateaued. This can also be seen in Figure 1 for stationary shear viscosity at the shear rate of 0.7 s^{-1}. This disadvantage of the model is easily overcome by the transition to the multi-mode approximation [5]. Herewith several relaxation processesare introduced for consideration, their time scales can differ by 4 orders. If it is not difficult in the viscometric flows modeling due to their simplicity, in the three-dimensional flows modeling this fact can not only lead to significant increase in calculation time but also makes the calculations impossible because of the increased "hardness" of the solved system. Also it should be noted that the paper [3] the vortex area size has fallen to zero when the initial relaxation time changes from 2 s to 0.2 s. In other words, errors in the definition of the relaxation time can change the flow pattern essentially. Thus the computation accuracy, obtained in Figures 2 and 3 can be considered acceptable. Anyway this issue needs comprehensive study both in terms of the convergence for numerical schemes and adequate description for the studied processes.

Therefore, model (1) describes experiments mentioned here in qualitatively, and the differences between experimental and theoretical curves can be attributed to both inaccuracy in defining τ_0 parameter, the single-mode nature of model (1) and the oscillatory character of the solutions obtained. Perhaps this is a consequence of instability inherent in the examined flows. Velocity vector oscillations are possible both in the absolute value and in the direction that allows to average the observed values over the period of oscillation. Despite this, it should be noted that the studied flow is not turbulent, in other words, the observed oscillations are regular and not chaotic. Anyway, these issues need further thorough research to be carried out in the following works.

4. Conclusion.

Thus, the paper has compared the polymer melt flows with different rheological characteristics in a convergent channel with a rectangular cross section. It has shown that an increase in the relaxation time of the polymer sample gives rise to the eddy flow which is detected in the experiments.

Therefore, it confirms the applicability of a modified rheological Vinogradov and Pokrovskii model for describing the dynamics of polymer melts in areas with complex geometry. Also, the results obtained in this paper prove the effective application of parallel computing CUDA technology at unsteady calculations for three-dimensional flows of nonlinear viscoelastic media with rheological law of behavior in a differential form.

However, this paper does not discuss many issues related to flows in converging channels. For example, it doesn't state that the channel geometry, the molecular structure of the polymer or the temperature can affect the flow pattern. These and other issues can be explored in subsequent papers on the basis of the developed mathematical models and numerical methods tested here.

5. References

[1] Altukhov Yu A, Golovicheva I E, Pyshnograi G V 2000 *Fluid Dynamics* **35** 1
[2] Altukhov Yu A, Pokrovskii V N, Pyshnograi G V 2004 *J. Non-Newtonian Fluid Mech.* **121** 73
[3] Koshelev K B, Pyshnograi G V, Tolstykh M Yu 2015 *Fluid Dynamics* **50** 16
[4] Hertel D, Valette ., Münstedt H 2008 *J. Non-Newtonian Fluid Mech.* **153** 82
[5] Merzlikina D A, Pyshnograi G V, Pivokonskii R, Filip P 2016 *Journal of Engineering Physics and Thermophysics* **89** 652

IOP Conf. Series: Journal of Physics: Conf. Series **1045** (2018) 012024 doi:10.1088/1742-6596/1045/1/012024

Research into Dry Turning of Welded Surface by Replaceable Cutting Insert with Closed Loop Heat Removal

V P Kuznetsov, A S Skorobogatov, V V Voropayev, A A Petunin

Institute of New Materials and Technologies, Ural Federal University, 28 Mira St., 620078 Yekaterinburg, Russia

E-mail: v.p.kuznetcov@urfu.ru

Abstract. A special tool has been devised incorporating a closed loop system based on Peltier thermoelectric couples to cool replaceable hard alloy inserts. Experimental research has been done into dry turning of a X20Cr13 steel welded surface with a SECO cubic boron nitride circular plate at the cutting speed increasing from 100 to 300 m/min. It has been determined that Ra surface roughness decreases by 22% while Rz surface roughness goes down by 26%; in addition, the wear flat of tool tips diminishes by 17%.

1. Introduction

Turning is accompanied by intensive heat generation due to friction interaction and deformation of the material that is being removed; it is made obvious by high temperatures of the chip, the cutter and the work-piece [1-5]. Growing temperature of the machined material produces an adverse effect on the quality of machining; heating of the cutter brings about lower hardness of the tool tip, its quick dulling, which significantly reduces tool durability [2]. The intensity of heat generation is influenced by the properties of the machined material, that of the tool, the geometry of the cutter, and cutting modes. The biggest impact is made by cutting speeds while the feed and the cutting depth produce a lesser influence [3].

The heat generated in turning is distributed between the chip, the cutter and the work-piece. The biggest share of heat goes into the chip; the rest of it is distributed between the cutter and the work-piece. According to [4-5] the amount of heat diverted into the tool falls from ~60 to 18% when cutting speeds applied to AISI 316L and AISI 1045 steels are increased from 10 to 80 m/min. As the speed is further increased, the correlation of thermal flows does not change. As far as the quality of machining, tool life and production rate are concerned, most favorable conditions occur at a cutting speed of 50 m/min when the intensity of heat generation is relatively low and the heat transfer into the tool amounts to ~30%. Optimal cutting speeds recommended by SECO in regard to machining chromium steels with PCBN inserts range within 50…100 m/min.

In order to provide for high-quality machining and long tool life at higher cutting speeds it is necessary that the fraction of heat relieved from the cutting zone into the tool should be increased if the cutting speeds are be increased. As soon as external application of a lubricant for dry cutting is impossible and cooling of the work-piece is not efficient, the only remaining approach offering prospects for the future is to internally remove heat from the replaceable cutting insert.

A number of approaches toward producing turning tools with internal cooling of replaceable turning plates (RTP) are described in the papers [6-8]. The operational principle of these systems is aimed at supplying a cooling liquid to the base of the RTP through channels provided inside the tool holder.

The paper [6] reports on an open loop system through which a liquid heat medium is pumped. The system seems easy to fabricate: at the same time, it is in need of an efficient pumping unit, its application is associated with high consumption of cooling liquid. The papers [7, 8] present a two-stage system with a closed loop, its operation based on evaporation of a liquid coolant under a replaceable turning plate. The closed loop system features more efficient cooling of the RTP; somehow, due to the large dimensions of its condenser its application for mass production seems difficult.

A more efficient cooling system is presented in the papers [9-10]; it is designed to increase production rates of nanostructuring burnishing. An open loop system [9] employing high-pressure pumping of a coolant in the interior of the tool provides for a possibility to increase the rate of machining of X20Cr13 (43…45 HRC) steel by 65%. The closed loop cooling system built on Peltier thermoelectric couples has proven to be very efficient in removing heat from the tool, making it possible to raise the rate of burnishing by 2.2 [11].

The purpose of the present research is to determine the impact of internal heat removal, relieving heat from RTP's of the PCBN type, on the smoothness of machined surfaces and tool life in the case of dry turning of welded material at a cutting speed 3 times higher than the recommended value.

2. Experimental setup

To do the experimental study a turning tool was devised equipped with an internal closed loop for cooling its RTP (Fig. 1). High intensity of its heat removal is achieved by decreasing the temperature of the coolant in the closed loop which comprises a heat exchanger with four integrated Peltier thermoelectric couples. The cooling performance of the Peltier thermoelectric unit is 400 W; it enabled to maintain the temperature of the coolant at 9 °C when the turning speed was increased to 300 m/min.

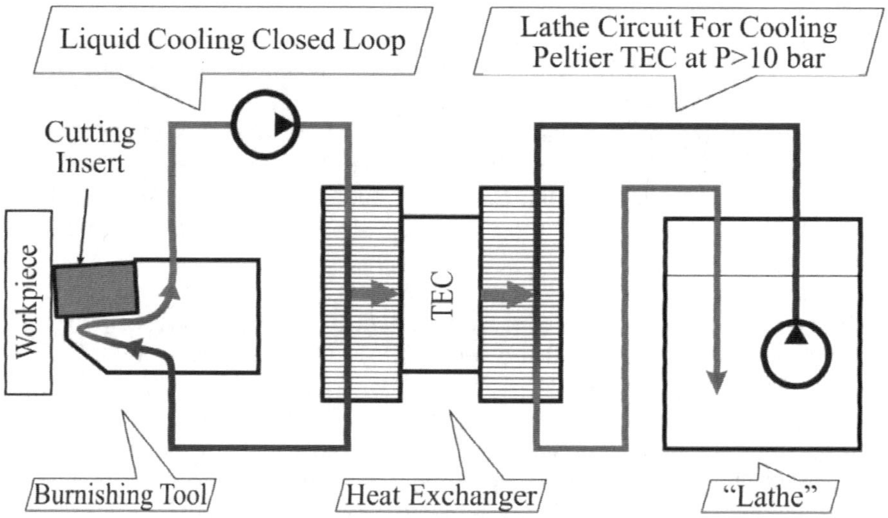

Figure 1. Liquid cooling closed loop system for cooling the replaceable insert based on Peltier thermoelectric couple

A disk-shaped sample used in the experiment was made from X20Cr13 steel (as delivered); a X20Cr13 steel layer 5 mm thick was welded to it, its hardness measuring from 45 to 48 HRC. Turning was done on an Okuma Multus B300 lathe (Fig. 2).

Figure 2. Turning tool with a closed loop system for cooling a replaceable cutting insert on an Okuma Multus B300 lathe: 1 – tool holder with the RCT; 2 – cooling system; 3 – thermovision camera; 4 – workpiece

Preliminary turning was performed with a Sandvik WNMG 080408-SM hard alloy cutting plate at a cutting speed of 80 m/min and a feed of 0.6 mm/rev.

Hard finishing turning of circular areas on the face was done with a SECO RNMM190400S-10020 round plate manufactured from CBN500 cubic boron nitride, its diameter being 19 mm. The cutting depth was set at t=0.15 mm, the feed was f=0.15 mm/rev. The cutting speed was set at 100, 200, 250 and 300 m/min separately for each path.

In the process of machining the temperature on the surface of the replaceable turning plate was measured with a Testo 875 thermovision camera. The roughness of the machined circular areas was measured by means of the contact method on a Calibr 170623 profilometer. The arithmetic mean value resulting from ten measurements of roughness was assumed to be the final result

3. Results and Discussion

Equally, when turning with and without heat removal, the dependence of the temperature of the RTP on the cutting speed turns out to be identical (Fig. 3). When the speed is raised from 100 to 250 m/min, the temperature grows linearly. Alternatively, increasing the speed from 250 to 300 m/min results a significant decline of the temperature of the RTP. In the case of machining without heat removal the maximum and the minimum temperature values of the RTP were 263 °C and 69 °C respectively. It was discovered that application of the cooling system leads to lowering maximum and minimum temperatures by 12% to 235 °C and by 32 % to 47 °C respectively. The significant temperature drop at 300 m/min may be attributed to lower friction in this speed range.

IOP Conf. Series: Journal of Physics: Conf. Series **1045** (2018) 012024 doi:10.1088/1742-6596/1045/1/012024

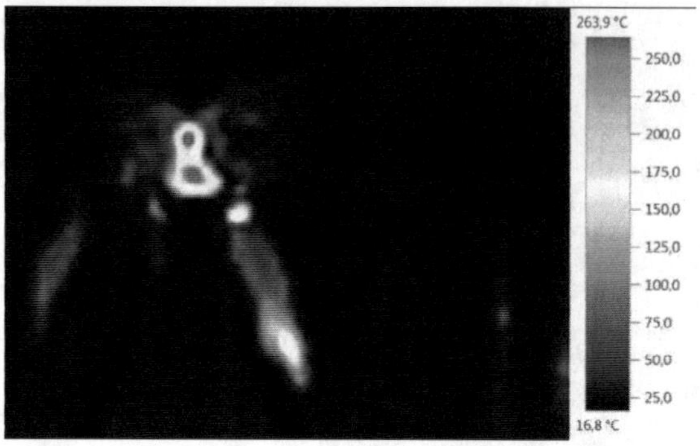

(a)

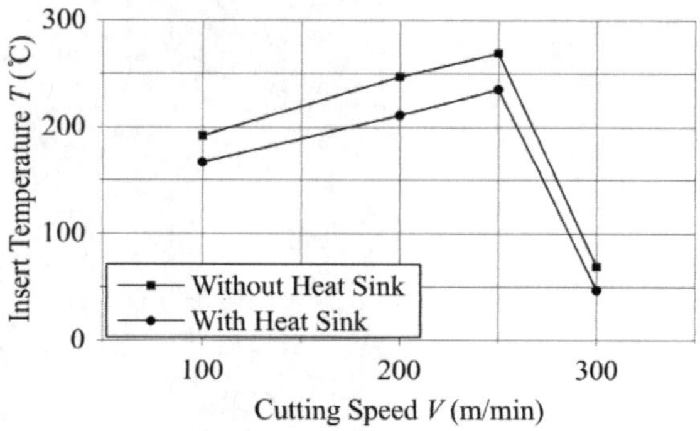

(b)

Figure 3. Thermograph of the cutting zone when machining without heat removal at 300 m/min (a) and dependence of the RTP temperature on the cutting speed (b)

Dependences of changes in the surface roughness as after turning exhibit an excellent correlation with the RTP temperature. Cooling the RTP is beneficial for increasing the surface smoothness; it is specifically evident for turning at 300 m/min. The roughness Ra goes down by 22% from 0.23 μm to 0.18 μm and the roughness Rz – by 26% from 1.43 μm to 1.06 μm.

Usage of the cooling system also contributes to reducing the wear of the RTP. At the optimal speed of 300 m/min the width of the wear flat in the tool tip decreases by 17% from 2.3 mm to 1.9 mm (Fig. 4).

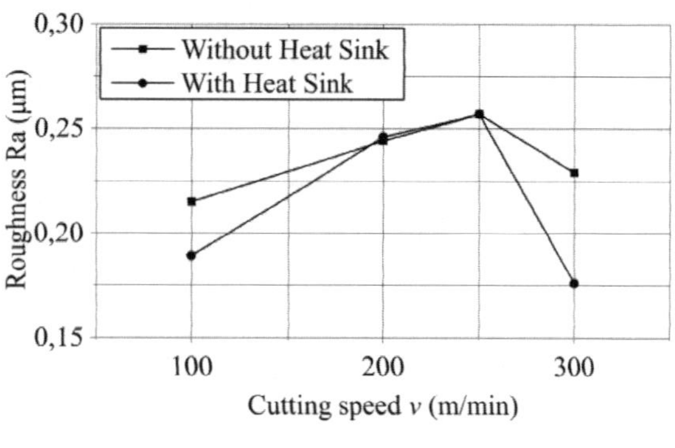

(a)

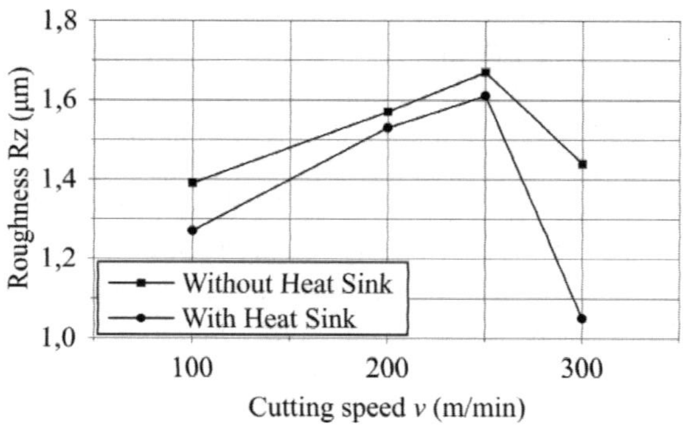

(b)

Figure 4. Dependence of roughness Ra (a) and Rz (b) on cutting speed

4. Conclusions

The research has proven that application of the closed loop cooling system results in both better smoothness of machined surfaces and longer tool lives of RTP's. It has been determined that the most favorable conditions for dry turning of a X20Cr13 steel welded layer exist at a speed of 300 m/min.

Application of the cooling system at the optimal speed leads to lowering the roughness Ra by 22% from 0.23 to 0.18 μm and the roughness Rz by 26% from 1.43 to 1.06 μm in comparison to machining with no heat removal. The width of the wear flat in the tool tip of the RTP goes down by 17%. Therefore, application of a closed loop cooling system is an efficient way for improving the quality of machining and the tool life of RTP's.

Acknowledgements
The work was supported by Ministry of Education and Science of the Russian Federation, contract № 02.G25.31.0148 with Sverdlovsk Instrumental Plant within the framework of Research and Development Project No H979.210.007/15, July 28, 2015 for the federal autonomous educational institution of higher education "Ural Federal University".

References

[1] Wang X, Da Z J, Balaji A K, Jawahir I S 2007 *J. of Manuf. Proc.* **9–1** 61
[2] Che-haron C H 2001 *J. Mater. Process. Tech.* **118** (1–3) 231
[3] Venkata R R, Kalyankar V D 2012 *Eng. Appl. of Artif. Intell.* **26–1** 524
[4] Bonnet C, Valiorgue F, Rech J, Claudin C, Hamdi H, Bergheau J M, Gilles P 2008 *Int. J. of Mach. Tools and Manuf.* **48** 1211
[5] Ben Abdelali H, Claudin C, Rech J, Ben Salem W, Kapsa P, Dogui A 2012 *Tribol. in Manuf. Proc.* **286–287** 108
[6] Ferri C, Minton T, Ghani S B C, Cheng K 2014 *J. of Mech. Eng. Sci.* **228–1** 135
[7] Sanchez L E A, Scalon V L, Abreu G G C 2011 *Proc. of Third Int. Workshop on Adv. in Cleaner Prod.* 18
[8] Vicentin G C, Sanchez L E A, Scalon V L, Abreu G G C 2011 *Clean Technol. and Environ. Policy* **13–6** 831
[9] Kuznetsov V P 2013 *High Technol. in Manuf. Eng.* **11** (29) 19
[10] Kuznetsov V P, Skorobogatov A S, Gorgots V G 2015 *Pap. of IV Int. Sci. and Tech. Conf. (Reznik's Readings)* 34
[11] Kuznetsov V P, Skorobogatov A S, Gorgots V G, Yurovskikh A S 2016 *IOP Conf. Ser.: Mater. Sci. and Eng.* **124** 012127 https://doi.org/10.1088/1757-899X/124/1/012127

IOP Conf. Series: Journal of Physics: Conf. Series **1045** (2018) 012025 doi:10.1088/1742-6596/1045/1/012025

Control of the effects of the concentration on the viscoelastic behavior of the aqueous Polyethylene-Oxide solution

N Lahlou[1], M Riahi[1], S Aniss[1], M Ouazzani Touhami[1]

[1]Laboratory of Mechanics, Faculty of Sciences Aïn-Chock, B.P.5366 Mâarif, Casablanca, Morocco.

E-mail: nouhalahlou10@gmail.com

Abstract. It is established that the rheological properties of polyethylene oxide solutions are related to the concentration, molecular weight, nature of the solvent, external stresses such as temperature and pressure and microstructural interactions. The main objective of this study is to carry out the effect of the concentration of aqueous PEO solutions on their rheological behavior. In this context, we consider quasi-stationary, transient and oscillatory flow regimes for three values of concentrations (250 ppm, 500 ppm and 1000 ppm). Except the case where the solution is Binghamian (250 ppm), we denote that for quasi-stationary mode, the observed behaviors correspond to a rheofluidifying fluids with yield stress. The model adapted from Rheowin Software is Herschel-Bulkley for whicha yield stress and consistency increases and a flow indexdecreases when the concentration increases. The studies of these solutions under transient and subsequently under oscillating conditions are made to specify their viscoelastic properties. For a concentration less than 500 ppm the viscoelastic nature is described by the Maxwell model and by the Jeffrey model for a concentration of 1000 ppm. Otherwise, and for the oscillating test, the determination of the evolution of the critical frequency fc (for which the conservation factor G 'and the loss factor G' 'are comparable) as a function of the concentration consolidates the actual control protocols of PEO aqueous solutions and allows to optimize the processes of their use in industrial environment.

1. Introduction

POLYOX Water-Soluble Resins are non-ionic poly (ethylene oxide) polymers. They are hydrophilic polymers available in a wide range of molecular weights and are supplied as white, free flowing powders. POLYOX offers a history of successful use in extended release applications, including osmotic pump technologies, packaging, construction, heavy industry, electrical and electronics, automotive and agriculture. Other applications include hydrophilic matrices, gastro-retentive dosage forms, tamper-resistant technologies and other drug delivery systems such as transdermal and muco-adhesive technologies [1]. POLYOX water-soluble resins provide a number of benefits including:

- Wide range of molecular weights for formulation flexibility
- Versatile application in direct compression and granulation
- Rapid hydration and swelling for use in osmotic pump technologies
- Fast hydration and gel formation for use in hydrophilic matrices
- Extreme hardness and virtually crush-proof for tamper-resistant technologies

Most of these solutions exhibit a complex rheological behavior associating their viscoelastic and thermal properties to their mechanical history as well as memory effects that result in thixotropic behaviors. An earlier study [2], the viscosity of the solutions increases with the concentration of PEO and the flow curves obtained don't exhibit a yield stress. Otherwise, the concentrations are considered to verify that G 'and G" modules increase with the frequency [3]. The rheofluidifying character was confirmed by Bahlouli at al. [4] at a different concentrations and the rheological parameters have been calculated from the Cross's model. These conclusions are the results of many studies carried out [5].

In this work, we make a contribution to this study in order to better understand the rheological behavior of PEO aqueous solutions at low concentrations, mainly through the diversification of flow regimes.

2. Materials and Experimental Procedure

The tests are carried out with a Rheostress 1 Rheometer from Haake. It is a rotating rheometer operating at imposed speed and imposed stress, provided with several measuring cells. For our study, the coaxial cylinders cell called Z-41 is retained, and the samples used are solutions based on polyethylene oxide POE incorporated as a powder in distilled water. Three concentrations were prepared: 1000 ppm, 500 ppm and 250 ppm. The mixture is stirred through a rod rotating at a constant angular velocity for 24 hours.

It should be noted that each test is the result of the implementation of a procedure specified in the Rheowin software. This procedure is used to determine the nature and duration of the excitations imposed.

Firstly, the quasi-stationary tests are done to determine the yield stress of the prepared samples as well as their flow curves. The fluid is subjected to a constrained flow ramp imposed in a given range for a long time.

To carry out the viscoelastic properties, the solution is solicited by a sufficiently low stress (less than the yield stress to respect the linear behavior) for 5 min, this test allows to have a creep function of these solutions.

To complete the characterization of the viscoelastic behavior [3] (determination of the loss and conservation modules), oscillating mode tests were conducted.

3. Results and Discussion

3.1. In quasi-stationary mode

From the procedure mentioned above concerning the determination of the yield stress [5], Figure 1 shows the evolution of the deformation as a function of the shear stress, and Figure 2 represents two curves; (a) and (b) concerning respectively therheograms of the PEO solution for different concentrations and the Variation of the viscosity in function of the shear rate.

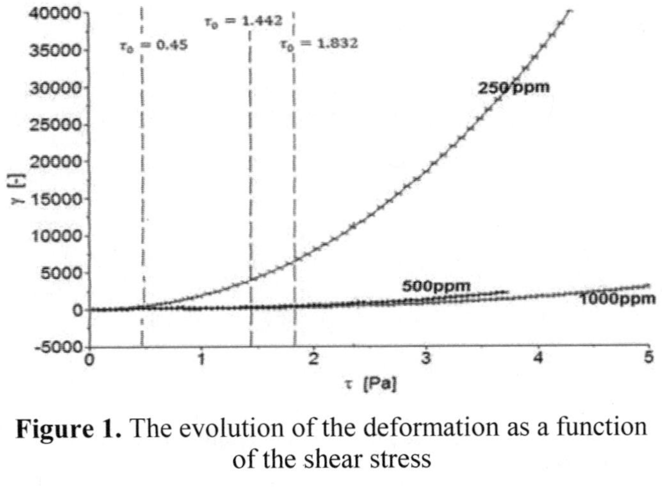

Figure 1. The evolution of the deformation as a function of the shear stress

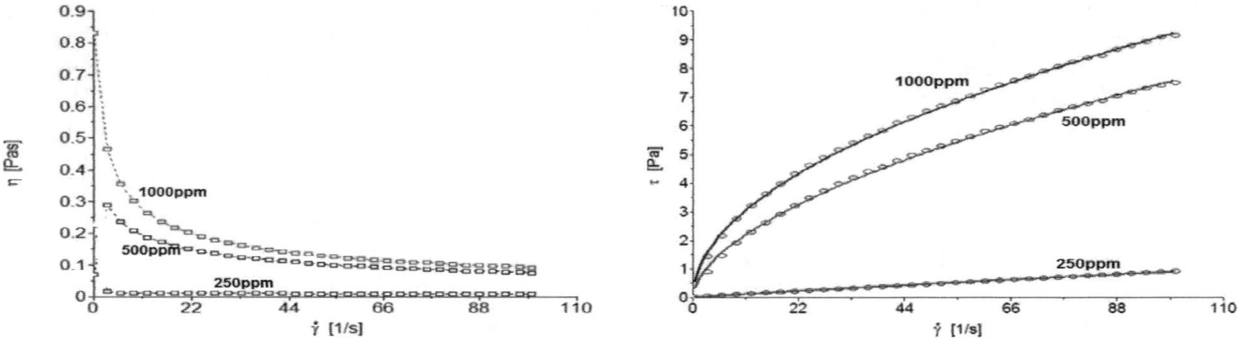

(a) (b)

Figure 2. Rheograms of the PEO solution for different concentrations (a), the variation of the viscosity in function of the shear rate (b)

According of these curves, the yields stress are other than zero for the three cases and they increasewith the concentrations (Figure 1). Moreover, for 250 ppm, the fluid has a constant viscosity, and for 500 ppm and 1000 ppm, the viscosities decrease with shear rate (Figure 2 (b)).

On the following table, we present the behavior law's corresponding of these evolutions; it's about a Bingham model for 250 ppm and Herschel-bulkley model for the other cases (Figure 2 (a)).

Concentration	Rheological models	Model settings
1%	$\tau = \tau_o + K\,\dot{\gamma}^n$	$\tau_o = 1.832$ K=0.9262 n=0.4996
0.5%	$\tau = \tau_o + K\,\dot{\gamma}^n$	$\tau_o = 1.441$ K=0.578 n=0.5593
0.25%	$\tau = \tau_o + \eta\,\dot{\gamma}$	$\tau_o = 0.45$ η =0.009

3.2. In transient mode

The viscoelastic properties are determined from the different expressions of their creep function.For 250ppm, this function is an affine one (Figure 3 (a)) similar to that of a Maxwell fluid:

$$f(t) = J_0 + \frac{1}{\eta_0}\,t\,,$$

Which has an instantaneous elasticity $J_0 = 3.404$ Pa^{-1}, a viscosity $\eta_0 = 0.009$ Pa.s, close to that obtained in quasi-stationary mode ($\eta_0 = 0.01$ Pa.s), and a relaxation time $\lambda = J_0\,\eta_0 = 0.034$ s.

A behavior identical to this one is observed for a concentration of 500ppm where $J_0 = 0,\ 1789$Pa^{-1}, η_0=0.38 Pa.s and $\lambda = 0.0696$ s.

For 1000 ppm (Figure 3 (b)), the creep function is approximated by:

$$f(t) = J_0 + \frac{1}{\eta_0}\,t + J_1\left(1 - e^{-t/\lambda_1}\right),$$

which corresponds to that of a fluid of Jeffrey. This has an instantaneous elasticity J_0, a delayed elasticity J_1, a delay time λ_1 and a relaxation time $\lambda_2 = J_1\,\eta_1$.

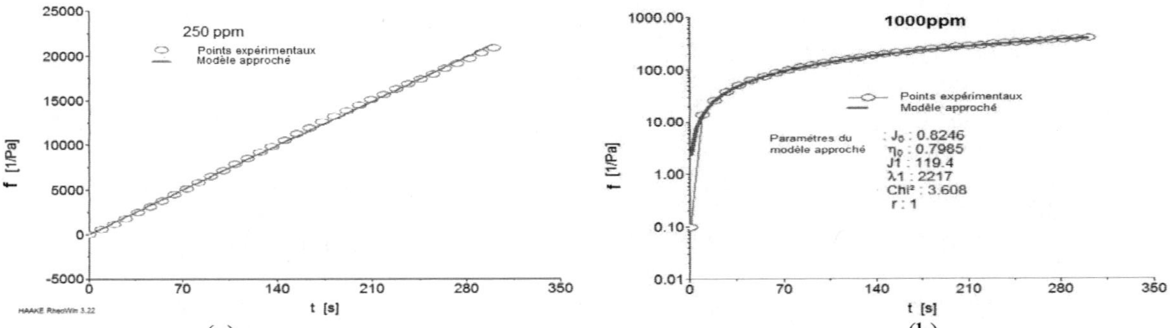

(a) (b)

Figure 3. The creep function for a concentration of 250 ppm (a), 1000 ppm (b).

3.3. In Oscillation *mode*

To complete the description of the viscoelastic behavior, the evolutions of conservation modules G 'and loss G" are presented for 500 ppm and 1000 ppm, andfor a frequency range varying from 1 to 10 Hz (Figure 4).

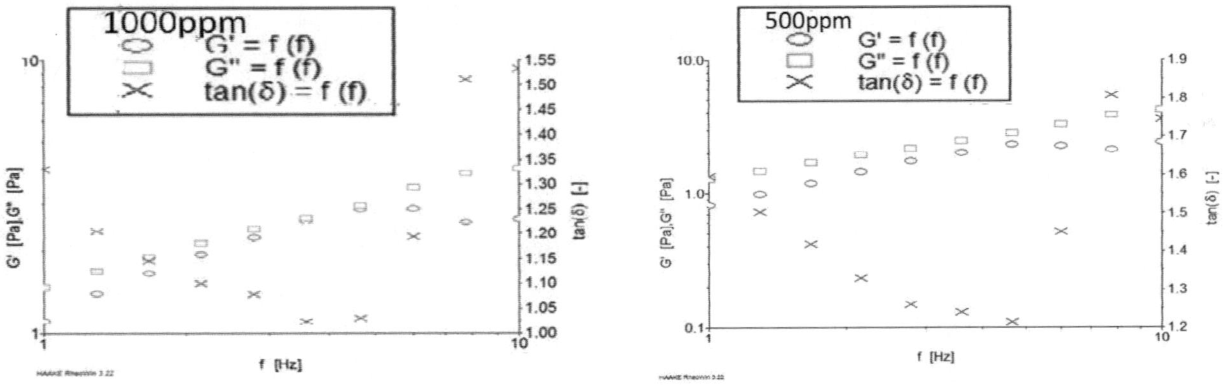

(a) (b)

Figure 4. The viscoelastic propertiesfor a concentration of 1000 ppm (a), 500 ppm (b).

According to the figure 4 (a), the loss factor is greater than the conservation factor across the frequency range. The solution then has a rather viscous character which becomes even more accentuated when the frequency is greater than a critical value, $f_c \approx 3.6$ Hz.

A behavior similar to this one is observed for a concentration of 500ppm where $f_c \approx 4.6$ Hz as shown in Figure 4 (b).

4. Conclusion

In conclusion, we first confirm that these solutions behave as threshold fluids, as already noted [4] even at low concentrations furthermore all the rheological parameters are related to the concentrations. Otherwise, the evolution of the value of f_c, which corresponds to the case where G' and G" are comparable, is inversely proportional to the concentration. This last result consolidates the control protocols of aqueous solutions of Polyox oscillating regime and optimizes the processes of their use in industrial environment.

Acknowledgement
The research was conducted as a part of a PhD thesis at the laboratory of mechanics- Faculty of Science Aïn Chock University Hassan II Casablanca.

5. References

[1] Niedzwiedz K, et al. 2008 *Macromolecules* **41** (13) 4866
[2] Ebagninin K W, Benchabane A, Bekkour K 2009 *Journal of Colloid and Interface Science* **336** 360
[3] Niedzwiedz K, Wischnewski, A, Pyckhout-Hintzen, W, Allgaier J, Richter D, Faraone A, 2008 *Macromolecules* **41** 4866
[4] Bahlouli M I, Bekkour K, Benchabane A, Nemdili A 2011 Comportement thermo-rhéologique des solutions aqueuses de Polyéthylène Oxyde Institut de Mécanique des Fluides et des Solides. Université de Strasbourg-CNRS, Strasbourg, France
[5] Ortiz M, Kee D De, and Carreau P J, 1994 *Journal of Rheology* **38** (3) 519

IOP Conf. Series: Journal of Physics: Conf. Series **1045** (2018) 012026 doi:10.1088/1742-6596/1045/1/012026

Rheological characteristics of bitumen mastic depending on composition and filler dispersity

M S Lebedev[1] and N I Kozhukhova[2]

[1]High Technology Center, Belgorod State Technological University named after V.G. Shoukhov, Kostukov St, 46, Belgorod, Russia
[2]Department of material science and material technology, Belgorod State Technological University named after V.G. Shoukhov, Kostukov St, 46, Belgorod, Russia

E-mail: michaelL1987@yandex.ru

Abstract. As fillers in asphalt concrete are used for bitumen viscosity increase, structure modification it is necessary to estimate the filler structuring effect in bitumen binder. For this purpose the influence of phase and size parameters of fine-dispersed powdered fillers on rheological characteristics of bitumen mastics (bitumen–filler composition) is studied. Maximal effect for bitumen properties is observed for such fillers as kaolin and chalk-stone. These two fillers have the highest specific surface area (SSA) and the highest void content. So, they are able to absorb a higher content of bitumen vs. fillers with lower SSA and void content values. Variations in Delta Ring and Ball values for kaolin- and chalk stone-based mastics at 83°C and 24°C, respectively are significant vs. other compositions (up to 8–10°C). It can be associated with structuring effect under high temperature. Significant change in bitumen physical and mechanical properties can be occurred when kaolin application is connected with thin flake form of its particles. Study of rheological characteristics of bitumen mastics by the dynamic shift remoter under high temperatures demonstrates lower values of complex modulus for bitumen mastics containing lime stone and sand fillers with medium and high dispersity vs. chalk-stone- and kaolin based analogues.

1. Introduction

Fillers added to asphalt concrete are fine-dispersed powders, produced by grinding of rocks or powdered industrial by-products. They are used as modifier for bitumen to increase the binder viscosity (i.e. structuring effect) as well as to expand of operating temperature range. The resulting structured disperse system acts as binding component in bitumen concrete. It leads to reducing of creep deformation (plastic yield) and increasing of bitumen admixture. It is generally believed, that structuring capacity of filler is associated with highly-developed surface.

However, it should be noted that not all disperse materials similarly effect on bitumen due to the following reasons: mineral and chemical compositions, dispersity, particle shape and surface reactivity of them. These parameters are interacted, so it difficult to determine particular effect of each parameter on general structuring effect of asphalt-based materials.

Degree of structuring of organic binder containing filler is associated with particle distribution in bulk i.e. packing density. Also this parameter could be determined by particle-size distribution i.e. filler dispersity as well as particle shape [1, 2]. In case of multimodal particle distribution it is the number of interpartial contacts, packing density of them and strength of system [3]. Due to complex interdependencies between such parameters as particle-size distribution, particle shape and packing density, it is easier to measure packing density than to calculate it. Packing density is characterized by interpartial pores presence, that generally determined by Rigden voids method [4].

According to Rigden theory filler can keep some quantity of bitumen in pore space [5] that initiates increasing of bitumen viscosity. Extra bitumen content is called «free bitumen». This quantity depends on type of filler. So, fillers those are differed in composition and dispersity have different content of air pores in «dry» packing density and different amount of «fixed» bitumen in admixture. That leads to difference in structuring capacity that should be determined according to rheological characteristics.

2. Experimental Part

2.1. Materials

Fillers, applied in this work, were obtained by grinding of different types of rocks such as carbonates (limestone and chalk-stone), silicates (hydrothermal quartz, kaoline). According to XRD-analysis, carbonates consists of calcite $CaCO_3$, hydrothermal quartz – quartz SiO_2, and kaoline – kaolinite ($Al_4[Si_4O_{10}](OH)_8$) as well as nacrite, dickite, halloysite. The above materials are «pure» or monomineral. Construction bitumen BND 60/90 (Russia) was used as binding component.

2.2. Methods.

Real density was measured with helium pycnometer Pycnomatic ATC.

Dispersity of fillers is determined by following parameters:

– Particle-size distribution and specific surface area measured with laser particle analyzer FRITSCH Analysette 22 NanoTec plus;

– Specific surface area measured by Blaine method as well as by BET method based on nitrogen adsorption using SORBI-M equipment.

Mastics for test procedure were prepared by joint mixing of bitumen and fillers at 150–160 °C, stirring rate is 100 r/min with following component ratio (%): filler/bitumen – 37.5 /62.5 (Table 1).

Table 1. Composition of bitumen mastics with different fillers

Filler		Limestone medium-dispersed LM	Limestone fine-dispersed LF	Chalk-stone C	Hydrothermal quartz medium-dispersed QM	Hydrothermal quartz fine-dispersed QF	Kaoline K
bitumen/filler ratio	vol. %,	62.5/37.5					
	wt. %,	38.3/61.7	38.3/61.7	38.2/61.8	38.8/61.2	38.6/61.4	39.2/60.8

Rheological characteristics of mastics, obtained by mixing of bitumen and filler were measured with standard methods (penetration test, melting point-test). Variation in penetration and melting point was fixed as difference between bitumen and mastics values. Flow behavior of bitumen and based mastics were studied in wide temperature range using rotational viscometer Rheotest RN4.1.

Standard test system of rheometer consists of two parallel metal plates (upper moving plate with diameter of 12 mm), temperature monitoring sensor. Choice of measuring system is based on gypothesis of higher viscosity of mastics vs. free-of-filler bitumen. Thickness of test fused beads for bitumen and mastics was 1 mm. Rheological characteristics were studied in temperature range of 45–80°C with step size of 5°C. So, controlled parameters were followings: temperature, applied stress (i.e. torque moment was 100 μmN·m) and loading period with frequency of 1.59 Hz. Test was realized in CSS regime with controlled shearing stress. Output parameter was deformation (taking into account that applied stress was fixed) that was used for calculation of complex modulus G* as «amplitude stress – deformation» ratio [6].

Microstructure of bitumen mastics was studied with laboratory microscope AXIO SCOPE A1 (Carl Zeiss) in transmission regime. Test samples were thin layers of mastics at glass slides.

IOP Conf. Series: Journal of Physics: Conf. Series **1045** (2018) 012026 doi:10.1088/1742-6596/1045/1/012026

3. Results and Discussions

3.1. Characteristics of fillers

According to grain size analysis, all fillers are characterized by polydispersed particle size distribution (Fig.1). They can be graded by size in following sequence: QM (size range is 0.01–80 μm) → K (size range is 0.1–50 μm) → LM (size range is 0.01–35 μm) → QF (size range is 0.01– 30 μm) → LF (size range is 0.01–30 μm) → C (size range is 0.01–20 μm). It should be noted that high dispersity of C with average particle size of 1.52 μm and content of particle smaller than 100 nm is 2.73 % takes place.

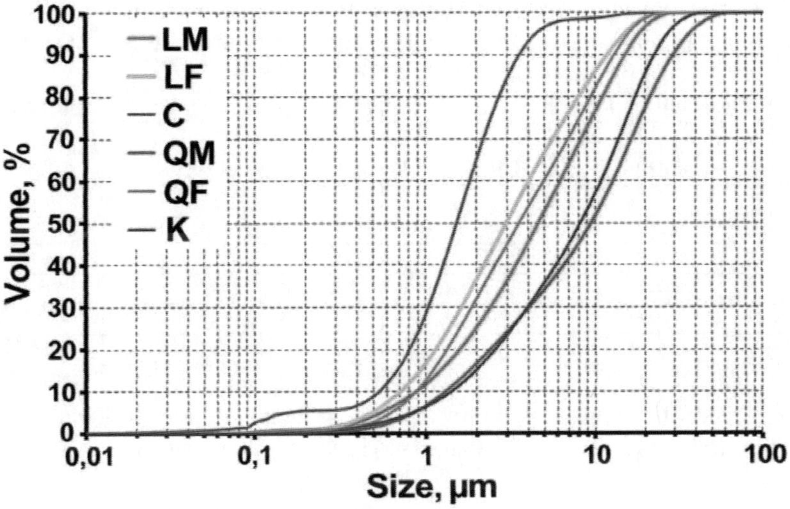

Figure 1. Integral curves of particle size distribution for the fillers

Specific surface area (SSA) is also the basic parameter especially for characterization of fillers in bitumen (Table 2). This parameter, measured by Blain method is in agreement with data of particle size distribution analysis (Figure 1, Table 2) besides for K.

Table 2. Characteristics of fillers

Filler	Real density, g/cm^3	Specific surface area, m^2/kg			Void content, vol. %
		by Blaine method	by BET method	Particle Size Distribution	
LM	2.713	703.8	5500±1200	1041.9	58.3
LF	2.710	978.8	4100±600	1412.9	55.6
C	2.718	1520.7	9700±300	2760.4	70.5
QM	2.658	329.7	1600±200	711.1	44.8
QF	2.675	885.4	3000±400	1244.6	61.3
K	2.613	1357.1	10200±1100	714.5	74.8

SSA for kaolin by Blain method is 1357.1 m^2/kg and by BET method is also high (10200 m^2/kg). It is associated with morphology of the filler: separate slides of kaolinite with high surface area those are collected in «packs». That explains an appearance of large particles (Fig. 1). Reduction of BET-measured SSA when increasing of grinding period for Limestone (Table 2) can be associated with particle aggregation due to extra energy. Actually, both two methods of SSA measurement characterize air pore content in powdered material that explains data of SSA and void content. For exp., void content, calculated for fillers is similar to Rigden porosity, applied in other countries [4, 5]. So, void content will be strongly effect on rheology of mastics, produced by bitumen and filler mixing.

3.2. Standard rheological tests for bitumen mastic

Due to small variations in real density of fillers, the «bitumen-filler» (B/F) ratio in mastics is varied from 38.2/61.8 to 39.2/60.8 (Table 1). I.e. bulk and weight concentrations of filler in bitumen are comparable in all mastics compositions.

According to data obtained, K and C maximally influent on bitumen properties (Table 3). Two both fillers are characterized by the highest SSA and void content values. So they can absorb more of bitumen vs. others fillers. This parameter significantly effects on melting point gradient (for exp., 24°C for C-based mastics and 83°C for K-based mastics and 8–10°C for other mastics compositions). Relative viscosity (penetration) gradient is less visible, but also takes place at 25 °C and at 0 °C (Table 3).

Table 3. Physical and mechanical characteristics of mastics with different fillers

Parameters	Filler					
	LM	LF	C	QM	QF	K
Penetration gradient						
At 25 °C	42	43	54	45	48	56
At 0 °C	9	7	8	8	7	12
Delta Ring and Ball	10	8	24	10	10	83
Penetration index	-0,71	-1,13	0,57	-0,98	-1,06	6,27

High correlation between fillers porosity, melting points gradients (values of Delta Ring and Ball) (Table 2) and penetration index (Table 3) at 25°C for two types of fillers is observed (Figure 2).

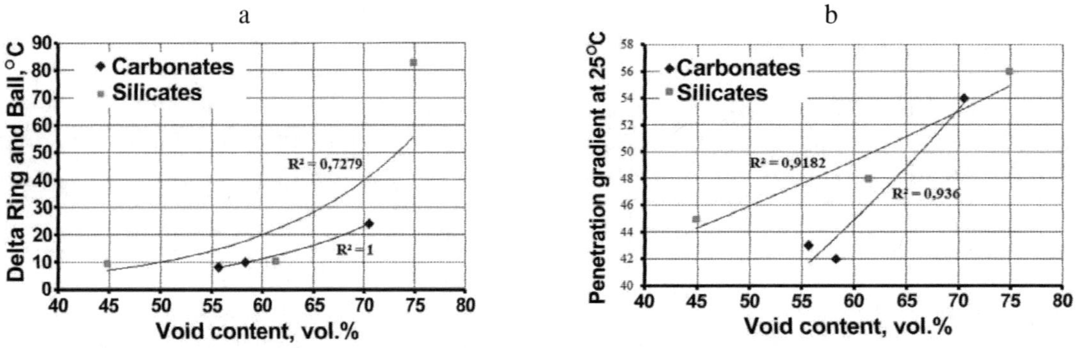

Figure 2. Correlation between fillers porosity, melting points gradients obtained (a) and penetration gradient at 25°C (b) for filler-containing bitumen mastics

In case of melting point as a relative viscosity factor at high temperature vs. penetration test we can see, that growing of temperature leads to increasing of gradient in mastics viscosity. In can be associated with intensification of structuring effect at higher temperatures due to higher void content. This relationship is clearly represented by penetration index values (Figure 3): introduction of carbonate and silicate fillers effect on this parameter vs. reference bitumen in range of ±0.24 (penetration index =-0.95). At the same time C and K initiate penetration index up to 0.57 and 6.27, respectively. Penetration index demonstrates thermal sensibility of bitumen and based mastics. Increasing of penetration index leads to reducing of sensibility of bitumen viscosity of temperature and increasing of heat-resistance of material. These results are confirmed by others studies [7, 8].

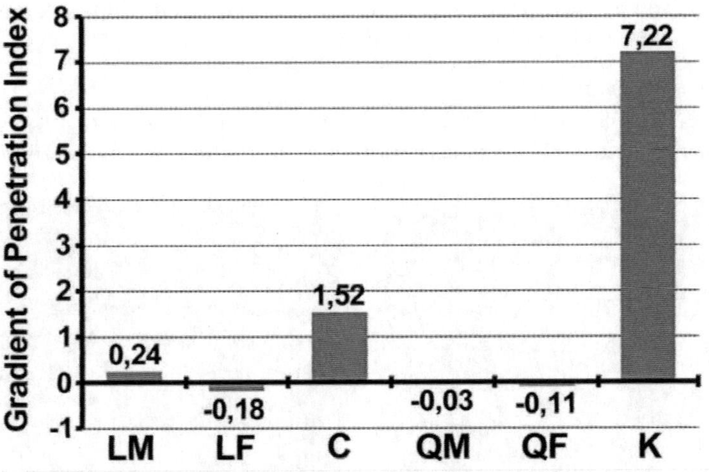

Figure 3. Dependence of penetration index for bitumen on fillers with different phase and size heterogeneity. Carbonates: LM, LF, C. Silicates: QM, QF, K

Significant variation in physical and mechanical properties of bitumen when kaolin (K) introduction is associated with plate-like particle shape. The results obtained are in agreement with earlier studies [1], devoted to effect of particle shape of fillers on rheological properties of bitumen mastics.

3.3. Rheological tests of bitumen mastics by vibrational method

According to results of the above studies, physical and mechanical characteristics of bitumen mastics should be investigated more deeply at temperature conditions.

Data from Figure 4 demonstrates, that complex shear modulus for mastics based on limestone and quartz mineral fillers is significantly lower vs. C- and K based ones but higher than for bitumen. It is confirmed by data of penetration gradient at 25°C, melting points and penetration index.

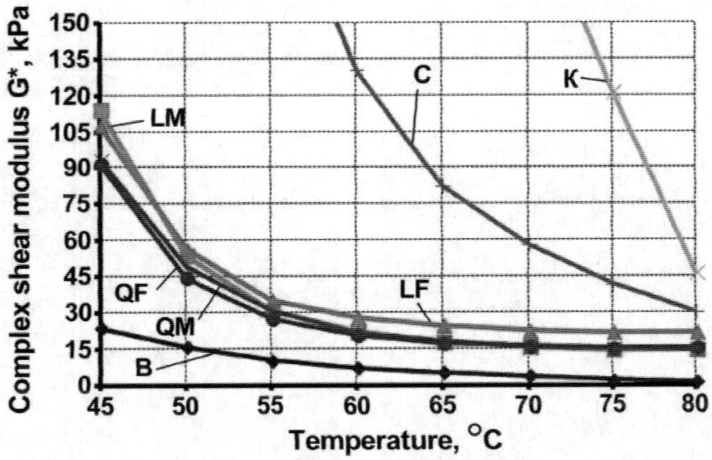

Figure 4. Effect of temperature of complex shear modulus G* for bitumen (B) and mastics with different fillers

C and K based mastics demonstrate the max reduction of temperature sensitivity vs. bitumen. It can be initiated by small-size particles (for C) and high SSA (for C and K as well) and leads to dramatic

IOP Conf. Series: Journal of Physics: Conf. Series **1045** (2018) 012026 doi:10.1088/1742-6596/1045/1/012026

structuring effect to bitumen. The most clearly this effect is realized at high temperature (Figure 4), especially, for limestone and less, for hydrothermal quartz.

3.4. Microstructure of bitumen mastics

Different structuring effect of fillers on bitumen is confirmed by images in Figure 5. Microstructure for C based mastics is presented at higher resolution due to small-size particles (up to 5 μm).

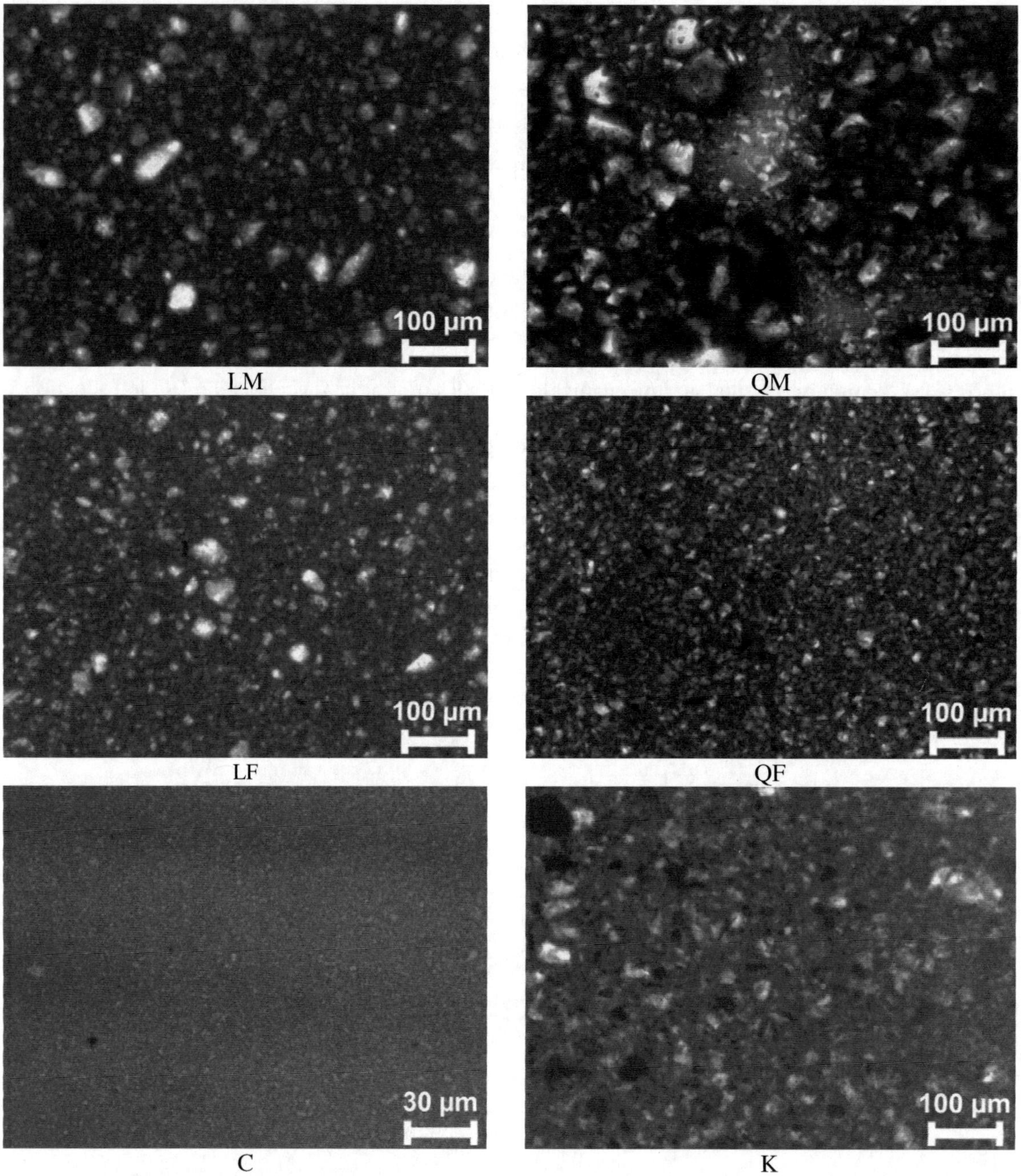

Figure 5. Microstructure bitumen-mineral mastics

According to images in Figure 5, structure of all experimental mastics is characterized by homogeneous distribution of filler in bitumen. Microstructure of limestone- and hydrothermal quartz-based mastics demonstrate a presence of large particles having a little effect on rheological characteristics as great part of bitumen is unbonded. For C and K based mastics thickness of intergrain bitumen layer is significantly lower leading to formation of more compact structure. For K based mastics filtration of bitumen into interlayer space of kaolin particles as well as distribution of separate layers in bitumen matrix takes place. But due to small-size particles of fillers, it is difficult to study structure formation with optical microscopy.

4. Conclusions

Results of study of effect of composition and dispersity of fillers on rheological characteristics of bitumen mastics demonstrate strong correlation. For example, variation in dispersity leads to transformation of composite structure, in generally. The rheological results demonstrate that the max reducing of temperature sensitivity is characterized for mastics contained fine-dispersed fillers with high SSA due to small-size particle (for C) and shape of particle (for K). These fillers are characterized by the highest concentration of air pores initiating a dramatical structuring effect on bitumen, especially, at high temperature.

Acknowledgement

This research work is financially supported by Ministry of Education and Science of Russian Federation in framework of State Assignment №11.9329.2017/8.9, Russian Federation Ministry of Education and Science in the framework of the Presidential Scholarship № PS-2099.2015.1, using equipment of High Technology Center at BSTU named after V.G. Shoukhov.

References

[1] Barnes H A 2000 A handbook of elementary rheology, Institute of Non-Newtonian Fluid Mechanics, University of Wales

[2] Wypych G 1999 Handbook of Fillers, *Chem Tec Publishing*, Toronto, Canada

[3] Petropavlovskaya V B, Novichenkova T B, Belov V V and Buryanov A F 2013 *Construction-materials* **1** 64

[4] BS EN 1097-4: 2008 Tests for mechanical and physical properties of aggregates. Determination of the voids of dry compacted filler (BSI)

[5] Rigden D J 1947 *Journal of the Society of Chemical Industry* **66** 299

[6] AASHTO T315 2010 Standard Method of Test for Determining the Rheological Properties of Asphalt Binder Using a Dynamic Shear Rheometer (American Association of State Highway and Transportation Officials)

[7] Delaporte B, Di Benedetto H, Chaverot P and Gauthier G 2008 *Transportation Research Record: Journal of the Transportation Research Board* **2051** 41

[8] Lebedev M Sand Chulkova I L 2016 Bulletin of Belgorod State Technol. University named after V.G. Shoukhov **11** 47

IOP Conf. Series: Journal of Physics: Conf. Series **1045** (2018) 012027 doi:10.1088/1742-6596/1045/1/012027

Ceramics with a hierarchical porous structure based on aluminum oxide

R V Levkov[1, 2], N V Pigaleva[3], M V Korobenkov[3] and S N Kulkov[1, 2, 3]
[1]Tomsk Polytechnic University, avenue Lenin, 30, 634034 Tomsk, Russia
[2]Institute of Strength Physics and Materials Science SB RAS, pr. Akademicheskii, 2/4, 634055, Tomsk, Russia
[3]National research Tomsk State University, avenue Lenin, 36, 634050, Tomsk, Russia

E-mail: levkov.r.v@mail.ru

Abstract. It have been studied porous ceramic Al_2O_3 obtained via decomposition of aluminum hydroxide. This method allows obtaining ceramics without any impurities on the inner surfaces of porous material. Initial powders were subjected to mechanical treatment in a planetary ball mill for 48 hours and next cold molding samples were sintered in air at 1450-1650 °C for a one hour. It was shown that the porosity of the ceramic increases with increasing content of aluminum hydroxide in the powder mixture and was equal 64% for samples contains 100% $Al(OH)_3$ sintered at 1450 °C. It was found that the compressive strength increases with increasing sintering temperature and concentration of aluminum hydroxide affects on the compaction rate of ceramic samples. The maximum strength of these ceramics was 180 MPa for samples without of hydroxide. It has been estimate that activation energy of the sintering process was 40 ± 5 kJ / mol.

1. Introduction

It is well known, what ceramics has a special place among the materials. This is due to a wide range of applications and high characteristics of ceramics. Oxide ceramics due to its corrosion resistance, the ability to work at high temperatures, high physics-mechanical characteristics has found its application in many areas of industry Porous ceramics can be used in the manufacture of filter elements, in catalytic carriers, for fine gas purification. The main characteristics of these ceramics are porosity and its morphology [1-3].

There are many methods of pore formation in ceramic materials one of which is a method based on decomposition of aluminum hydroxide to alumina and H_2O accompanied by gassing and as a consequence, the formation of the required porous structure. In this case, a different concentration of hydroxide in the powder mixture will affect the porosity in the sample. This method allows forming a multi-level pore structure without blowing agent method [4-6].

The aim of this work was to study the influence of the concentration aluminum hydroxide in hydroxide-oxide mixture and sintering temperature on formation of the pore structure in the alumina and its mechanical properties of sintered ceramic.

2. Materials and Experimental Procedure

Gibbsite modification of aluminum hydroxide powder and aluminum oxide powder obtained by heat treatment aluminum hydroxide powder were used as initial components. The volume change of the pore space in the ceramic was provided by varying the sintering temperature and the concentration of aluminum hydroxide in a mixture.

Alumina powder was obtained by annealing the aluminum hydroxide powder at a temperature of 1100 °C 1h. The alumina powder and aluminum hydroxide powder were mechanically activated in a

IOP Conf. Series: Journal of Physics: Conf. Series **1045** (2018) 012027 doi:10.1088/1742-6596/1045/1/012027

ball mill for 48 hours. After mechanical activation alumina powder was mixed with aluminum hydroxide powder (in concentrations of 0-100%) in a ball mill during 10 h. The mixture was cold pressed at 200 MPa. Sintering of compacts was carried out in air at temperatures 1450 - 1650 °C. Compression tests of sintered samples were carried out on a universal test machine at a strain rate 10^{-2} c^{-1}.

3. Results and Discussion

SEM images of powders and distribution of particle size are shown on Figure 1. As one can see the particle shape is nearly spherical. The average particle size was 50.3 µm; standard deviation was 22.7 µm.

(a) (b) (c)

Figure 1. (a) SEM-image of alumina powder after annealing, (b) SEM-image of aluminum hydroxide powder and (c) its particle size distribution before mechanical activation.

SEM image of the powder mixture from a mechanically activated aluminum hydroxide and aluminum oxide in Fig. 2. Mechanical activation made it possible to break up particles with <D> = 50.3 µm to <D> = 2.2 µm (powdered alumina and aluminum hydroxide).

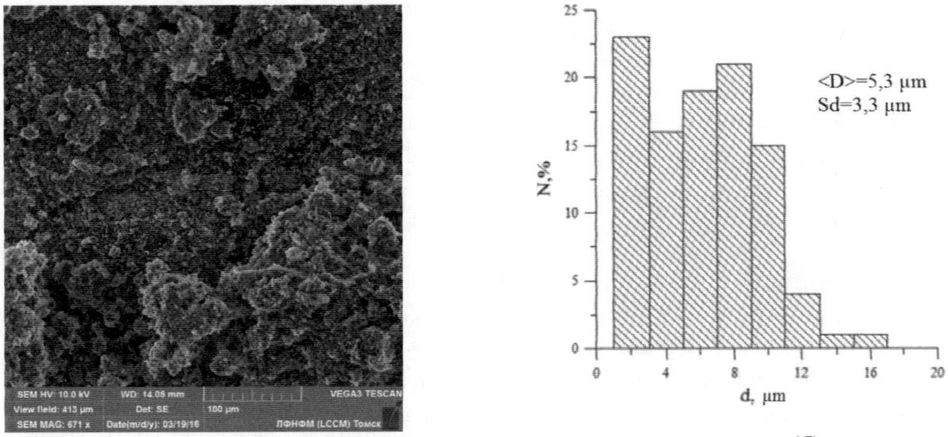

(a) (б)

Figure 2. SEM image of the powder mixture from a mechanically activated aluminum hydroxide and aluminum oxide.

SEM images of the fracture surface of ceramics synthesized at various sintering temperatures are shown on Fig.3. One can see that the particles of ceramics sintered at 1450 °C have a fractured shape, while the particles of the ceramic sintered at 1650 °C have a rounded shape.

IOP Conf. Series: Journal of Physics: Conf. Series **1045** (2018) 012027 doi:10.1088/1742-6596/1045/1/012027

(a) (б)

Figure 3. SEM images of fracture surface of porous ceramic samples obtained from alumina powder with addition of a hydroxide powder at different sintering temperatures: (a) 5% $Al(OH)_3$, 1450 °C, (b) 75% $Al(OH)_3$, 1650 °C.

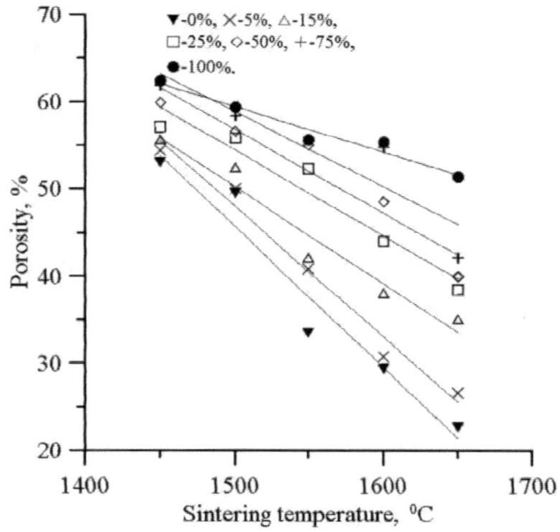

Figure 4. Dependence of porosity from sintering temperature with different concentrations of $Al(OH)_3$.

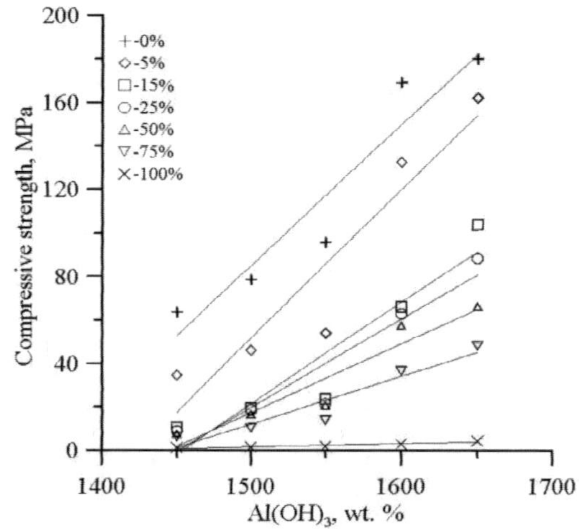

Figure 5. Dependence of compressive strength from sintering temperature with different concentrations of $Al(OH)_3$.

The dependence of porosity vs. sintering temperature is shown in Fig.4. With increasing sintering temperature, the porosity of ceramic samples decreases. Aluminum hydroxide affects the porosity dependence on the sintering temperature. Decreasing the aluminum hydroxide amount in the powder mixture, the greater is an effect of sintering temperature on the porosity. One can see that the effect of aluminum hydroxide on porosity at a sintering temperature of 1450 °C is lower in too time of aluminum hydroxide on the porosity at a sintering temperature of 1650 °C.

When the sintering temperature increases, the pore volume in the ceramic samples decreases, and porosity in $Al(OH)_3$ ceramics decreases less than in case of ceramics made of pure alumina. The increase of $Al(OH)_3$ content leads to more noticeable decrease in pore volume with increasing sintering temperature from 1450 to 1650 °C. Density of ceramic after sintering at 1450 °C varies slightly and is equal to 2.2± 0.2 g/cm3. After sintering at higher temperatures, an increase in density is observed. The highest density 3.3 g/cm^3 was obtained for samples sintered at 1650 °C. The dependence of the sample density on the sintering temperature allowed to estimate the activation energy of the compaction process during sintering, Q = 45 ± 5 kJ / mol, which corresponds to the literature data for highly disperse powders [7, 8] and corresponds to surface diffusion.

Figure 5 shows the dependence of compressive strength on sintering temperature for ceramics with different aluminum hydroxide content. One can see that the compressive strength increases with increasing sintering temperature, while the tensile strength differs for ceramics with different aluminum hydroxide content sintered at the same temperature. The ultimate strength of ceramics without aluminum hydroxide is maximal and increases almost 3 times with sintering temperature increasing.

Base on the data obtained, one can conclude that with increasing sintering temperature the porosity decreases. This is due to the fact that during the sintering, the formation and growth of contacts between particles, this means a decrease in the pore size, which is accompanied by an increase in their size.

Dependence of porosity of materials based on alumina and aluminum hydroxide powders at different sintering temperatures determine the possibility of synthesizing strength ceramic materials with a porosity of 30-50% changing the corresponding technological parameters.

4. Conclusion

It was found that the decomposition of aluminum hydroxide during sintering leads to the formation of porosity in the ceramic. It was shown that with increasing sintering temperature density and shrinkage also increase. It was shown that the porosity in ceramic samples increases with the content of aluminum hydroxide in the powder mixture. It has been found that the compressive strength increases with increasing of sintering temperature and the concentration of aluminum hydroxide affects on the compaction rate of ceramic samples. It was shown that the porosity of the ceramic samples increases with increasing content of aluminum hydroxide in the powder mixture and was equal 64% for samples contains 100% $Al(OH)_3$ sintered at 1450 °C. The maximum strength of these ceramics was 180 MPa for samples without of hydroxide. The average activation energy of the sintering process is 40±5 kJ/mol.

Acknowledgement

The work presented was carried out in accordance with Tomsk State University's competitiveness program and with partial support of project 14.584.21.0026 (RFMEFI58417X0026).

5. References

[1] Zhen-Yan Deng, Takayuki Fukasawa, Motohide Ando 2001 *J. Am. Ceram Soc.* **84** (11) 2638
[2] Zhen-Yan Deng, Takayuki Fukasawa, Motohide Ando 2001 *J. Am. Ceram. Soc.* **84** (3) 485
[3] Korobenkov M V, Kulkov S N 2016 *AIP Conf. Proc.* 1760. 020033. doi 10.1063/1.4960252
[4] Ostrowski T, Rodel J 1999 *J. Am. Ceram. Soc.* **82** (11) 3080
[5] Gömze A L, Kulkov N S, Shabalin L I, Gotman I, Fedraza F, Lecomte G L, Mayarova T, Kurovics E, Hamza A 2015 *Építőanyag-JSBCM* **67** (4) 143 http://dx.doi.org/10.14382/epitoanyag-jsbcm.2015.24
[6] Egesz A, Gömze A L, 2015 *J. Phys.: Conf. Ser.* **602** 012023 https://doi.org/10.1088/1742-6596/602/1/012023
[7] Levkov R V, Kulkov S N 2017 *IOP Conf. Ser.: Mater. Sci. Eng.* **175** 012045 https://doi.org/10.1088/1757-899X/175/1/012045
[8] Savchenko N L, Sevostyanova I N, Sablina T U, Molnar L, Geber R, Gömze L A, Kulkov S N, Gomze L N 2014 *Építőanyag-JSBCM* **66** (2) 44 http://dx.doi.org/10.14382/epitoanyag-jsbcm.2014.9

IOP Conf. Series: Journal of Physics: Conf. Series **1045** (2018) 012028 doi:10.1088/1742-6596/1045/1/012028

On the mechanics of a fusiform cerebral aneurysm: Mooney-Rivlin mathematical model for the experimental data

A I Lipovka[2], K S Ovsyannikov[3], A V Dubovoy[3] and D V Parshin[1, 2]

[1]Lavrentyev Institute of Hydrodynamics SB RAS, Novosibirsk, Russia
[2]Novosibirsk State University, Novosibirsk, Russia
[3]Federal Neurosurgical Center, Novosibirsk, Russia

E-mail: lipovkaai@yandex.ru and parshin@hydro.nsc.ru

Abstract. In this work, mathematical modelling of the process of rupturing the wall of a human fusiform aneurysm is carried out. We consider 3-parameter and 5-parameter Mooney-Rivlin models. The comparison between the results of the modelling and the experimental data is made. Analyzing the relation of parameters of 3-parameters and 5-parameters Mooney-Rivlin models, the relation between the greatest (least) value and Cauchy strain was found. The field of application of these models for hemodynamics was shown.

1. Introduction

An intracranial aneurysm (IA) is a cerebrovascular disease, which occurs approximately in 1 out of 50 people [1, 2], but nevertheless is rarely diagnosed. To know the risk of rupture of given aneurysm is one of the main concerns of the neurosurgery. According to statistics approximately only 1-2% of unruptured aneurysms will rupture at some point. To calculate the risks of rupture the numerous factors are to be considered and it presents a demanding problem to examine all of them at the same time. However doing so provides the better understanding of the aneurysm structure and properties which increases the accuracy of prediction whether or not the aneurysm is going to rupture. The mechanism of the aneurysm formation is not entirely known yet, hence the studies of unruptured IAs, such as [3, 4] and [5], all use results from various experimental approaches.

Mathematical modelling of aneurysm tissue provides more profound understanding of its mechanics [6]. The model that sufficiently corresponds experimental and clinical data can be used in FSI calculations during the estimation of critical parameters of aneurysm rupture. Besides each model of that kind has its own field of application. Mooney-Rivlin kind of models are widely used in mathematical physics for modelling the mechanics of elastic materials and particularly in hemodynamics related field [7].

2. Methods

The experiments on investigating the mechanic properties of cerebral aneurysm tissue were first performed at studies [8]. And the technology was significantly developed by the group under the leadership of A.M. Robertson [9, 10].

After the neurosurgeries were performed, the obtained IA tissues were taken to laboratory, preserved with sodium 0.9% at +2°C-+5°C. According to [11] refrigerated aneurysm show slightly different results with relation to used temperature. At the laboratory the rupture machine (machine Zwick/Roell Z010, Germany) was used to perform a series of experiments on tissues. As the tissue is quickly deteriorating becoming unfit for the further investigation, only one series of experiments is possible for every sample. Speed of pulling equals 1mm per minute and was the same for all experiments.

Finding Mooney-Rivlin constants

As the conducted tension of aneurysm tissue in mechanic tests was uniaxial we will consider the mathematical model under that condition.

In that case:

$$S_{3p} = 2C_{10}\left(\lambda - \frac{1}{\lambda}\right) + 2C_{01}\left(1 - \frac{1}{\lambda^3}\right) + 6C_{11}\left(\lambda^2 - \lambda - 1 + \frac{1}{\lambda^2} + \frac{1}{\lambda^3} - \frac{1}{\lambda^4}\right) \quad (1)$$

$$S_{5p} = 2C_{10}\left(\lambda - \frac{1}{\lambda}\right) + 2C_{01}\left(1 - \frac{1}{\lambda^3}\right) + 6C_{11}\left(\lambda^2 - \lambda - 1 + \frac{1}{\lambda^2} + \frac{1}{\lambda^3} - \frac{1}{\lambda^4}\right)$$
$$+4C_{20}\lambda\left(1 - \frac{1}{\lambda^3}\right)\left(\lambda^2 + \frac{2}{\lambda} - 3\right) + 4C_{02}\left(2\lambda + \frac{1}{\lambda^2} - 3\right)\left(1 - \frac{1}{\lambda^3}\right) \quad (2)$$

Where S_{3p} and S_{5p} are corresponding uniaxial stress for 3-parameters and 5-parameters models, λ is a stretch ratio and C_{10}, C_{01}, C_{11} are parameters for 3-parameters model while C_{10}, C_{01}, C_{11}, C_{02}, C_{20} are parameters for 5-parameters model.

To find the Mooney-Rivlin constants that correspond best to the data of each experiment we used Wolfram Mathematica (commercial software, License of NSU). The approximate function for strain-stress curve was built, using least squares method. After that, based on obtained function, the Mooney-Rivlin constants were found, independently for each experiment.

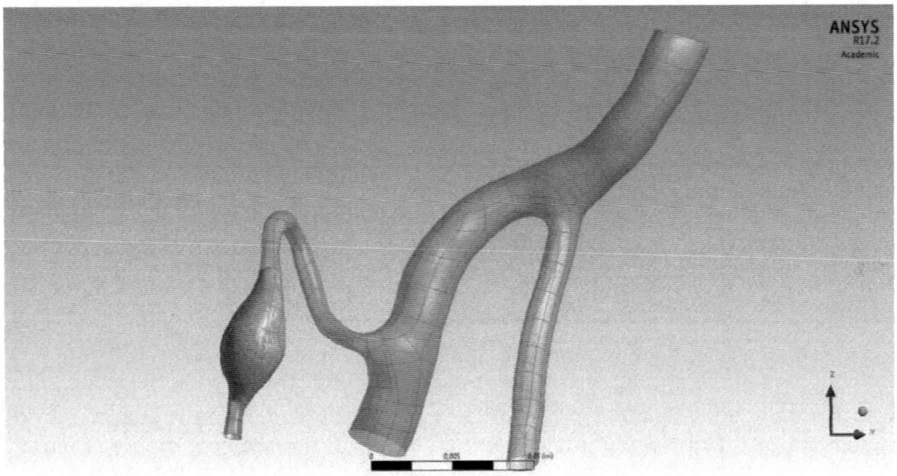

Figure 1. Reconstructed 3D model of vessel with aneurism

IOP Conf. Series: Journal of Physics: Conf. Series **1045** (2018) 012028 doi:10.1088/1742-6596/1045/1/012028

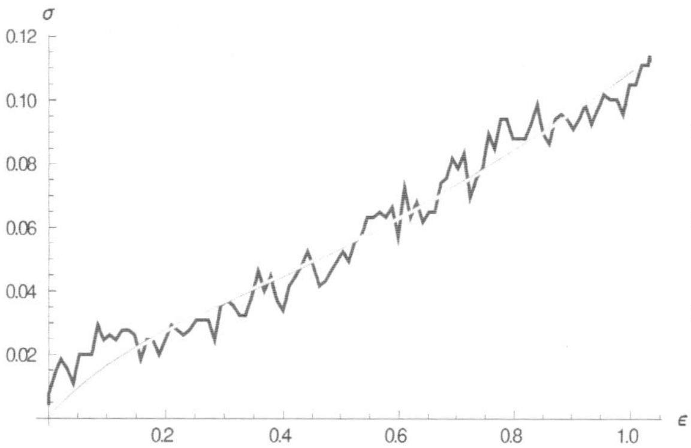

Figure 2. Results of the experiment №2 (blue) and mathematical modelling with 3-parameter (red) and 5-parameter (yellow) Mooney-Rivlin models

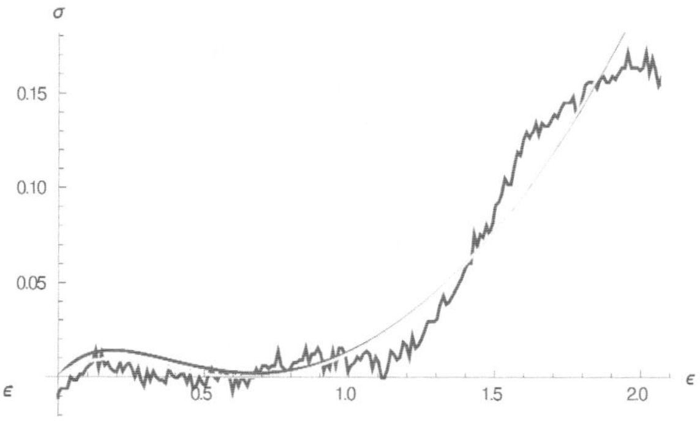

Figure 3. Results of the experiment №4 (blue) and mathematical modelling with 3-parameter (red) and 5-parameter (yellow) Mooney-Rivlin models

3. Results and discussion

We have examined the effect that growth of the relative elongations of the sample has on the changes in the parameters of the Mooney-Rivlin models under cyclic loading. In Figures 4-11 plot points correspond to largest Cauchy strain of previously described experiment. As it can be noted from the plots, in area of ε from 1.6 to 2.6 every one of parameters for each model reaches its extremum. From the Figure 3 it is clear that after given Cauchy stress value the error increases significantly (up to 13%) to in relation to experimental data.

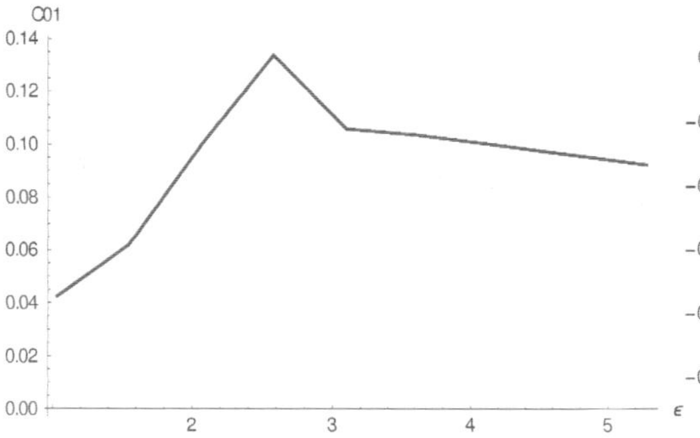

Figure 4. Coefficient C01 value for 3-parameters model (1) in relation to Cauchy strain

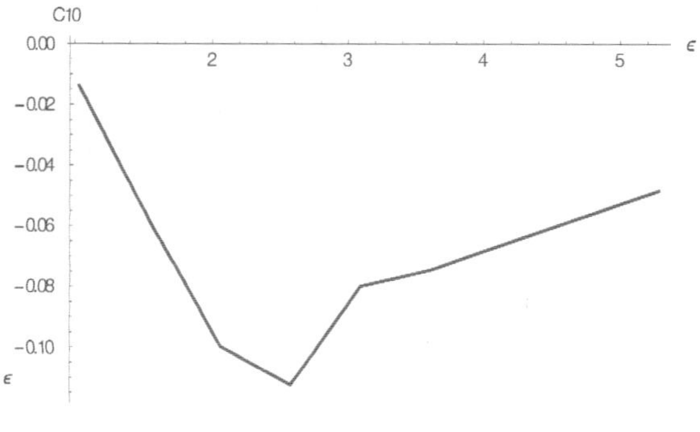

Figure 5. Coefficient C10 value for 3-parameters model (1) in relation to Cauchy strain

IOP Conf. Series: Journal of Physics: Conf. Series **1045** (2018) 012028 doi:10.1088/1742-6596/1045/1/012028

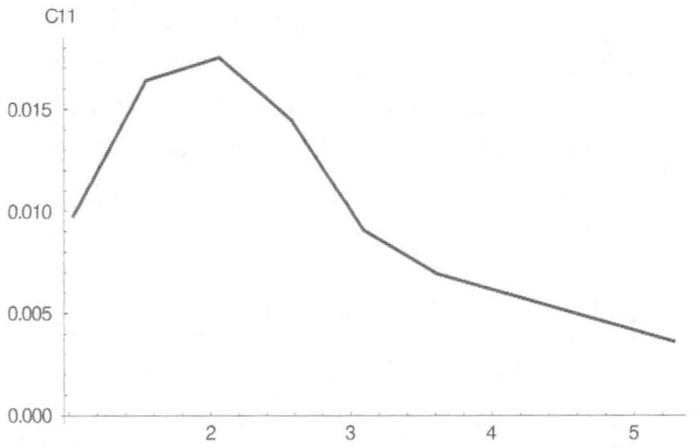

Figure 6. Coefficient C11 value for 3-parameters model (1) in relation to Cauchy strain

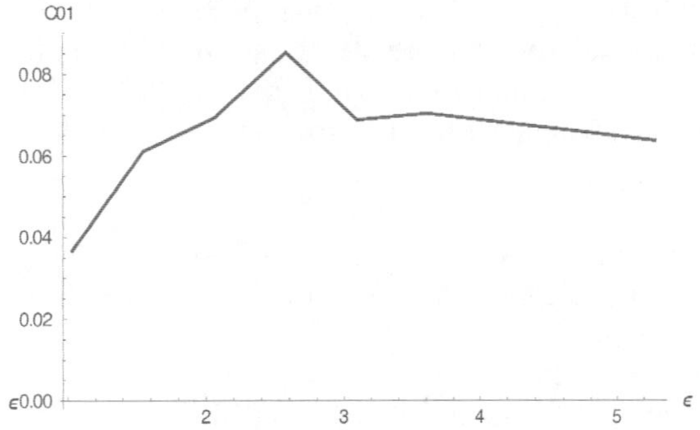

Figure 7. Coefficient C01 value for 5-parameters model (1) in relation to Cauchy strain

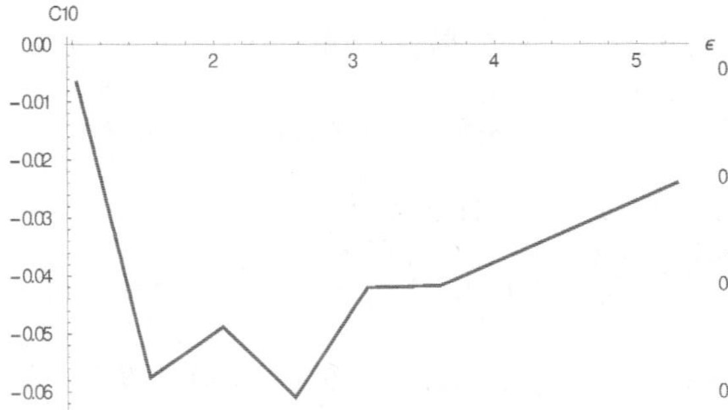

Figure 8. Coefficient C10 value for 3-parameters model (1) in relation to Cauchy strain

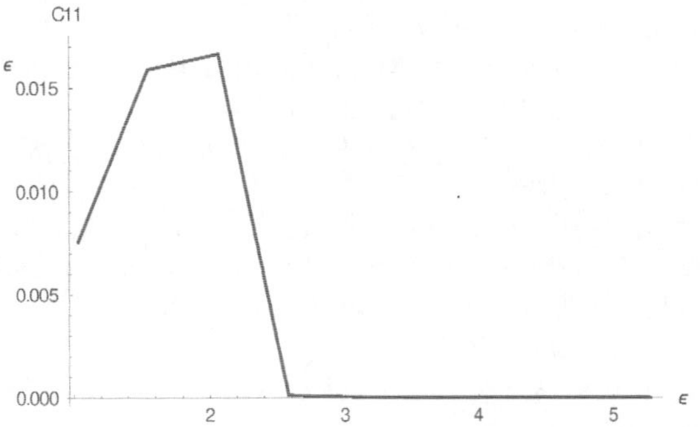

Figure 9. Coefficient C11 value for 5-parameters model (2) in relation to Cauchy strain

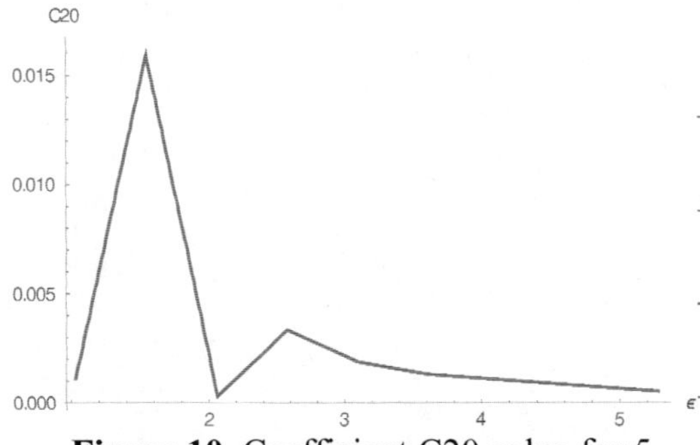

Figure 10. Coefficient C20 value for 5-parameters model (2) in relation to Cauchy strain

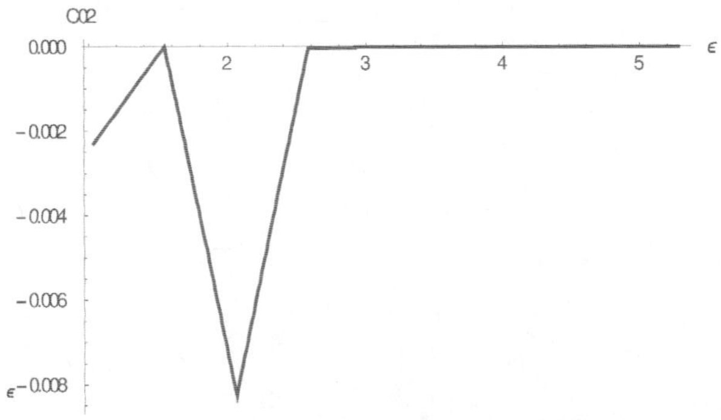

Figure 11. Coefficient C02 value for 5-parameters model (2) in relation to Cauchy strain

4. Conclusion

The satisfying correspondence of stress-strain curves were obtained during the mathematical modelling for both 3-parameters and 5-parameters Mooney-Rivlin models. It is shown that for the relative elongations in the range from 1.6 to 2.6 the coefficients of the above models reach their extremum. With further increase in the relative elongations of the aneurysm material, the error in the approximation of the models of experimental data of technologies increases (up to 13%).

This analysis suggests that the usage of 3-parameters and even 5-parameters Mooney-Rivlin models is justified for relatively small deformation of the material and particularly it is adequate to use such models in FSI calculations in cases where the area of aneurysm does not contain diverticula (protrusions) which are usually are characterized by relatively large displacements relative to the dome of the aneurysm.

Acknowledgements
This work was supported by a grant from Russian Science Foundation, No. 17-11-01156.

References

[1] Morita A, Kimura T, Shogima M, Sameshima T and Nishihara T 2010 *Neurol. Med. Chir.* **50** 777
[2] Rinkel G J E, Djibuti M, Algra A, van Gijn J 1998 *Stroke* **29** (1) 251
[3] Cebral J R, Mut F, Weir J and Putman C 2011 *AJNR Am. J. Neuroradiol.* **32** 145
[4] Khe A K, Chupakhin A P, Cherevko A A, Eliava S S and Pilipenko Yu V 2015 *Russ. J. Numer. Anal. Math. Modelling* **30** 277
[5] Larrabide I, Aguilar M L, Morales H G, Geers A J, Kulcsar Z, Rufenacht D and Frangi A F 2013 *AJNR Am. J.Neuroradiol.* **34** 816
[6] Humphrey J D 2002 *Cardiovascular Solid Mechanics: Cells, Tissues, and Organs. Springer: New York* **386**
[7] Fatma Gulden Simsek, Young W Kwon 2015 *J Biol Phys.* 2015 **41**(2) 173
[8] Scott S, Ferguson G G and Roach M R 1972 *Can. J. Physiol. Pharmacol.* **50** 328
[9] Cebral J R, Duan X, Chung B J, Putman C, Aziz K M and Robertson A M 2015 *AJNR Am. J.Neuroradiol.* **36** 1695
[10] Robertson A M, Duan X, Aziz K M, Hill M R, Watkins S C and Cebral J R 2015 *Ann. Biomed. Eng.* **43** 1502
[11] Stemper B D, Yoganandan N, Stineman M R, Gennarelli T A, Baisden J L and Pintar F A 2007 *J. Surg. Res.* **139** 236

IOP Conf. Series: Journal of Physics: Conf. Series **1045** (2018) 012029 doi:10.1088/1742-6596/1045/1/012029

Momentum coupling at laser ablation of liquids in a porous matrix

E Loktionov

Bauman Moscow State Technical University, Moscow, 105005, Russia

Email: stcpe@bmstu.ru

Abstract. The main problems at laser impact on liquids are energy dissipation and splashing, leading to poor performance even after application of special tricks like viscosity increase, thin films, cavities or droplets formation. We studied the laser ablation of liquids, including energetic ones, confined by porous matrix made of sintered titanium powder. We have compared the liquid-filled targets (ethanol, kerosene, nitromethane) momentum coupling coefficient dependency on laser fluence with the same unfilled target. Filling a porous target with a liquid may have both positive and negative effects depending on impact regime. Porous matrix prevents heat and mass losses characteristic to bulk solid and liquid targets. Momentum coupling coefficient doubled at porous target density increase by 6% only with ethanol filling, and at least tripled at ca. 22% density decrease as compared to a bulk target. Filling of porous matrices with energetic liquids leads to improvement of laser thrust generation due to combination of heat and splashing losses substantial reduction and chemical energy release.

1. Introduction

Laser plasma generation has been considered mainly for solids rather then liquids [1], since high energy density can be reached in a surface layer of the former, and plasma plume containing mainly high-speed light particles [2]. We have considered possible ways of efficiency increase at laser ablation of liquids recently; liquid media were considered because of easy desired contents preparation (properties variation) [3], well developed precise dosing (down to 10^{-15} L and 0.1% stability [4]) and feeding systems. The main problems at laser impact on liquids are energy dissipation [5] and droplets formation due to splashing [6], leading to poor performance [7] even after application of special tricks like viscosity increase [8], thin films [9], cavities or droplets [10] formation. Solidification (by UV curing) [11] and splashed droplets retrieval by magnetic field [12] have shown that efficient laser propulsion is possible with liquids.

Heat dissipation is restricted in porous media similar to thin films, so laser energy losses are decreased. Splashing and energy dissipation for liquids confined in pores is also suppressed. However, such target configuration has never been laser ablated before. The only work studying momentum coupling at laser ablation of liquids with porous matrices we managed to find was [13]. Liquids of 0.2 to 700 Pa*s viscosity have been considered at 10.6 μm/60 ns laser irradiation of 0.9 to 4.6 mm thick layers, so impact region was out of the 0.2 to 40 porosity medium. The latter was used as a permeable solid interface actually; its effect on momentum coupling was not investigated and is not fully used in such configuration.

Solid or liquid confining layer on either irradiated or back side (or both) of the target prevents energy losses and leads to plume compression [14]. Plume temperature rises then due to effective

IOP Conf. Series: Journal of Physics: Conf. Series **1045** (2018) 012029 doi:10.1088/1742-6596/1045/1/012029

absorption of laser radiation by more dense plasma and increased interaction time. Unlike flat target, in multi-layered ones, ablated medium is spread through a hole in the cover layer, in fact a nozzle is formed, so the plume is well collimated [15]. It was shown in [16, 17], that for such "exotic targets" required laser energy is reduced and momentum coupling is significantly improved. E.g., in aluminum target coated with 2 mm glass, momentum coupling coefficient increased by a factor of 12 due to detonation [18] (an increase of more than 100 was shown in [19] with an optimum shift of an order of magnitude towards lower energies).

Of all liquids, energetic ones [20, 21] are the most promising for propulsion because of energy release induced by laser irradiation. But for efficient output, those should stay compressed at high temperature. We have studied the laser ablation of liquids, including energetic ones, confined by sintered titanium powder. The aim of our study was to check if such a confinement could improve momentum coupling for a solid target, and to compare the results in terms of momentum coupling coefficient with non-impregnated Ti targets.

2. Experimental layout

The sketch of experimental setup is presented at Figure 1. We irradiated targets at ambient conditions in a single pulse mode using frequency doubled Nd:YAG laser (532 nm, 12 ns, up to 80 mJ, Lotis TII LS-2147), spot diameter at target was 224 μm. Incident laser energy was measured by control photodiode, calibrated following standard procedure, using energy meter (Ophir 30A-P-RP) placed just after lens instead of a target. Recoil momentum has been measured with a calibrated force sensor (531 mV/N, PCB Piezotronics 209C11). Electric signals were recorded by PC-coupled digital oscilloscope (Aktakom ADS-3114). Force signal was integrated over time to get recoil momentum.

The targets consisted of 0.9 mm thick porous target made of sintered titanium powder (fine gas filter manufactured by VMZ-Techno, 0.26 porosity = non-solid volume / total volume, average pore dia. 8 μm) impregnated with ethanol, kerosene or nitromethane. Liquids were supplied using a pipette until substrate saturation. To avoid target jumping, it was attached to the sensor by sticky tape.

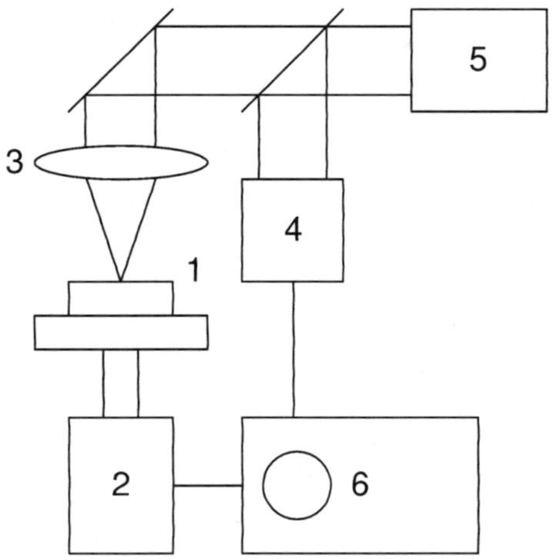

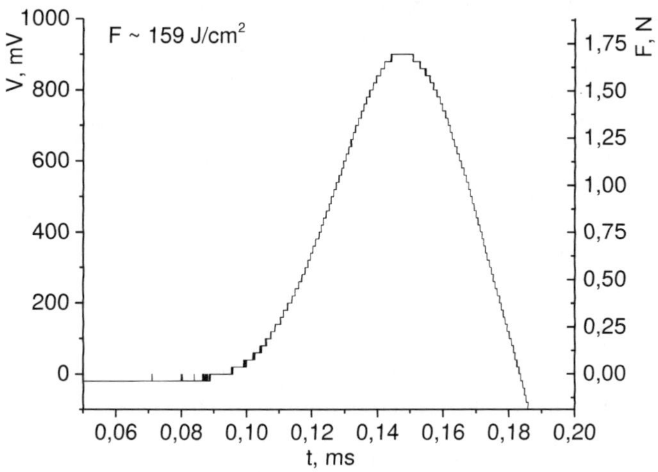

a b

Figure 1. Experimental setup (a) and typical force signal (b):
1 – target, 2 – force sensor, 3 – lens, 4 – control photodiode, 5 – laser, 6 – digital oscilloscope

3. Results and discussion

First, we have obtained reference data on momentum coupling for solid Ti targets: bulk, foil and porous (Figure 2a). Presence of heat transfer restrictions for the latter two has resulted in recoil momentum increase, expectedly. It's worth mentioning, that $C_m(F)$ dependency, which represents recoil momentum normalized by laser pulse energy [22], had a quite different pattern as compared to bulk titanium target. Similarity of values obtained for all targets at high intensities can be explained by reduction of residual energy coefficient at laser fluence increase [23] and by increase of ambient air input in recoil momentum, making target properties not so important. For foil and porous targets, momentum coupling (both absolute values and dependency pattern) at low fluences were similar to polymers (which are heat insulators) [24] and metals at femtosecond laser irradiation (when heat affected zone is very small) [25]. This represents the effect of heat restriction. Although considered liquids are not heat insulators, thermal conductivity of those is still 100 times lower than for bulk titanium.

After that, we have conducted analogous measurements for porous targets impregnated by liquids (Figure 2b). The most pronounced effect was discovered for ethanol: momentum coupling coefficient doubled from its minimum and than followed the unfilled pattern. For kerosene, negative effect was observed leading to further C_m decrease after unfilled minimum (almost halved), and then jump to unfilled values ($2*10^{-4}$ N*s/J) at the edge of the studied range. Nitromethane resulted in a combination of the previous patterns: 1.5 times decrease after minimum and than jump to ethanol filled values. These patterns might be caused by heat sink to a liquid – negative effect was proportional to the heat of evaporation. Momentum coupling jump obviously happened due to exothermic reactions, fluence needed for ignition was proportional to flash temperature. However, there was no evidence of liquid detonation within porous matrix, i.e., no macroscopic destruction or other visible difference to non-filled target. Visually, it looked like combustion happened in the plume.

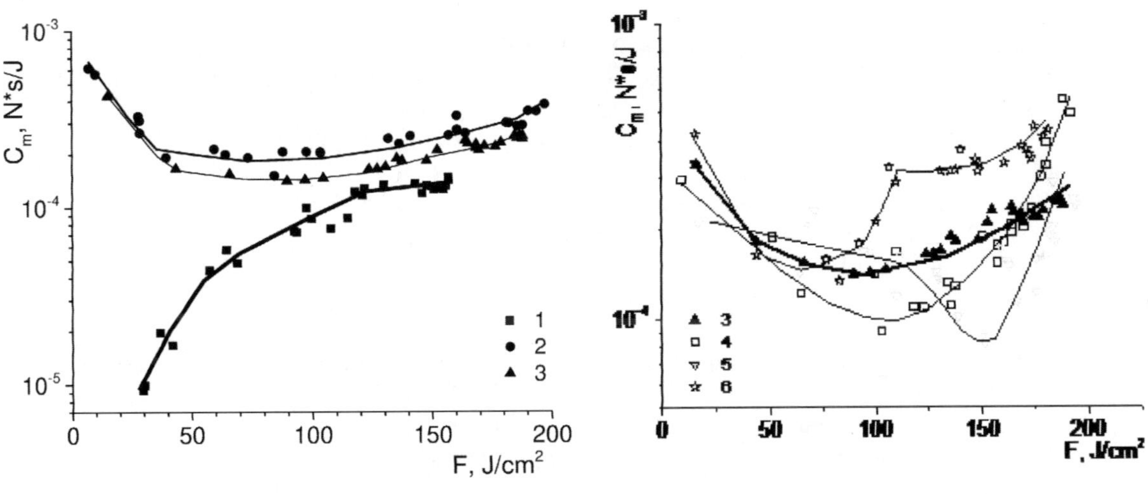

a b

Figure 2. Momentum coupling at solid (a) and filled (b) Ti targets irradiation:
1 – bulk (1.5 mm), 2 – foil (0.1 mm), 3 – porous, 4 – nitromethane filled,
5 – kerosene filled, 6 – ethanol filled

So, filling a porous target with a liquid may have both positive and negative effects depending on impact regime. Any way, porous matrix prevents heat and mass losses characteristic to bulk solid and liquid targets. C_m doubled at porous target density increase by 6% only with ethanol filling, and at least tripled at ca. 22% density decrease as compared to a bulk target. These relations mean that specific impulse might also increase in both cases, but it could not be stated for sure without target weighing. We did not weigh targets to evaluate mass loss because liquid evaporation rate could not be estimated properly. To answer this question closed pores targets filled with liquids or laminated open pored ones are needed. Full confinement in such pores is close to the conditions in multi-layered targets [15], so the effect might be even more pronounced. However, manufacturing of such targets is rather complicated, at least of metal ones.

4. Conclusions

Filling of porous matrices with energetic liquids leads to improvement of laser thrust generation due to combination of heat and splashing losses substantial reduction and chemical energy release. However, heat of evaporation and flash temperature of the filling liquid should be considered, and possibly decreased for better effect. Porous targets could be easily manufactured countinously by sintering (e.g., by laser), and powder supply system can be similar to liquid. For magnetisc powders, special supstrates and porous layer recovery could be used. In high performance pulsed periodic plasma generators refilling of pores with a liquid could be enhanced by ultrasound, electroosmosis and other effects. Impact on targets with energetic liquid filled closed pores could lead to even greater effect due to pressure increase.

Acknowledgements

This research has been performed at "Beam-M" facility, and supported by grant of the President of Russia No. MK-633.2018.8. We are grateful to V.A. Bolshakov for providing porous samples.

References

[1] Phipps C R, ed. 2007 Laser Ablation and its Applications *Springer,* New York
[2] Kuzenov V V and Ryzhkov S V 2017 *J. Phys.: Conf. Ser.* **830** 012124
[3] Loktionov E, Protasov Y and Telekh V 2017 *J. Phys.: Conf. Ser.* **830** 012061
[4] Li D, ed. 2015 Encyclopedia of Microfluidics and Nanofluidics *Springer*, New York
[5] Sinko J E and Pakhomov A V 2008 *Proc. AIP* **997** 195
[6] Fardel R, Urech L, Lippert T, Phipps C, Fitz-Gerald J and Wokaun A 2009 *Appl. Phys. A* **94** 657
[7] Sinko J 2005 Time resolved force and imaging study on the laser ablation of liquids. Thesis M. Sc. Huntsville
[8] Choi S, Han T-h, Gojani A and Yoh J 2010 *Appl. Phys. A* **98** 147
[9] Wang B, Li L, Tang Z-p and Cai J 2010 *Proc. AIP* **1230** 243
[10] Li X, Hong Y and He G 2010 *J. Propuls. Technol.* **31** 105
[11] Loktionov E Y 2014 *J. Appl. Spectrosc.* **81** 305
[12] Loktionov E Y 2015 *Appl. Phys.* 12
[13] Uchida S 2006 *Efficient laser ablation*. AOARD-044033. Institute for Laser Technol.: Osaka
[14] Zhang C, Yao J, Lan S, Trofimov V A and Lysak T M 2013 *Opt. Comm.* **308** 54
[15] Nehls M, Edwards D and Gray P. 2002 *Proc. AIAA* 2002-2154
[16] Phipps C R, Seibert Ii D B, Royse R W, King G and Campbell J W 2000 *Proc. SPIE*, **4065** 931
[17] Uchida S, Hashimoto K, Fujita K, Niino M, Ashizuka T and Kawasahima N 2002 *Proc. AIAA* 2002
[18] Zhi-Yuan Z, Jie Z, Zuo-Qiang H, Xiao-Hui Y, Zhe Z, Xin L, Zhao-Hua W and Zhi-Yi W 2006 *Chin. Phys.* **15** 580
[19] Fabbro R, Fournier J, Ballard P, Devaux D and Virmont J 1990 *J. Appl. Phys.* **68** 775
[20] Bouddu V and Redner P, eds. 2011 Energetic materials. Thermophysical properties, predictions and experimental measurements *CRC Press*, Boca Raton
[21] Shaw R W, Brill T B and Thompson D L, eds. 2005 Overview of recent research on energetic material*s* World Scientific: Singapore
[22] Loktionov E Y, Protasov Y S and Protasov Y Y 2013 *Opt. Spectrosc.* **115** 758

[23] Vorobyev A Y and Guo C 2007 *J. Phys.: Conf. Ser.* **59** 418
[24] Loktionov E Y, Ovchinnikov A V, Protasov Y S, Protasov Y Y and Sitnikov D S 2012 *Opt. Spectrosc.* **112** 631
[25] Loktionov E Y, Ovchinnikov A V, Protasov Y S, Protasov Y Y and Sitnikov D S 2013 *High Temp.* **51** 867

Measuring viscosity of polyethylene blends using a rotational rheometer

Dorottya Nagy[1] **and Károly Belina**[1]

[1] Department of Materials Technology, GAMF Faculty of Engineering and Computer Science, John von Neumann University, Kecskemét, Hungary

E-mail: nagy.dorottya@gamf.uni-neumann.hu

Abstract. Polyethylene (PE) is one of the most important and widely used polymer in packaging industry, film and tube forming technology [1]. Knowing the properties of polymer blends and the characterization of their rheological behavior is useful for industrial processes. For that very reason in our research PE blends were prepared and characterized by rheological method to find out the differences between the blends and the base materials. The base materials were TIPELIN FA 381-10 (high-density polyethylene, HDPE) and TIPOLEN FB 243-51 (low-density polyethylene, LDPE). The PE blends were made in the following proportions: 80/20, 60/40, 40/60 and 20/80. The first numbers mean the FA content percentage in the blend, while the second numbers mean the FB content percentage. Viscosity of the polymer melts were measured by rotational rheometer at low shear rate and the effect of temperature on viscosity were also measured. The structure viscous behavior of the polymer melt was confirmed, as the physical structure of the polymeric chain changes in flow. The fact that the number and the structure of the branches influence viscosity was revealed. FA material (HDPE) has the highest viscosity, while FB (LDPE) has the least. The blends show a good correlation between the two limits, although the complex viscosity does not show linear behavior in FA material content. The least viscosity was measured at 20% FA content. The modulus values plummet with temperature. The polydispersity index is 2.2 for FA material at 230°C. The temperature dependence of viscosity was revealed for blend (40-60). Moreover the activation energy was calculated at three frequencies. The fact that the activation energy decreases in function of angular frequency was demonstrated.

1. Introduction

Polyethylene (PE) is one of the most important and widely used polymers in packaging industry, film and tube forming technology [1]. Different types of PE are known according to the chain structure and the manufacturing technologies. Polyethylene with few branches is capable to reach high amount of crystalline, it is called high-density polyethylene (HDPE). Branched polyethylene is hardly form into crystalline fractions, so it is called low-density polyethylene (LDPE). The density influences the properties of the polymer. Hence polyethylene blends having different branches are worth to investigate [2].

Because of the polymer forming process, rheology of thermoplastic polymers is essential. Polymer forming is usually done in molten state, then the polymer melt becomes solid and is used in many fields of our life. The properties of the solid product depend on the forming process and the properties of the melt. Viscosity is generally measured to characterize the flow of polymer melts. The viscoelastic behavior of the polymer melt is known, but it depends on the structure of material. As viscosity is not a constant value namely it depends on deformation rate, it is worth to investigate materials at different ranges of flow. A novel measuring method examine materials at low shear rate range, so the beginning of the flow can be observed. There are some reports, where polymer melt is examined in oscillation mode, so even storage and loss modulus can be determined besides complex viscosity [3-10].

Since the characteristic values are temperature depending ones, measurements must be done at least three different temperatures above the melting temperature of the polymer. Temperature dependence of the viscosity of polymeric materials can be characterized by the activation energy of flow at a given frequency [11-12].

In our research PE blends were prepared and characterized by oscillation rheological method to find out the differences between the blends and the base materials. Moreover, our aim is to determine the flow activation energy for one chosen blend in function of frequency.

2. Materials and Experimental Procedure

The base materials were TIPELIN FA 381-10 (high-density polyethylene, HDPE) and TIPOLEN FB 243-51 (low-density polyethylene, LDPE). They were purchased by MOL Group, manufactured by TVK Nyrt. in Hungary. HDPE is a copolymer with hexen-1 comonomer [13]. LDPE does not contain any additives [14]. Both are appropriate for film blowing and can be processed by conventional extruder. The PE blends were made in the following proportions: 80/20, 60/40, 40/60 and 20/80. The first numbers mean the FA content percentage in the blend, while the second numbers mean the FB content percentage.

Polyethylene blends were made by a Collin Teach-line conventional extrusion line with an E20T single-screw extruder, as shown in Figure 1. The zone temperatures were 190-190-200-210°C. The extruder speed was 80 turn/min.

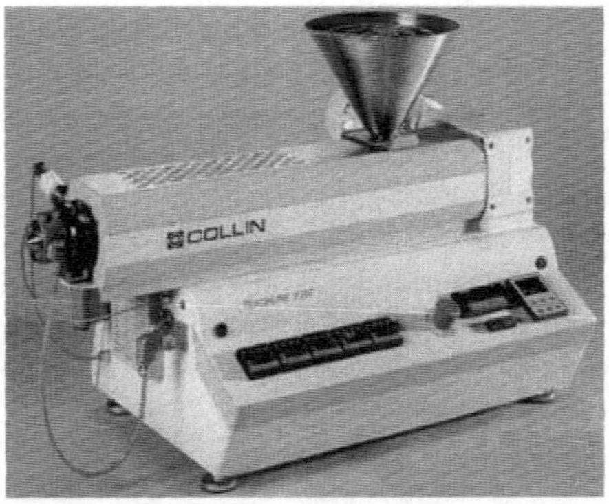

Figure 1. Extruder with standard screw

The blends and the extruded materials were investigated by a rotational rheometer, ARES G2 in oscillation mode with cone-plate geometry, as shown in Figure 2. ARES G2 contains SMT system, which means Separated Motor Transducer, and has a Forced Convention Oven (FCO).

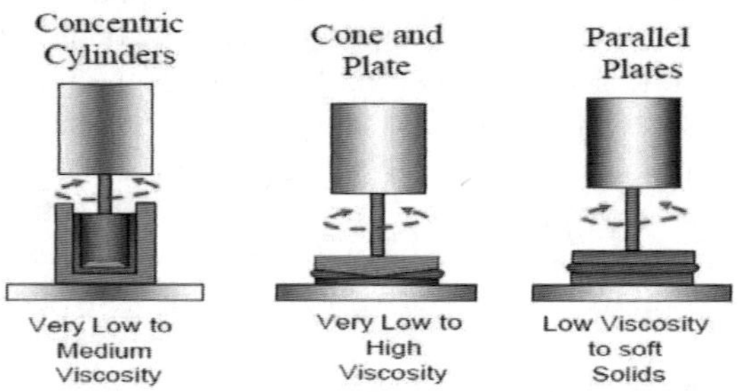

Figure 2. Possible measuring geometries [15]

Approximately 0.5g granulate was used for one measurement. The angular frequency was between 6.28 and 628 rad/s and complex viscosity, storage and loss modulus were determined. Each material was measured at 230°C. Moreover, 40/60 blend was measured at three different temperatures (220, 230 and 240°C). HDPE, which was marked as FA, was examined at 230 and 250°C.

3. Results and Discussions

Complex viscosity is shown in function of angular frequency in Figure 3. The blends and the extruded materials were examined at 230°C. The viscosity is decreasing with frequency, because polymer melts show non-Newtonian behavior. The physical structure of the macromolecules changes during the flow, therefore the internal resistance is not constant. In low frequency range, the viscosity is almost constant, due to the mechanism of flow (deformation of the physical network). FA material (HDPE) has the highest viscosity, while FB (LDPE) has the least. The number and the structure of the branches influence the viscosity. HDPE (copolymer) has the most branches therefore the inner resistance of the polymer melt is the highest. The viscosities of the materials show a good correlation with the ratio of the materials in the blend.

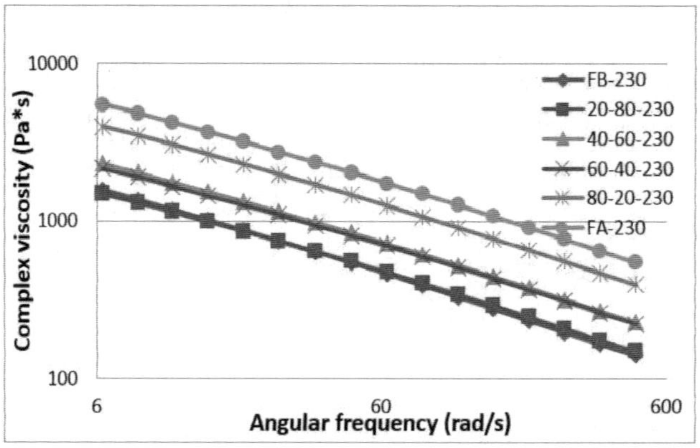

Figure 3. Complex viscosity in function of angular frequency

IOP Conf. Series: Journal of Physics: Conf. Series **1045** (2018) 012030 doi:10.1088/1742-6596/1045/1/012030

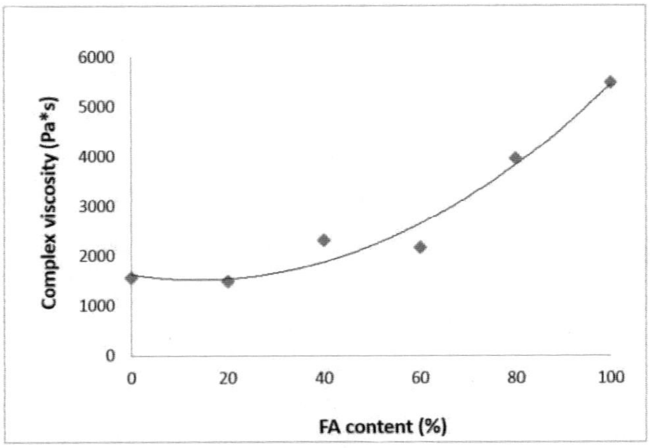

Figure 4. Complex viscosity (Pas) in function of FA content

The complex viscosity in function of FA material content at constant angular frequency (at 6.28 rad/s), does not show linear behavior, as shown in Figure 4. The viscosity change shows negative deviation from the linear. It is assumed, that this deviation due to the change of the free volume in the melt.

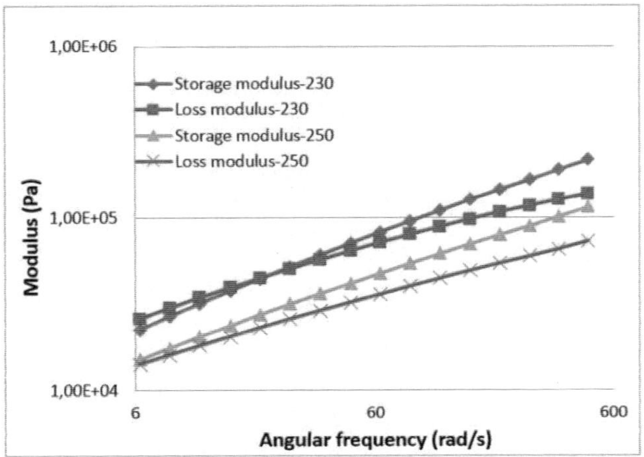

Figure 5. Modulus values of FA material

The storage and loss modulus values were measured for FA material at 230 and at 250°C, as shown in Figure 5. The modulus values plummet with temperature, because of the softening, and the steepness of the curves decreases in small extent. The intersection of the storage and loss modulus curves shows the polydispersity-index of the material, giving 2.2 for the FA material at 230°C.

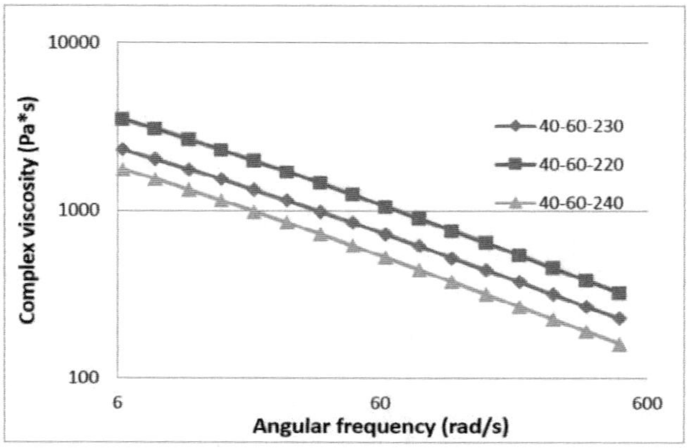

Figure 6. Complex viscosity of 40% FA+ 60% FB blend

One blend (40-60) was chosen to measure at three temperatures: at 220°C, at 230°C and at 240°C, as shown in Figure 6. In this way the temperature dependence of viscosity can be investigated. At higher temperature, complex viscosity is lower, the inner resistance against flow of the material decreases.

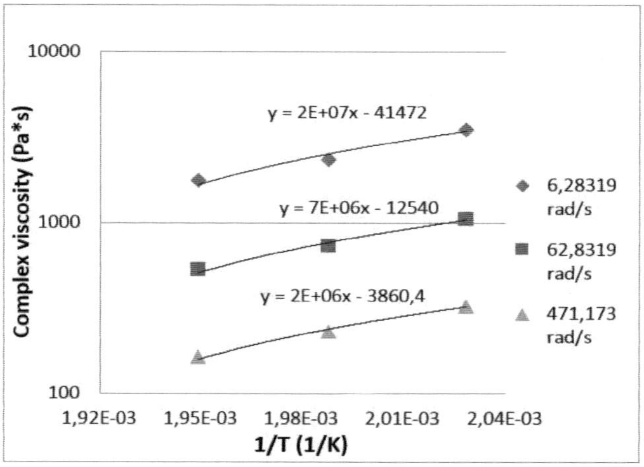

Figure 7. Complex viscosity (Pas) in function of 1/temperature

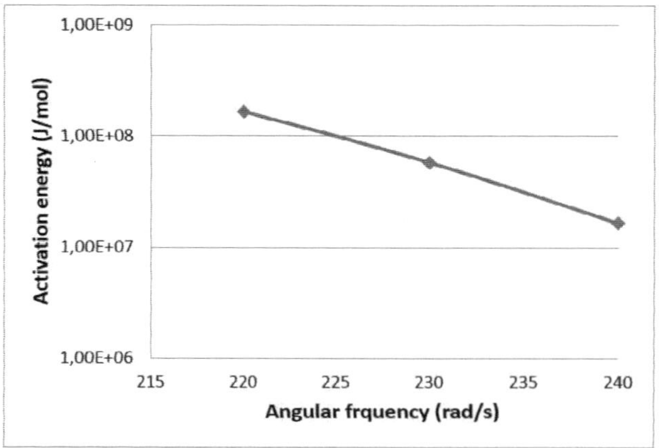

Figure 8. Activation energy in function of angular frequency

This blend (40/60) was appropriate to present the complex viscosity in function of 1/temperature at constant angular frequency (6.28; 62.8 and 471 rad/s), as shown in Figure 7. These curves are linear in logarithmic scale, which means that the viscosity change follows the Arrhenius law.

From the slope of the linear lines, activation energy can be calculated at different frequencies. The activation energy decreases with frequency, as shown in Figure 8. This result also emphasizes the structural change of the polymeric melt.

4. Conclusion

In our study, polyethylene blends were prepared and investigated by rheological method. The results of extruded materials and blends were compared. Complex viscosity, loss and storage modulus, polydispersity index and the activation energy were determined.

The structure viscous behavior of the polymer melt was confirmed, as the physical structure of the polymeric chain changes in flow. The fact that the number and the structure of the branches influence viscosity was revealed. FA material (HDPE) has the highest viscosity, while FB (LDPE) has the least. The blends show a good correlation between the two limits, although the complex viscosity does not show linear behavior in FA material content. The least viscosity was measured at 20% FA content.
The modulus values plummet with temperature. The polydispersity index is 2.2 for FA material at 230°C. The temperature dependence of viscosity was revealed for blend (40-60). Moreover the activation energy was calculated at three frequencies. The fact that the activation energy decreases in function of angular frequency was demonstrated.

Acknowledgement
This research is supported by EFOP-3.6.1-16-2016-00006 "The development and enhancement of the research potential at John von Neumann University" project. The Project is supported by the Hungarian Government and co-financed by the European Social Fund.

5. References

[1] Elvers B, Hawkins S, Schulz G 1992 Ullman's Encyclopedia of Industrial Chemistry 5th ed. 487.
[2] Litvinov V M, Ries M E, Baughman T W, Henke A, Matloka P P 2013 *Macromolecules* **46** 541 http://dx.doi.org/10.1021/ma302394j
[3] MohammadiM, Yousefi A A, Ehsani M 2012 *J Polym Res.* **19** 9798 http://dx.doi.org/10.1007/s10965-011-9798-9
[4] Stadler F J, Piel C, Kaschta J, Rulhoff S, Kaminsky W, Münstedt H 2006 *RheolActa* **45** 755 http://dx.doi.org/10.1007/s00397-005-0042-6
[5] Byutner O, Smith G D 2001 *J. of Polym Sci. Part B: Polym Phys.* **39** 3067
[6] Raju V R, Smith G G, marin G, Knox J R, Graessley W W 1979 *J. of Polym Sci.: Polym Phys* **17** 1183
[7] Lee J H, Archer L A 2002 *Macromolecules* **35** 6687 http://dx.doi.org/10.1021/ma020398e
[8] Mead D W 2013 *J. of Non-Newtonian Fluid Mechanics* **195** 99 http://dx.doi.org/10.1016/j.jnnfm.2013.02.001
[9] Mavridis H, Shroff R N 1992 *Polym Eng. and Sci* .**32** (23) 1778
[10] Shen H-W, Xie B-H, Yang W, Yang M-B 2013 *J. of Appl Polym Sci* http://dx.doi.org/10.1002/app.38850
[11] Ye Z, AlObaidi F, Zhu S 2004 *Macromol. Chem. Phys.* **205** 897http://dx.doi.org/10.1002/macp.200300128
[12] Gu S-Y, Zhang K, Ren J, Zhan H 2008 Carbohydrate Polymers **74** 79 http://dx.doi.org/10.1016/j.carbpol.2008.01.017
[13] https://mol.hu/images/pdf/.../polimer.../hdpe.../fa-381-10_eng.pdf
[14] https://www.granulat.com.pl/media/.../tipolen%20fb%20243%2051.p...
[15] https://polymerinnovationblog.com/rheology-thermosets-part-2-rheometers/

Flow structures in a swirl flow - vortex breakdown condition

P Novotny[1, 2], B Weigand[2], F Marsik[1], C Biegger[2] and M Tomas[1]

[1] New Technologies - Research Centre, University of West Bohemia, Univerzitni 8, 30614 Plzen, Czech Republic
[2] Institute of Aerospace Thermodynamics, University of Stuttgart, Pfaffenwaldring 31, 70569 Stuttgart, Germany

E-mail: novotnyp@ntc.zcu.cz

Abstract. This paper presents an experimental and numerical analysis of swirl flow in a circular tube. Swirl flow is important for technical and natural processes, where high heat transfer and good fluid mixing are needed. Flow and vortex structures respecting redistribution of momentum and possible occurrence of vortex breakdown are taken into account. The swirl flow is analysed experimentally by the measurement of the velocity field using Particle Image Velocimetry (PIV) and numerically via the commercial CFD code ANSYS CFX. Three cases are investigated: laminar flow regime (Re = 1,000), intermediate flow regime (Re = 2,000), and turbulent flow regime (Re = 5,000). The redistribution of the velocity field and the decrease of the swirl strength towards the outlet are shown. This redistribution affects the Reynolds number. Concerning the Rossby number, the occurrence of vortex breakdown in the swirl flow is determined. It is shown that the vortex breakdown takes place in the flow with higher Reynolds numbers, where an axial backflow may occur. Changes of the Reynolds numbers Re_ϕ and Re_z along the tube length also confirm this statement. Furthermore, a thermodynamic perspective of vortex breakdown phenomena is presented.

1. Introduction

Swirl flow has an impact in several technical applications due to its influence on fluid mixing and heat transfer. Historically, swirl flow in a tube was investigated by Ranque in 1933 who discovered a temperature separation induced in this kind of flow [1]. Then, Hilsch [2] followed his work and studied the effects of geometry and operational parameters of the tube proposed by Ranque. Hence, this type of tube is nowadays known as Ranque - Hilsch vortex tube. Their general work was followed by many researchers and the problem of a swirl flow in a tube was investigated from different views.

Many authors were interested in a flow in an annulus with through flow and a rotating wall and its effect on heat transfer, e.g. the works [3-5]. They focused on the influence of co - and counter - rotating tubes on heat transfer in the case of heated outer tube. They found out the positive influence of an inner rotating tube on the heat transfer. Reich et al. [6] studied also the effect of rotation on laminar flow in a heated pipe resulting in the statement that free vortices occurring due to the heated pipe wall disappeared with increasing pipe rotation. Weigand and Beer [7] analysed the heat transfer in rotating tubes with respect to the flow pattern inside the tube. They observed a laminarisation of the flow due to tube rotation, which caused a decrease of the heat transfer. They extended the work to a case where a fluid entered a rotating tube in turbulent regime, i.e. the axial flow in the pipe was initially turbulent, a flow relaminarisation appeared by increasing rotation rate of the tube [8].

Hartnett and Eckert [9] analysed the effect of a rotating flow on an axial flow and heat transfer. They focused on measurements of the flow fields and the temperature in a vortex flow generated at nozzle cross - section area and moved in one direction to the end of the tube. Ligrani et al. [10] investigated flow phenomena in swirl chambers, which models a cooling passage located at the leading edge of a turbine blade, with significant axial and tangential velocity components. They carried out flow visualisations for different Reynolds numbers. They stated that the swirl flow could be very useful in different engineering applications, including biomedical applications, heat exchangers and automobile engines. Glezer et al. [11] focused on the internal swirling flow and its influence on the heat transfer. They used a test rig simulating a rotating leading edge internal passage of a blade with heated walls and a screw - shaped cooling swirl resulted from flow generated by tangential slots. They stated that Coriolis forces appearing during blades rotation had a positive influence on the internal heat transfer in case of the same direction of the Coriolis forces with the tangential velocity component. In contrary, there was observed the opposite phenomenon in case of the direction of the Coriolis forces acting in opposite direction of the swirl flow (tangential velocity).

The abrupt change in the flow structure resulting in a backflow is, in literature, referred as vortex breakdown [12]. This phenomenon might be observed due to flow over a delta wing influencing the lift and stability of the wing, and in an axisymmetric swirling flow in tubes. Vortex breakdown is usually characterised by a deformation and deceleration of the vortex core, changes of velocity distribution (with flow reversal), and changes of pressure distribution in the swirling flow. Thus, the occurrence of vortex breakdown has a strong influence on the performance of the studied system, such as turbines, trailing vortices behind a wing, cyclone separators, vortex tubes, swirl chambers, etc. The phenomenon of vortex breakdown was studied by researchers for decades and, as mentioned by Lucca - Negro and O'Doherty [13], there might be an association of flow instability to vortex breakdown. Moreover, some theories describing and explaining vortex breakdown were developed. Thus, there are three basic theories that are associated with vortex breakdown as stated by Escudier [14] or by Hall [15]

- instability, i.e. axisymmetric disturbances, spiral disturbances, non - linear interactions,
- stagnation, i.e. separation analogy, failure of slender core (quasi - cylindrical approximation),
- wave phenomena, i.e. inertia waves, solitary waves, shock or hydraulic jump analogy.

A criterion that may be used to predict the onset of vortex breakdown was proposed by Spall et al. [16]. This criterion was derived based upon previous experimental, numerical, and theoretical studies and followed work by Squire [17] and Benjamin [18]. The criterion is based on the Rossby number defined as the ratio of the axial velocity to the tangential velocity. A threshold for the occurrence of vortex breakdown, below which it might appear was, by Spall et al. [16], determined as 0.65. Furthermore, Spall et al. also concluded that this criterion is due to characteristic scales applicable to confined and unconfined vertical flows.

This contribution follows the above mentioned principles. Firstly, the experimental and numerical works are shortly introduced. Then, methodology used for flow field analysis is described. Thus, the Rossby number is used to predict vortex breakdown. To predict vortex breakdown, changes of tangential Reynolds number (Re_{ϕ}) and axial Reynolds number (Re_z) towards tube outlet are used. Moreover, the decay of the swirl number, i.e. the decrease of the swirl strength of the flow, reflecting the redistribution of the flow inside the tube is represented also by the tangential Reynolds number (Re_{ϕ}). Thermodynamic perspective of the vortex breakdown is discussed respecting the dissipation and destabilising processes during flow redistribution. The paper is closed by a summary and a conclusion.

2. Procedures

2.1. Experimental setup

Flow measurements are conducted on a model swirl tube schematically depicted in Figure 1. A vacuum pump sucks air through the swirl tube and drives the open loop device. A laminar flow element is used to measure the mass flow rate into the swirl tube. Before the air approaches the swirl generator, the air is seeded with oil particles for the PIV measurements. Then, the air flows through the swirl generator consisted of two tangential inlet slots (nozzles) into the swirl tube. The end of the tube represents the tube outlet connected via an outlet plenum guiding the flow to the vacuum pump.

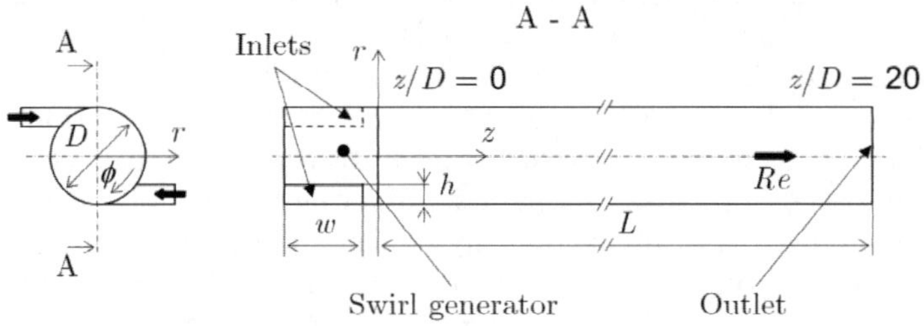

Figure 1. Swirl tube geometry including coordinate system.

The dimension of the swirl tube (Figure 1) are as follows: The inner diameter of the tube is $D = 50$ mm, length of the tube is $L = 20D$. The swirl generator consists of two inlets with the height of $h/D = 0.1$ and the width of $w/D = 0.67$. This geometry results in a geometrical swirl number $S_{IG} = 5.3$ given by equation (1).

$$S_{IG} = \frac{\pi R^2 (R - h/2)}{Rnwh}, \tag{1}$$

where n denotes the number of tangential slots. The investigated Reynolds numbers range from 1,000 to 5,000 and the definition of the Reynolds number is given as

$$Re = \frac{\overline{U^z} D}{\nu} = \frac{4\dot{m}}{\pi D \mu}. \tag{2}$$

Here, $\overline{U^z}$ represents the exit bulk velocity, D is the tube diameter, ν denotes the kinematic viscosity, $\dot{m}$ is the mass flow and μ is the dynamic viscosity. Table 1 summarises the range of investigated Reynolds numbers, corresponding bulk velocities and angular velocity taken from the experimentally obtained tangential velocity profile near the tangential inlets ($z/D = 0$), where the tangential velocity profile corresponds to a solid body vortex. The axial bulk velocity and angular velocity will be used as parameters to plot results in dimensionless form. The axial velocity profile is measured in the tube axis over the whole length. Furthermore, the tangential velocity profile is measured at the positions $z/D = 0$ - 19, which are one tube diameter (D) apart from each other.

Table 1. Investigated Reynolds numbers and corresponding axial bulk velocity, mass flow rate, angular velocity of solid body rotation and inlet velocities for numerical model.

Re (-)	$\overline{U^z}$ (m s^{-1})	$\dot{m}$ (kg s^{-1})	Ω (s^{-1})	U_{inlet} (m s^{-1})
1,000	0.3	0.0007	100	1.74
2,000	0.6	0.0014	200	3.48
5,000	1.5	0.0035	500	8.69

2.2. Numerical model

The numerical simulations are carried out for an isothermal turbulent flow with the commercial code ANSYS CFX, for modelling turbulence the BSLEARSM (Baseline Explicit Algebraic Reynolds Stress Model) model is employed [19]. In case of the lowest Reynolds number Re = 1,000, the Gamma model is used to model transition to turbulence with default values of the onset Reynolds number Re = 260 and the low turbulent intensity and eddy viscosity ratio. Usage of the Gamma model in case of the Reynolds number Re = 2,000 was also tested, but no positive effect was observed. So that, in this case the Gamma model is not utilised. Thus, in the cases of the Reynolds numbers Re = 2,000 and

Re = 5,000 only the medium turbulent intensity and eddy viscosity ratio and the high turbulent intensity and eddy viscosity ratio are set, respectively.

The domain for the numerical simulations is designed to be close as possible to the swirl tube depicted in figure 1. The tube diameter and dimensionless length are D = 50 mm and L/D = 20. The swirl generator consists of two inlets with the height of h/D = 0.1 and the width of w/D = 0.67. To correctly model the swirl generator, the total width of the swirl generator is large by the parameter 0.8D. So, that the total length of the domain is 21.47D.

The wall boundary conditions are set as no - slip conditions. Other details of these simulations respecting the different Reynolds numbers are listed in table 1. The swirl tube geometry is for the three Reynolds numbers meshed via hexahedral O - grids with a total mesh size of 1.1 million cells. The wall mesh is chosen to provide a dimensionless wall distance of $y_1^+ \leq 1$ for the first grid point near the wall. The time step is adjusted to comply the limit given by the Courant - Friedrichs - Lewy number (CFL) ≤ 1, which may ensure stability of the solution. In ANSYS CFX, the chosen time discretisation is expresses by a second order backward Euler scheme, and convective and viscous fluxes are approximated with high accuracy enabling the software usage of the second order upwind differencing scheme with a blending function to the first order upwind differencing scheme for better convergence. The numerical simulations are run for $3\Delta t_{domain}$ to ensure that the whole domain has been calculated and that simulations have already converged. Here, the time domain is $\Delta t_{domain} = L / \overline{U^z}$.

2.3. Flow analysis

2.3.1. Velocity approximation

The velocity field given by measurements and numerical simulations are analysed with respect to vortex breakdown, the swirl decay, and thermodynamic stability condition. According to these analyses smooth functions continuously distributed along the radial coordinate are desired. Thus, the obtained flow field is approximated (fitted) to get smooth curves. A model that is able to be used for the swirl flow in the tube is developed for an axisymmetrical helical vortex [20], which is, for our purposes, extended to capture the wall region of the swirl tube.

$$u^{\phi} = \frac{\Gamma}{2\pi r}\left[1-\exp\left(-\frac{r^2}{\varepsilon^2}\right)\right]\left\{1-\frac{1}{\exp\left[k\left(R-r\right)\right]}\right\}, \tag{3}$$

$$u^{z} = u^{0} - \frac{\Gamma}{2\pi l}\left[1-\exp\left(-\frac{r^2}{\varepsilon^2}\right)\right]\left\{1-\frac{1}{\exp\left[k\left(R-r\right)\right]}\right\}, \tag{4}$$

In these approximations, Γ denotes the circulation, ε is the vortex core size, l is the pitch of helical symmetry, u^0 is the velocity at the vortex axis and k is the wall parameter. These parameters are obtained via the free Toolbox EzyFit in Matlab [21].

2.3.2. Rossby number

Spall et al. [16] proposed a criterion based on the Rossby number that is used to predict the onset of vortex breakdown. Local values of the Rossby number are computed by the ratio of the axial velocity to the tangential velocity

$$Ro = \frac{u^{z*}}{r^*\Omega}. \tag{5}$$

Here, r^* defines the radial position where the swirl velocity reaches its maximum, $r^*\Omega$ is the swirl velocity where Ω denotes the angular velocity of the solid body rotation, and u^{z*} is the axial velocity at the position r^*.

2.3.3. *Axial and tangential Reynolds number*

The axial Reynolds number and the tangential Reynolds number, equation (6), take into account mean values of the axial velocity and mean values of the tangential velocity measured and/or computed at the defined axial position of the tube.

$$Re_z = \frac{\overline{u^z} D}{\nu}, \; Re_\phi = \frac{\overline{u^\phi} D}{\nu} \tag{6}$$

where $\overline{u^z}$ and $\overline{u^\phi}$ are the mean value of the experimentally and/or numerically obtained axial velocity and tangential velocity, D represents the tube diameter and ν is the kinematic viscosity.

2.3.4. *Thermodynamic stability condition*

The stability concept based on thermodynamics follows the work by Marsik [22]. The thermodynamic stability condition (TSC function) can be, in general, written as

$$\bar{\pi} = T\sigma(S) \geq 0, \tag{7}$$

where T is the thermodynamic temperature and σ (S) represents a term covering the production of entropy. Equation (7) is valid for each point x in the body with the volume V. Taking into account the balance of the total specific enthalpy h_t, the TSC function for cases without heat transfer can be rearranged to read

$$\pi = t_{dis}^{ik} \frac{\partial u_i}{\partial x^k} + u_i \frac{\partial t_{dis}^{ik}}{\partial x^k} \geq 0, \tag{8}$$

where t_{dis}^{ik} is the Cauchy stress tensor for an incompressible fluid. The preliminary outcomes of this thermodynamic concept can be found in the work by Marsik et al. [23]. The flow is from a thermodynamic point of view stable if equation (8) is satisfied. A global value of the TSC function is computed by volume integration of equation (8).

However, to obtain values of the global TSC function in dimensionless form, the volume integral is normalised by the inner volume of the tube V_t, the material parameter μ (dynamic viscosity), and the angular velocity Ω taken from table 1.

$$\frac{\Pi}{V_t \Omega^2 \mu} = \frac{1}{V_t} \int_V \frac{\pi}{\Omega^2 \mu} \mathrm{d}\overline{V} \geq 0, \text{ for } \mathrm{d}\overline{V} = \bar{r}\mathrm{d}\bar{\phi}\mathrm{d}\bar{r}\mathrm{d}\bar{z}\,. \tag{9}$$

The axial coordinate ($\mathrm{d}\bar{z}$) in the integral varies from zero to 1 mm and represents the thickness of the slice with a radius r changing from 0 to R, for which a value of the global TSC function at a position z/D is calculated. Moreover, there is a huge dissipation near the tube wall resulting in positive values of the global TSC function. On the other hand, due to the velocity redistribution in the tube, there are processes in the flow that the global TSC function should cover. Due to this hypothesis, the values of the global TSC function computed at a position (z/D) without the influence of the wall are used for the analysis of the flow field to present their development towards tube outlet. These values are computed by equation (9), where integration over the radial coordinate is taken from 0 to 0.9R.

3. Results

Data obtained from numerical results are extended by the profile near the tangential inlets (position z/D = -1) where measurements are not possible. These data are shown in order to better comprehend the redistribution of the velocity field from the tube inlets to tube outlet.

Vortex breakdown can be predicted by a change of the two different Reynolds numbers, the axial Reynolds number (Re_z) and the tangential Reynolds number (Re_ϕ), equation (6), respectively. In contrast to the analysis based on the Rossby number (equation 5), these Reynolds numbers are used to study the effect of the mean velocities. The decreasing trend of the calculated tangential Reynolds numbers, given mainly by a drop of the swirl number characterising the swirl strength, and an almost constant trend of the axial Reynolds numbers at different positions are obvious from Figure 2. The axial and tangential Reynolds numbers are determined for all investigated Reynolds numbers.

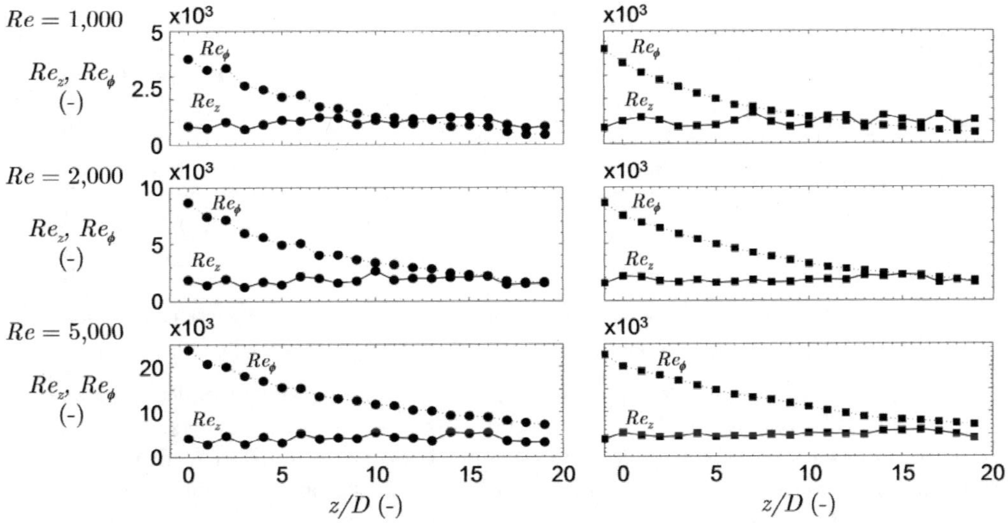

Figure 2. Evolution of the axial (Re_z) and tangential (Re_ϕ) Reynolds number towards tube outlet. Bullets and squares represent experiment and numerics, respectively.

In case of Re = 1,000, the tangential Reynolds number is larger than the axial Reynolds number in the region near the tangential inlets. This indicates that in this region, vortex breakdown may occur. As the fluid approaches the tube outlet, the flow becomes axial dominant without axial backflow indicating no vortex breakdown. This fact is represented by intersection of the tangential and the axial Reynolds number followed by a drop of the tangential Reynolds number under the axial Reynolds number. With an increase in the Reynolds number to Re = 2,000, the state of the fluid near the inlet region remains unchanged. The tangential Reynolds number is larger than the axial Reynolds number indicating the occurrence of vortex breakdown. Towards tube outlet, the tangential Reynolds number approaches, but it does not intersect, the axial Reynolds number. This means that near the tube outlet, the flow state approaches the onset of vortex breakdown as indicated by the Rossby number in figure 3. For the highest Reynolds number (Re = 5,000), there is obviously the widest gap between the tangential Reynolds number and the axial Reynolds number near the tangential inlets. Vortex breakdown in this case takes place over the entire tube length, because the tangential and axial Reynolds number do not intersect.

A connection between the Rossby number (Ro), equation (5), indicating vortex breakdown and the global TSC function (Π), equation (9), characterising processes due to the velocity field redistribution on the swirl flow as shown in Figure 3, where the black straight line indicates the critical value of the Rossby number of 0.65. All Rossby numbers larger than the critical Rossby number indicate no vortex breakdown. Figure 3 supports the results from figure 2, so vortex breakdown occurs over the entire tube length for Re = 5,000. For Re = 2,000, there is a difference between numerics and experiment. Analysis of the experiments show the occurrence of vortex breakdown over the entire tube length, but analysis of numerics predicts no vortex breakdown in the region near the tube outlet. This disagreement may be assigned to a slight inaccuracy of the numerical prediction of the axial velocity and to the difference in values of the global TSC function near the tangential inlets. In case of Re = 1,000, vortex breakdown is predicted in the tube region where the flow is swirl dominant. Near the tube outlet, no vortex breakdown is assumed.

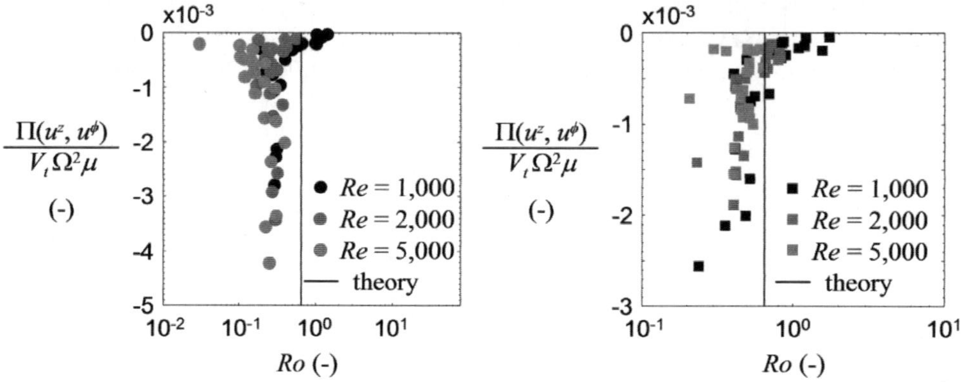

Figure 3. Values of the global TSC function against Rossby number. Bullets and squares represent experiment and numerics, respectively.

Values of the global TSC function, computed for the axial and tangential velocity, decrease towards tube outlet. Thus, for all investigated Reynolds numbers, as the fluid flows towards tube outlet, the global TSC function approach less negative values and the flow is stabilised from point of view of thermodynamics. The local Rossby number increases in region near the tube outlets for *Re* = 1,000, where the flow is axial dominant. In the case of *Re* = 2,000, the local Rossby number also increases in the region near the tube outlet, but this increase is not so significant as in the case of *Re* = 1,000. This is due to the fact that in this case the flow is neither axial nor swirl dominant, so that the Rossby numbers are very close to the limiting value in this region. For swirl dominated flow, i.e. in case of *Re* = 5,000, the local Rossby number decreases in the region near tube outlet. This decrease corresponds to a decrease of the swirl strength as the fluid moves towards tube outlet.

4. Summary and conclusions

An analysis of a swirl flow in a tube has been shown. The flow shows velocity redistribution. A prediction of vortex breakdown together with thermodynamic aspects of the flow, were presented.

We provided a description of experimental and numerical work carried out to obtain the velocity field of the swirl flow in the tube at several positions that are necessary for an analysis of vortex breakdown. Three Reynolds number were investigated, i.e. *Re* = 1,000, 2,000 and 5,000, to cover laminar, intermediate and turbulent flow regimes. To predict vortex breakdown, the methodology based on the Rossby number was employed. Moreover, vortex breakdown and its prediction were investigated via the change of the axial and tangential Reynolds numbers showing that vortex breakdown may be expected when the flow is swirl dominated, i.e. $Re_\phi > Re_z$, and when an axial backflow occurs. This condition is fully satisfied only in case of the investigated Reynolds number *Re* = 5,000. In the other two cases, vortex breakdown covered only part of the tube length (*Re* = 1,000) or almost the whole tube length except a region near the tube outlet (*Re* = 2,000).

Furthermore, a thermodynamic perspective, i.e. the TSC function introduced by equation (9), of the swirl flow together with the occurrence of vortex breakdown was presented. This analysis confirmed outcomes from the analysis based on the Rossby number and the axial and tangential Reynolds numbers. Moreover, the TSC function provided information on the redistribution of momentum and its impact on the dissipation processes taking place in the flow. It was shown that towards tube outlet, the swirl flow became more stable indicating a more pronounced potential character of the flow. This behaviour was observed for all three investigated cases.

So, the swirl flow can be analysed by the prediction of vortex breakdown based on the Rossby number. Furthermore, a decrease of the swirl strength towards tube outlet is able to be figured out by the analysis of a change in trend of the axial and tangential Reynolds number. Moreover, decrease of the swirl strength results in a redistribution of momentum accompanied by dissipation processes in the flow field. This redistribution causes a stabilisation of the flow from thermodynamic point of view.

Acknowledgments
The result was developed within the CENTEM project, reg. no. CZ.1.05/2.1.00/03.0088, co - funded by the ERDF as part of the Ministry of Education, Youth and Sports OP RDI programme, and, in the follow - up sustainability stage, supported through the CENTEM PLUS (LO1402) by financial means from the Ministry of Education, Youth and Sports under the National Sustainability Programme I. The authors, B. Weigand and C. Biegger, would like to acknowledge financial support by the Deutsche Forschungsgemeinschaft (DFG) for Project WE2549/38-1.

References
[1] Ranque G J 1933 *J. Phys. Radium* **4** (7) 112
[2] Hilsch R 1947 *Rev. Sci. Instrum.* **18** (2) 108
[3] Pfitzer H and Beer H 1992 *Int. J. Heat Mass. Transf.* **35** (3) 623
[4] Pfitzer H 1992 Konvektiver Wärmetransport im axial durchströmten Ringspalt zwischen rotierenden Hohlwehlen, PhD thesis (Technische Hochschule Darmstadt: Institut für Technische Thermodynamik) 132
[5] Rothe T 1994 Der Einfluß der Rotation af die Strömung und den Wärmetransport im turbulent durchströmten Ringspalt zwischen zwei rotierenden Wellen, PhD thesis (Technische Hochschule Darmstadt: Institut für Technische Thermodynamik) 156
[6] Reich G, Weigand B and Beer H 1989 *Int. J. Heat Mass. Transf.* **32** (3) 563
[7] Weigand B and Beer H 1992 *Int. J. Heat Mass. Transf.* **35** (7) 1803
[8] Weigand B and Beer H 1994 *Appl. Sci. Res.* **52** (2) 115
[9] Hartnett J P and Eckert E R G 1957 *J. Heat Transfer* **79** 751
[10] Ligrani P M, Hedlund C R, Babinchak B T, Thambu R, Moon H - K and Glezer B 1998 *Exp. Fluids* **24** (3) 254
[11] Glezer B, Moon H - K, Kerrebrock J, Bons J and Guenette G 1998 Heat Transfer; Electric Power; Industrial and Cogeneration (Stockholm) **4** (New York: *American Society of Mechanical Engineers (ASME)*)
[12] Sarpkaya T 1971 *AIAA J.* **9** (9) 1792
[13] Lucca - Negro O and O'Doherty T 2001 *Prog. Energy Combust. Sci.* **27** (4) 431
[14] Escudier M 1988 *Prog. Aerosp. Sci.* **25** (2) 189
[15] Hall M G 1972 *Annu. Rev. Fluid Mech.* **4** 195
[16] Spall R E, Gatski T B and Grosch C E 1987 *Phys. Fluids* **30** (11) 3434
[17] Squire H B 1960 Analysis of the "vortex breakdown" phenomenon part 1 *Tech. rep.* (102) London: Imperial College London, Department of Aeronautics
[18] Benjamin T B 1965 *J. Basic Eng.* **87** (2) 518
[19] *ANSYS CFX - Solver Theory Guide* Release 12.0 2009 Canonsburg: ANSYS 261
[20] Alekseenko S V, Kuibin P A and Okulov V L 2007 Theory of Concentrated Vortices Heidelberg: *Springer* 494
[21] Moisy F 2012 Ezyfit: A free curve fitting toolbox for Matlab v 2.41 http://www.fast.u-psud.fr/ezyfit/
[22] Marsik F 1989 *Acta Phys. Hung.* **66** (1) 195
[23] Marsik F, Kobiela B, Novotny P and Weigand B 2010 *Proc. 21st Int. Symp. on Transport Phenomena (Kaohsiung)* (Kaohsiung: National Kaohsiung University of Applied Sciences)

IOP Conf. Series: Journal of Physics: Conf. Series **1045** (2018) 012032 doi:10.1088/1742-6596/1045/1/012032

In situ testing of rail damages in accordance with Industry 4.0

Z Nyikes[1], D Tokody[2] and T Kovacs[3]

[1,2]Obuda University, Bánki Donát Faculty of Mechanical and Safety Engineering, Doctoral School of Safety and Security Science, Népszínház u. 8. Budapest H1081, Hungary
[3]Obuda University, Bánki Donát Faculty of Mechanical and Safety Engineering, Material Science Department, Népszínház u. 8. Budapest H1081, Hungary

E-mail: kovacs.tunde@bgk.uni-obuda.hu

Abstract. The railway is a very important segment of the public transport and therefore its part of the critical infrastructure system. The speed of the trains depends on the rail conditions. The state of the rails needs continuous control to determine the train speed and continuous traffic and transport. The goal of the Industry 4.0 is to build a smart industry and industrial environment [1]. Part of this smart system is the traffic and the transportation on road and railway too. The continuous railway transportation needs quality railroads and trains. The rails have complex load during their lifetime. In this research, we wanted to find a test method and innovate an experimental setup for the continuous state control of rails. In our work, we wanted to find a relationship between propagations of the rail surface damage and the loads. Also based on failure propagation rules we wanted innovate a special installation for in situ monitoring the process of the rail surface damage, predict the defects and crack propagations. This work is in accordance with the aim of the Industry 4.0.

1. Introduction

Nowadays the railway transport innovation is very important part of the engineering profession in Hungary and in the world. When we talk about railroad, we can separate the problems in two parts, one is the rail substrate and the other is the rail material. Nowadays for rail we usually use hot rolled rail steel in pearlite microstructure. Based on the knowledge of the wear process we know that the chemical composition and the microstructure are also relevant parameters. We also know the correlation between the wear resistance and the hardness, which depends on the heat-treating process in case of steels [2].

We focused in this article on the load of the wheel rail contact and the surface of the rail (Rolling Contact Fatigue; RCF) [3, 4]. We studied the Hertzian stress between the wheel and rail surface that depend on the surface roughness and surface failure. The stress depends on the real contact surface area and the size of the contact surface depends on the size of the failure. This theory can help us understand the causes of the surface cracks propagation and the damage process. These surface failures limit the train's maximum useable speed. The rail surface quality control is very important to assure the continuous traffic in railway system and prognosticate the maintenance time of the railroad. By monitoring the surface failure, we get the real rail quality data from which, we can calculate the real Hertzian stress, and based on those results we can determine the useable train speed on the measured rail parts (GPS coordinates). This information is very important for the next trains that are using the same railroad. The railway transport is a member of the critical infrastructure, which means we cannot transport the measured data by a traditional device, so it needs a safe way for the data transfer. It is required to use a safe data transfer device, which transfers the data processing results like determined usable maximal speed value attached with the GPS coordinates of the used trains.

2. Wheel-Rails Contact

Between the wheel and rail, we can find a rolling and sliding contact that increase wear and fatigue behaviour. The value of the contact surfaces size very important because the Hertzian stress depends on the size of the contact area. In the literature, we found different models for the train wheel and rail contact [3-5]. In our project, we used a modified model for the wheel-rail contact [3]. The used model is the roller-roller contact in case of railway (shown the Fig. 1.).

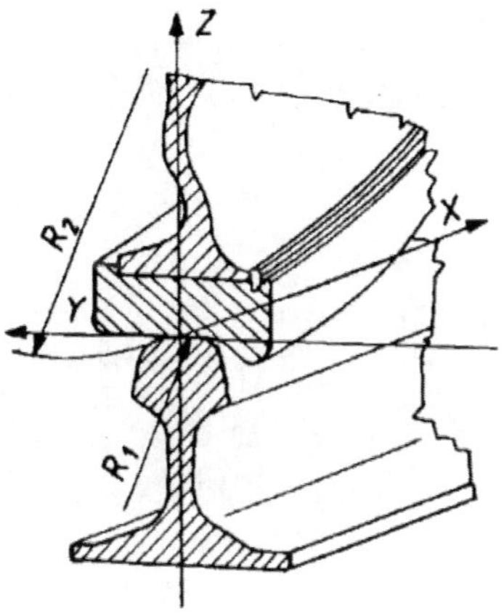

Figure 1. Wheel rail contact on base of Hertz model [3]

The maximum Hertzian stress in case of roller-roller contact model equation (1), where p is the average contact stress [N/mm^2], Z is the wheel load [kN], contact surface area (ellipse) [mm^2], 2a contact ellipse long axis [mm], 2b contact ellipse short axis [mm] (Fig.2.):

$$\sigma = \frac{3}{2}\bar{p} = \frac{3}{2}\frac{Z}{A} = \frac{3}{2}\frac{Z}{\pi ab}\left[\frac{kN}{mm^2}\right] \tag{1}$$

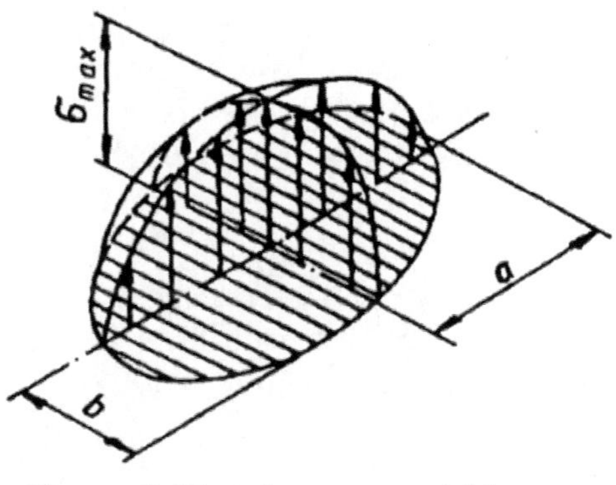

Figure 2. Hertz's stress model in case of wheel rail contact [3]

The Hertz stress calculation gives an adequate result in case of regular roller-roller geometry contact [6]. If we suppose any surface failure, crack, roughness error or other geometrical problem we need to modify the introduced equation. The real contact surface size is smaller than the calculated. The real Hertzian stress σ_{real} we can determine by equation (2) when e [mm^2] is the size of surface failure area.

$$\sigma_{real} = \frac{3}{2}\frac{Z}{(\pi ab) - e}\left[\frac{kN}{mm^2}\right] \quad (2)$$

The determination of the real Hertzian stress is very important because based on this result we can predict the rail lifetime and the defects propagations. The failure is detected with our device by visual testing, which is supported by a rapid camera and then it evaluates the failure size (e) by an imaging system.

Figure 3. Elliptical wheel-rail contact area and the failure size on the surface

Fig. 3. shows a welded joint failure, a cracked defect and some deformation on the rail. Cause of the welding joint defect the Hertzian stress is extremely high in this part of the rail. It is assumable, that the high Hertzian stress caused the detected failures (cracks, plastic deformation) in the roller-roller contact area.

3. Surface visual testing

The failure size measuring is the key of our project's success. The origin of the word diagnostic is the Greek diagnostics word, which means distinguish or identify. In rail diagnostics, we define the traditional visual inspection and the instrumental inspections and tests. The non-destructive testing methods usage is the way to assure the safe railway transportation. The diagnostic investigation means geometrical and material inspection. Nowadays the modern technology offers new material and geometrical testing methods. We built our device based on the surface geometry testing. On whole railway network (rail and parts of railroad) needs the continuous inspection [7-9]. Based on an adequate and successful monitoring system the railway transport can work safely and economically and it can predict the recovery works.

3.1. Monitoring by camera system

In our monitoring system, we needed to use high-speed camera because of the train speed [10]. That way we can take photos of the whole surface systematically. These pictures will contain the surface failures. Also for every picture, we need to attach a GPS coordinate. To custom this process, we need clear and clean rail surface.

3.2. Analysis of surface data

This imaging system is able to evaluate the failure size and based on that value the computer process based on the modified equation (2) calculate the real-time Hertzian stress. From the Hertzian stress value it can indicate the dangerous failures (and it's GPS location) and calculate a train speed limitation for the dangerous rail part or send an alarm sign [11, 12].

The data analysis processed by mathematical processing software, based on the data classified and the used modified Hertz stress model.

4. Investigated monitoring system

Fig. 4 shows the innovated monitoring system. The data transfer and the monitored rail local coordinates (GPS) registration is also very important part of our device.

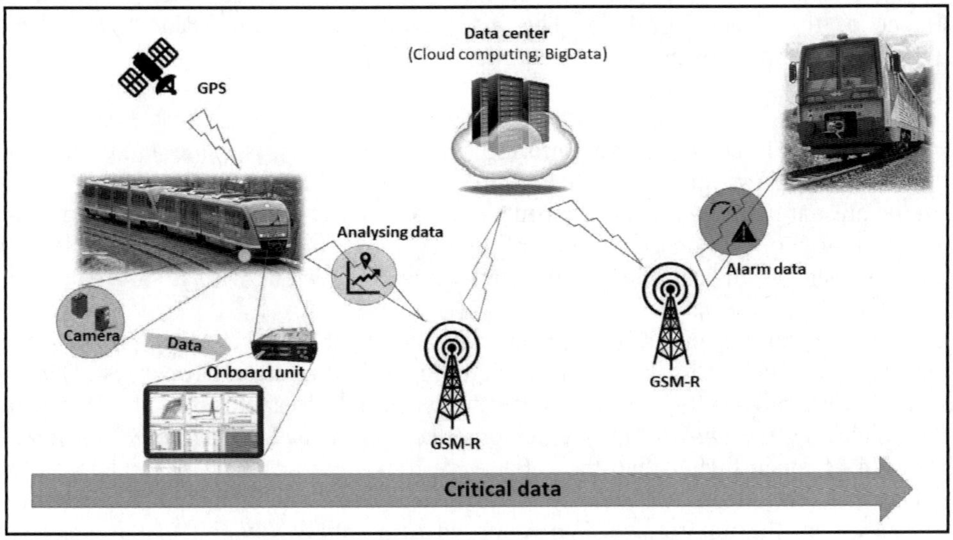

Figure 4. Innovated monitoring system

4.1. Classification of data and transfer type

Data classification

Primer data: The visual inspection gives numerous unprocessed data these are the primer data. It accumulates stores and transfers them in an adequate way for analysis [13]. To store temporary numerous data, we use a high-speed reader-writer puffer [14].

Seconder data: The storage and transfer of the local coordinate data (identify the tested rail segment) attached with the rail surface primer data have to be together [15, 16]. We use for these data the seconder data designation [17-19]. It is required to store and transfer the different origins and types involved data (data fusion).

Transfer types

The different data is classified in three different classes based on of the failure priority and the defect size.

Critical data: These are the measured critical sized failures data. These values are over the limit. They require a direct, real time transfer of this data to support the safety of the railway traffic. The critical data transferred immediately by GSM-R system to the datacentre.

Important data: These data are from the measured important, but not critical defect sizes. The measuring device saves these data and while it is near the station, it transferred to the station server by a radio transfer system. From the station server to the datacentre the data transfer to the cloud supported by optical or cable network.

General data: These are the data of the measured defects under critical or important limit size. These data are saved in the device memory and transferred to the data centre with radio transfer system when the device is arrives to the train terminal. These data are the bases of the prediction for the GPS coordinate identified railroad repair works.

4.2. Data critical infrastructure protection
The electronics information protection to support the transfer safety and the final point network protection is required to assure the measured and analysed data moving to the data centre and back transferring to the users. The execution of the data transfer required with accordance observing of the information safety rules in case of railway because this part of the state economy is a member of the critical infrastructure.
Nowadays cyber-attacks serially happen again the on critical infrastructure members which manifested that importance of the protecting them. The information safety rules adherence is required, to guarantee the safety operation of the introduced monitoring system.

5. Conclusion
In our work, we wanted to construct a monitoring system to support the safe and controlled railway infrastructure. We focused on one of the most important and difficult problem of the railroad defect detection. In the present traditional visual inspection used to detect the railroad surface defects that are outworn. We proposed a new complex monitoring system that is able to detect the failures and after the data analysis, our system can control the railway traffic system with a safe data transfer and protected from the cyber-attacks.

a) Railroads have failures on the surface. The railroad metal has mechanical properties and wears resistance. The surface load capacity is determines by the metal properties. The load is not constant in the railway system. Based on the wheel-rail contact model the Hertzian stress is determinable by the known Hertz equation. Because of the surface defect the real Hertzian stress can be much higher than the calculated from the contact stress model conception. To evaluate the real stress, it needed to modify the Hertz equation and decrease the contact surface size by the failure size. We proposed a modified equation that is the basis of our monitoring system data analysis.
b) The continuous surface visual inspection is realizable by the proposed monitoring system. The detected failure size data attached with the location coordinates and data classification transferred by a safe data transport way. Our system localises, detects and saves the railroad failures with local coordinates (GPS). The data transferring is done in a safe way to protect the system against cyber-attacks.
c) The innovated monitoring system realized in accordance with Industry 4.0.

References
[1] Kovács, T., Nyikes, Z., Tokody, D.: *Using a Complex Monitoring System for Railway Structure Investigation in Accordance with 4.0*, XVII. Műszaki Tudományos Ülésszak, Műszaki tudományos közlemények **6**., Kolozsvár/Cluj, **2016**. pp**151–162**.
[2] Prettner, L.: *Technical Service reducing technical development in the rail production cause of the Head Check failures*, Sínekvilága, **2014/6**. www.sinekvilaga.hu
[3] Kazinczy, L.: *Wheel rail Hertzian contact investigation in case of vehicular train structure*, BME **2017**. Budapest, www.epa.uz.ua/00000/00028/00005/pdf/2.pdf
[4] Bhushan, B.: *Introduction To Tribology*, A John Wiley & Sons, Ltd., Publication, **2013**. New York p. **95-119**
[5] Bagyinszki, Gy., Bitay, E.: *Surface Trating*, Műszaki Tudományos Füzetek **5**., EME, Kolozsvár/Cluj, **2009**, p. **359**; http://eda.eme.ro/handle/10598/15436
[6] Totten, G., Howes, M., Inoue, T.: *Handbook of Residual Stress and Deformation of Steel*, ASM International **2002**. p.**424-436**
[7] Flammini, F., at all: *Towards Wireless Sensor Networks for railway infrastructure monitoring, Electrical Systems for Aircraft*, Railway and Ship Propulsion, **2010**, pp. **1–6**.
[8] Flammini, F., Pragliola, C., Smarra, G.: *Railway infrastructure monitoring by drones*, **2016** International Conference on Electrical Systems for Aircraft, Railway, Ship Propulsion and Road Vehicles International Transportation Electrification Conference (ESARS-ITEC), **2016**, pp. **1–**

6.

[9] Boronahin, A. M., and all: *Optical profilometers for rail track diagnostics*, **2016** IEEE NW Russia Young Researchers in Electrical and Electronic Engineering Conference (EIConRusNW), **2016**, pp. **404–407**.

[10] Santur, Y., Karaköse, M., Akin, E.: *A new rail inspection method based on deep learning using laser cameras,* **2017** International Artificial Intelligence and Data Processing Symposium (IDAP), **2017**, pp. **1–6.**

[11] Li, Y., at all: *Rail Component Detection, Optimization, and Assessment for Automatic Rail Track Inspection*, IEEE Trans. Intell. Transp. Syst., vol. **15**, no. **2**, pp. **760–770, 2014**.

[12] Santur, Y., Karaköse, M., Akin, E.: *Big data framework for rail inspection*, **2017** International Artificial Intelligence and Data Processing Symposium (IDAP), **2017**, pp. **1–4.**

[13] Mester, Gy., at all: *Adaptive Control of Robot Manipulators with Fuzzy Supervisor, Using Genetic Algorithms*, Proceedings of International Conference on Recent Advances in Mechatronics, ICRAM'95, O. Kaynak (ed.), Vol. **2**, pp. **661–666**, ISBN **975-518-063**-X, Istanbul, Turkey, August **14-16, 1995**.

[14] Mester, Gy., at all: *Flexible Planetary Gear Drives in Robotics*, Proceedings of the **1992** International Conference on Industrial Electronics, Control, Instrumentation and Automation - Robotics, CIM and Automation, Emerging Technologies, IEEE IECON '92, Vol. **2**, pp. **646-649,** ISBN **0-7803-0582-5,** DOI: **10.1109**/IECON. **1992.254556**, San Diego, California, USA, November **9-13, 1992**.

[15] Iantovics, L.B.: *A New Intelligent Mobile Multiagent System*, in Proc. of the IEEE Int. W. on Soft Computing Applications (SOFA). Szeged-Hungary and Arad-Romania, IEEE Computer Society Press, IEEE Hungarian Section, **2005**, pp.**153-159**.

[16] Brassai, S.T., Iantovics, L.B., Enăchescu, C.: *Artificial Intelligence in the Path Planning Optimization of Mobile Agent Navigation*, Procedia Economics and Finance, **3**, **2012**, pp. **243-250** ISSN **2212-5671**

[17] Bánáti, A., Kacsuk, P., Kozlovszky, M.: *Reproducibility Analysis of Scientific Workflows*, Acta Polytechnica Hungarica Vol. **14**, No. **2**, Budapest, **2017**, pp**201-2017**, DOI: **10.12700**/APH.**14.2.2017.2.11**, http://uni-obuda.hu/journal/Banati_Kacsuk_Kozlovszky_73.pdf,

[18] Jocic, M., at all: *Managing Big Data Using Fuzzy Sets by Directed Graph Node Similarity*, Acta Polytechnica Hungarica Vol. **14**, No.**2**, Budapest, **2017**, pp**183-200**, DOI: **10.12700**/APH.**14.2.2017.2.10,** http://uni-obuda.hu/journal/Jocic_Pap_Szakal_Obradovic_Konjovic_73.pdf,

[19] Yun Lim, S., Mat Kiah, M.L., Fong Ang, T.: *Security Issues and Future Challenges of Cloud Service Authentication*, Acta Polytechnica Hungarica Vol. **14**, No.**2**, Budapest, **2017**, pp**69-89,** DOI: **10.12700**/APH.**14.2.2017.2.4,** http://uni-obuda.hu/journal/Lim_MatKiah_Ang_73.pdf

IOP Conf. Series: Journal of Physics: Conf. Series **1045** (2018) 012033 doi:10.1088/1742-6596/1045/1/012033

Rheological study of concentrated dispersions. Application to the drilling fluid.

H Ouaer[1], M Gareche[1] and A Allal[2]

[1]Laboratory of Hydrocarbons Physical Engineering, Faculty of Hydrocarbons and Chemistry, University of M'Hamed Bougara -Boumerdes-, Independence Avenue 35000 Boumerdes, Algeria
[2]IPREM-EPCP, University of Pau and of Pays of Adour, Hélioparc, 2 av. Pierre Angot, Pau 64053, France

E-mail: ouaer-91@hotmail.com

Abstract. In order to understand the rheological behavior of concentrated aqueous Algerian bentonite dispersions of drilling (sodium bentonite of Mostaganem "m'zila"), rheological tests were carried out. By varying the concentration of bentonite, flow tests have allowed to estimate the yield stress and apparent viscosity for each concentration and to see their influence on the rheological behavior of these dispersions. In addition dynamic tests (oscillatory) are used to define the linear region of our samples, the state of our fluid (elastic solid or viscous liquid) and understanding the mechanisms of structuring of the particles constituting the material. In parallel, other tests coupled with rheological measurements such as x-rays diffraction to know the mineralogical composition and granulometry to estimate the bentonite particle size.

1. Introduction

During oil drilling, many problems can be encountered due to the interactions between the various geological formations crossed and the drilling fluid used; this latter is formulated on the surface by using various components where bentonite is the main one for water based mud. The composition of bentonite used and the study of its rheological behavior in dispersion in water represent a very important step to avoid problems during implementation.

The bentonite used in this study comes from Mostaganem (m'zila, west of Algeria). It currently indicates the name of smectites which consist essentially of montmorillonite where are found three levels of organization which are: the sheet, the primary particle and the aggregate [1]. The sheets are composed essentially of atoms of silicium, oxygen, aluminum and magnesium [2].

In the literature, little research has been carried out on understanding the mechanisms of structuring of bentonite dispersions often used in drilling fluids in Algerian oil fields. It is with this objective that this study contributes by studying these mechanisms through rheological tests coupled with other methods such as X-ray diffraction (XRD) and particle size to better understand and reinforce the results obtained by identifying the origin of the observed properties.

2. Materials and methods

2.1. Materials and equipment

The bentonite used is primarily made up of sodium montmorillonite. This type of clay consists of an assembly of aggregates with a crystalline structure. It belongs to the phyllosilicate family 2: 1, i.e. it is

IOP Conf. Series: Journal of Physics: Conf. Series **1045** (2018) 012033 doi:10.1088/1742-6596/1045/1/012033

formed of an octahedral layer located between two tetrahedral layers [2-9]. The Si^{4+} ions are positioned in the tetrahedral layers while the octahedral layer contains Al^{3+} ions. Compensating cations (Na^{+}, Ca^{2+}) are often found between the sheets. These minerals are more sensitive to hydration which causes partial swelling, and consequently changes in rheological properties [2, 10].

All the rheological tests were carried out in an imposed stress rheometer (Physica MCR301) having a geometry of coaxial cylinders (Re=14.464 mm, Ri=13.325 mm, Re Ri=1.088, h=39.997 mm) and were maintained at a constant temperature at 25±0.1°C.

The mineralogical composition of bentonite was determined by x-rays diffraction. The measurement of bentonite particle size was made using a granulometry test in laser instrument (Malvern Mastersizer 2000).

2.2. Experimental protocol
Since the protocol of preparation of such suspensions has a great effect on the final state of the suspension and consequently, on its rheological behavior [3], the samples are prepared as follows: initially a quantity of given demineralised water is stirred, then a mass of bentonite of variable concentrations (3, 4, 5, 6 or 7%) are versed in water in way closely to avoid the formation of aggregates of clay particles. In order to homogenize the obtained system, the suspensions are left under magnetic agitation during 24h.

Before each rheological test, the suspensions are subjected to a pre-shear then a rest time in the measurement geometry to have the same structural state of reference. In the case of a long-term test (dynamic tests), the surface of the sample is covered with a film layer of paraffin oil to avoid evaporation problem.

The pH of our suspensions is checked systematically, it is between 9.5 and 10.5, which shows that the faces and edges of the clay particles are negatively charged, giving a first indication on the microstructure of our suspensions [11].

3. Results and Discussion

3.1. Rheological tests

3.1.1. Flow tests
This test consists of applying a stress along a rising ramp of 20s for each stage. It will inform us about the type of the rheological behavior of the fluid by estimating parameters such as the yield stress of flow, the dynamic viscosity of our sample, as well as the evolution of these parameters as a function of the bentonite concentration.

From the flow curves obtained (Fig. 1), it can be seen that our bentonite suspensions exhibit a non-Newtonian behavior, most often a shear thinning (pseudoplastic) behavior, as that is illustrated in figure 1. Similar results were obtained by [4].

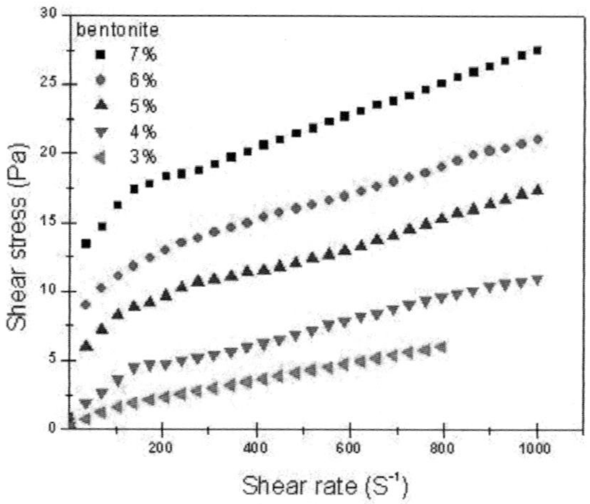

Figure 1. Rheograms of bentonite suspensions (3 to 7%) immediately after preparation (aged 24h).

The observed shear thinning behavior results in a decrease in the apparent viscosity of the suspensions with increasing the shear rate (Fig. 2) and this can be explained by the deformation or disaggregation of the suspended structural units under the effects of hydrodynamic forces.

All the suspensions have a yield stress which increases with the concentration (Fig. 3), below this yield stress the particles are linked together and the applied stress is insufficient to break these bonds and cause the flow of the sample which opposes resistance to the latter by its intact microstructure translating a solid behavior. As soon as this yield stress is exceeded, the sample is destroyed and there will be a flow, so it passes from a solid state to a liquid state.

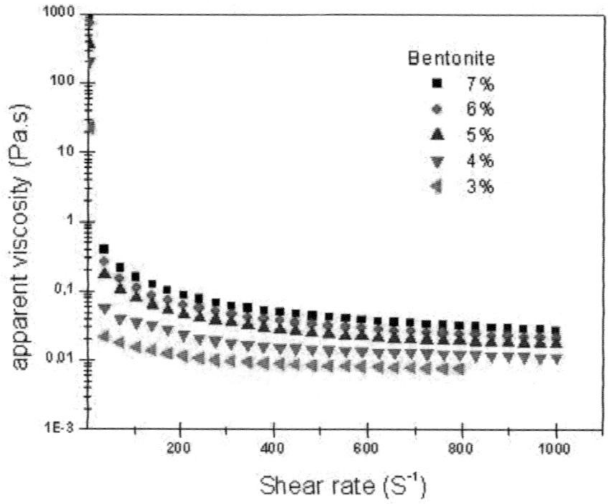

Figure 2. Variation of apparent viscosity as a function of shear rate of bentonite suspensions (3 to 7%).

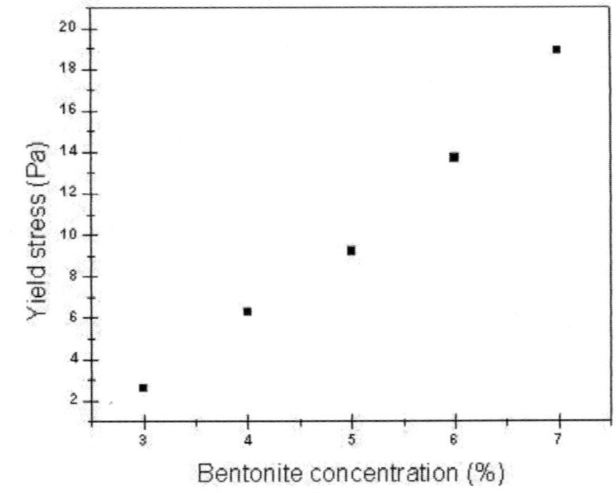

Figure 3. Variation of the yield stress of bentonite suspensions as a function of concentration.

Concerning the viscosity of the suspensions, it increases with increasing concentration (Fig. 4).

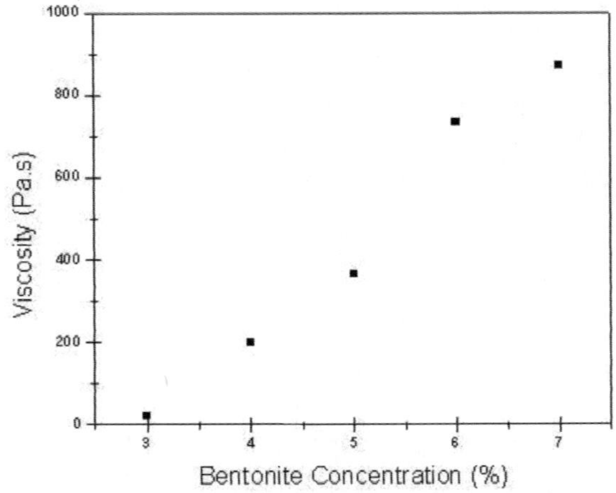

Figure 4. Variation of dynamic viscosity of bentonite suspensions as a function of concentration.

3.1.2. Dynamic tests

Rheological tests in dynamic regime consist in measuring the viscoelastic properties of given samples by studying the elastic G' and viscous G'' modulus as a function of stress, frequency and time.

a. Sweeping in stress

In order to know the linear viscoelasticity domain of the studied bentonite suspensions, i.e. measuring the evolution of the two modulus storage (G') and loss (G'') using a sweeping in stress for a given frequency (in our case 10 rad/s). The test consists in determining a critical stress beyond it there will be the rupture of the three-dimensional structure or of the network formed by the particles.

According to the Fig. 5 which represent the variation of G' and G'' as a function of stress at a frequency of 10 rad/s, it is noted that the values of the two modulus G' and G'' remain constant as long as the stress is less than a limiting value and beyond this value there is a collapse of the structure. That means the transition from the solid behavior to the liquid.

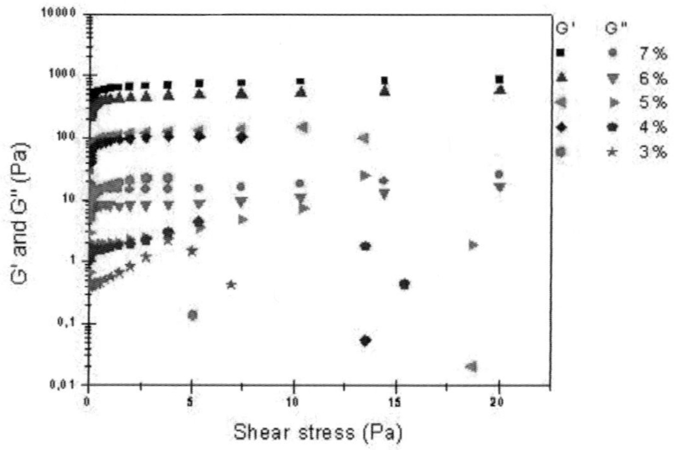

Figure 5. Evolution of viscoelastic modulus as a function of stress for different bentonite concentration.

b. Sweeping in frequency

Sweeping in frequency tests, which consist of measuring the evolution of the two modulus storage (G') and loss (G'') as a function of frequency, were carried out for various bentonite suspensions, choosing a frequency included between 0.01 and 100 rad/s and fixing the stress at a value belongs to the linear domain.

According to the results obtained (Fig. 6), it is noted that G' remains always higher than G'' for all the bentonite suspensions, this is a signature of an elastic solid behavior. In addition, the two viscoelastic moduli also increase according to the bentonite concentration. Similar results were obtained by other authors [5].

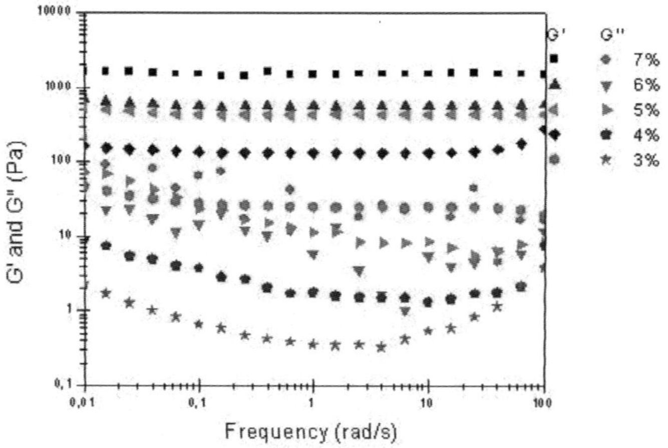

Figure 6. Evolution of viscoelastic modulus as a function of frequency for different bentonite concentration.

IOP Conf. Series: Journal of Physics: Conf. Series **1045** (2018) 012033 doi:10.1088/1742-6596/1045/1/012033

The results obtained in dynamic regime of Mostaganem bentonite confirm well those found previously in stationary regime. The behavior of these suspensions is an elastic solid even if for the weak mass concentrations (3%).

c. Sweeping in time (followed kinetic)

For better controlling the rheological properties of a material over time, firstly the structuring or the restructuring mechanisms of the particles and aggregates that make up this material must be understood [8]. These mechanisms were apprehended by a kinetic follow-up which consists in observing the evolution of the storage (G') and loss (G'') modulus by carrying out a sweeping in time. The results obtained for the concentrations (4, 5, 6 and 7%) are identical. Indeed, the suspensions present a behavior of a gel at the initial state (G' remains higher than G''). On the other hand for the suspensions 2, 2.5 and 3%, these present a time of gel approximately 76 min, 71min and 109 S respectively representing the moment where the crossing of the viscoelastic modulus is carried out. This time of gel physically represents the passage of the liquid state into a solid state and it decreases with increasing in bentonite concentration. For illustrating, just the figures 7, 8, 9 and 10 are represented which corresponding to the bentonite suspensions of 2, 2.5, 5 and 6%.

For the suspensions 4, 5, 6 and 7%, it can be noted that the loss modulus (G'') oscillates but tends to stabilize at an average value which is due to the acoustic resonance phenomenon [6, 8], while that the storage modulus evolves little and stabilizes. This shows that the structuring kinetics of these suspensions is very rapid. On the other hand, for dilute suspensions such as 2, 2.5 and 3%, the structuring kinetics is more significant and the liquid-solid transition can also be noticed. The authors Overlaz and Coussot [7] explained this effect through a percolation phenomenon: the initial increase of G'' indicates the progressive formation of solid aggregates in the fluid. When the volume fraction of the aggregates reaches a critical concentration, the aggregates then form a continuous solid network which is reinforced over time (inducing a sudden increase in G'). The material then becomes aging while G'' decreases.

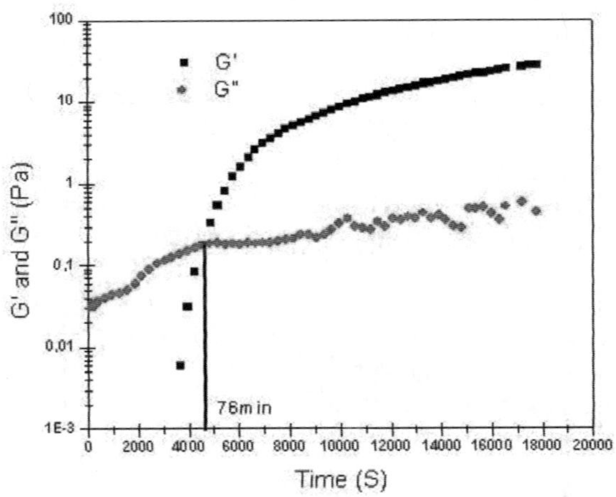

Figure 7. Variation of G' and G'' as function of time for the suspension of 2%.

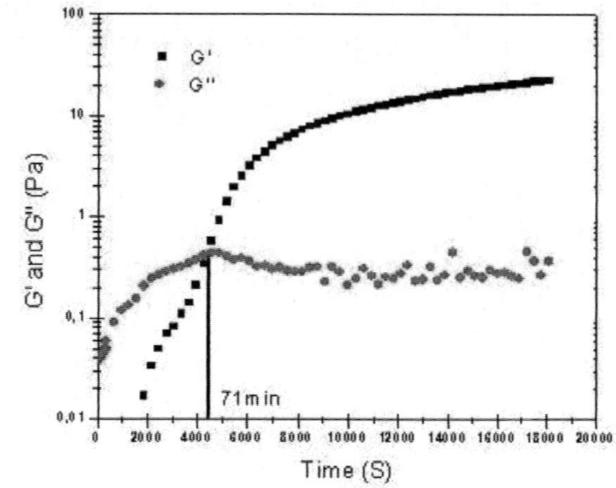

Figure 8. Variation of G' and G'' as function of time for the suspension of 2.5%.

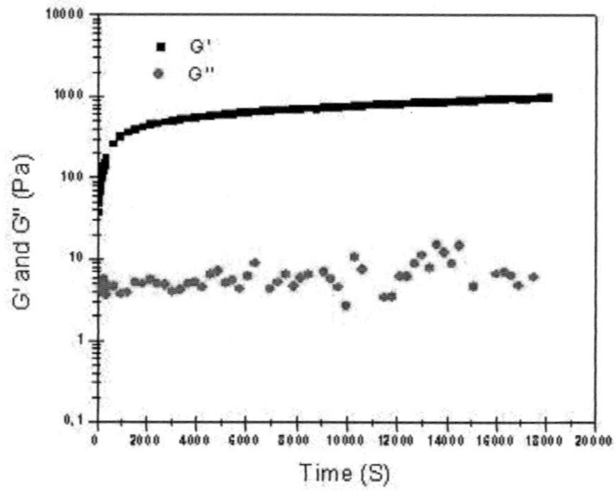

Figure 9. Variation of G' and G'' as function of time for the suspension of 5%.

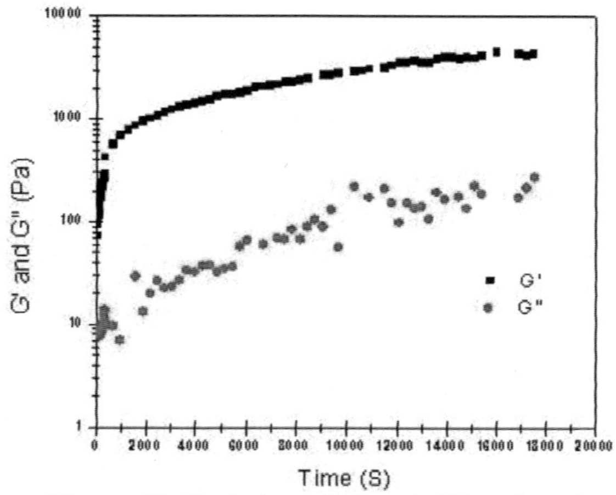

Figure 10. Variation of G' and G'' as function of time for the suspension of 6%.

IOP Conf. Series: Journal of Physics: Conf. Series **1045** (2018) 012033 doi:10.1088/1742-6596/1045/1/012033

3.2. Mineralogical analysis

In order to determine the various minerals constituting Mostaganem bentonite, a test of X-ray diffraction (DRX) is carried out.

The diffractogram (Fig. 11) was obtained by placing the sample of bentonite in the form of a powder in a sample holder. This figure shows several peaks in which the two principal ones correspond to montmorillonite which is the most encountered mineral in bentonite samples. The existence of other peaks shows that the sample is not pure and contains impurities which are: quartz at d = 4.16 Å, feldspath at d = 9.5 Å and d = 4.38 Å and calcite at d = 2.98 Å.

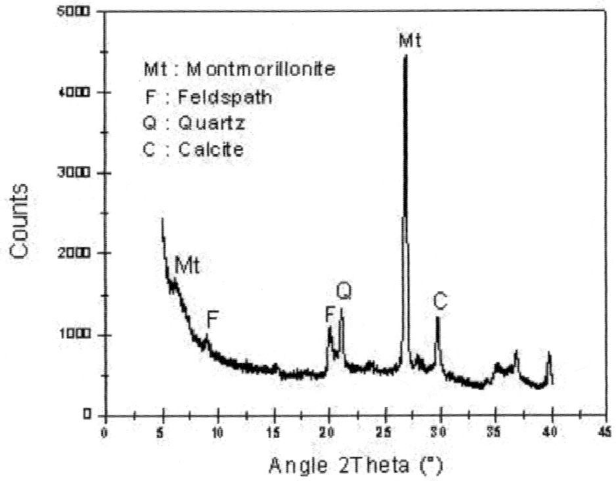

Figure 11. X-ray diffraction pattern of Mostaganem bentonite.

3.3. Granulometric analysis

The granulometric analysis of our bentonite illustrated in Fig. 12 revealed a particle distribution composed of 0.13% of sand (> 50 μm), 91.96% of silt (between 2 and 50 μm) and 7.86% of fine particles (< 2 μm).

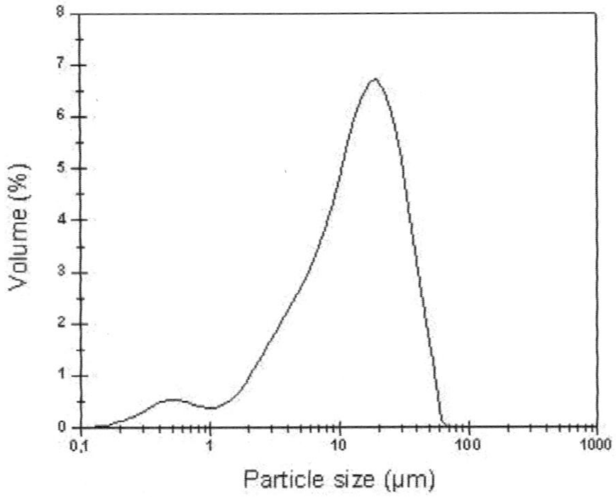

Figure 12. Granulometric distribution of Mostaganem bentonite particles (Algeria).

IOP Conf. Series: Journal of Physics: Conf. Series **1045** (2018) 012033 doi:10.1088/1742-6596/1045/1/012033

4. Conclusion

The study of the rheological behavior of Mostaganem bentonite showed, on the one hand, that the behavior is shear thinning with a yield stress of flow, having a dynamic viscosity which increases as a function of the concentration of the clay particles and, on the other hand, The presence of the viscoelastic properties, i.e. the liquid / solid transition for the low concentrations and a gel state for the high concentrations. These results are in good agreement with the results of the mineralogy and the granulometry which show that montmorillonite is the main mineral constituting Mostaganem bentonite and that it contains less sand (0.13%) which gives it more important rheological properties.

5. References

[1] Le pluart L, Duchet J, Sautereau H, Halley P, Gerard J F 2004 *Applied clay science,* **219** 25
[2] Luckham P F, Rossi S 1999 *Colloid Interface Sci.* **82** 43
[3] Benchabane A 2006 Etude du comportement rhéologique de mélanges argiles - polymères. Effets de l'ajout de polymères, *Thèse de doctorat, Université Louis Pasteur de Strasbourg*
[4] Paumier S, Monnet P, Pantet A 2007 Etude des propriétés viscoélastiques de suspensions de smectite homoioniques à bi-ioniques (Ca^{2+}, Na^{+}) : influence de la concentration (2-100 g/l), *18ème Congrès Français de Mécanique, Grenoble*
[5] Bekkour K, Kherfellah N 2002 *Applied Rheology,* **12** 234
[6] Kaldasch J, Laven J, Stein H N 1998 *J. Rheol,* **42** (6) 1285
[7] Ovarlez G, Coussot P 2007 *Physical Review E.* **76** (1) 11406
[8] Gareche M 2013 Formulation et caractérisation rhéologique des suspensions de bentonite en présence d'additifs polymériques, *Thèse de doctorat, Université M'Hamed Bougara de Boumerdes*
[9] Grandjean J 1997 *J Colloid Interface Sci.* **185** 554
[10] Van Olphen H 1977 An introduction to clay colloid chemistry, *Willey,* New York
[11] Durán J D G, Ramos-Tejada M M, Gonzalez-Caballero F 2000 *J. Colloid Interface Sci.* **229** 107

IOP Conf. Series: Journal of Physics: Conf. Series **1045** (2018) 012034 doi:10.1088/1742-6596/1045/1/012034

Wave Propagations in Metamaterial Based 2D Phononic Crystal: Finite Element Analysis

Z Özer[1], O Özer[2], M Karaaslan[2], A Mamedov[3] and E Özbay[3]

[1]Mersin University, Mersin Vocational High School, Department of Electronics and Automation, Mersin, 33335, Turkey

[2] Iskenderun Technical University, Department of Electrical-Electronics Engineering, Iskenderun, 31200, Turkey

[3] Bilkent University, Nanotechnology Research Center (NANOTAM), Ankara, 06800, Turkey

E-mail: zaferozer@hotmail.com

Abstract. In the present work, the acoustic band structure of a two-dimensional (2D) phononic crystal (PnC) containing composite material were investigated by the finite element method. Two-dimensional PC with triangular and honeycomb lattices composed of composite cylindrical rods are in the air and liquid matrix. The existence of stop bands are investigated for the waves of certain frequency ranges. This phononic band gap - forbidden frequency range - allows sound to be controlled in many useful ways in structures. These structures can be used as sonic filters, waveguides or resonant cavities. Phononic band diagrams $\omega=\omega(k)$ for a 2D PnC were plotted versus the wavevector k along the Γ-X-M- Γ path in the first Brillouin zone. The calculated phonon dispersion results indicate the existence of full acoustic modes in the proposed structure along the high symmetry points.

1. Introduction

The PnC's have been classified according to their physical structure of inclusion and matrix materials. The solid/solid, fluid/fluid, and the mixed solid/fluid composite systems have been studied theoretically and experimentally. These composite media have stop bands in their transmission spectra. In that case, the propagation of the acoustic or elastic waves is strictly forbidden [1, 8, 21]. As in photonic crystal, highly confined waveguides can also be created for elastic and acoustic waves by adding a line defect in PnC [9-14].

The unusual properties of the dispersion relation in the propagation band results in negative refraction. A flat slab with a negative refraction index behaves as a lens [23]. PnC can be designed as new component in acoustic and ultrasonic range to control sound.

It has been shown that, the width of the complete band gap can be increased by reducing the lattice symmetry [3, 9, 15]. In this work, we consider two dimensional PnC consist of composite cylinders in air. We compare the band structure and transmission spectra of triangular and honeycomb lattice both numerically and experimentally for sonic frequency and numerically for ultrasonic frequency. For ultrasonic frequency, we choice tungsten cylinder inclusion in water. Since, tungsten has high contrast of density and speed of sound [2, 16-19]. We investigate whether band gaps could be identified and obtained by using periodic-boundary finite element method (FEM) for the unit cell of infinite structure. This is investigated by comparing transmission magnitudes with the conventional finite structure [9, 17-18, 20-22].

IOP Conf. Series: Journal of Physics: Conf. Series **1045** (2018) 012034 doi:10.1088/1742-6596/1045/1/012034

2. The model and calculation method

The destructive interferences lead to forbidden bands in the periodic structures via wave scattering. If the frequency of incident wave is within the band gap region, total reflection is observed. PnCs are generally divided into two main categories; acoustic and mechanic PnCs. In this study, solid-fluid PnCs are investigated by applying acoustic waves. We assumed that cylindrical rod inclusions are infinitely long along z direction. The periodicity is assigned throughout *xy* directions and acoustic properties of the PnC are also periodic along *xy* direction. We consider that a longitudinal wave propagates in *xy* direction.
Finite element implemented calculation is used for $\Gamma - X - \mathrm{M} - \Gamma$ -directions to obtain eigenvalues and eigenvectors of PnC.

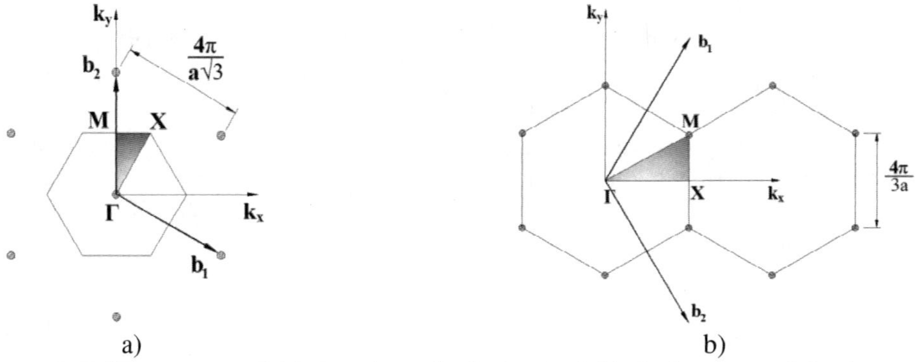

Figure1. Brillouin zone of (a) the triangular lattice and (b) the honeycomb lattice.

PnC with triangular and hexagonal lattices are considered. These lattices consist of composite circular cylinders. They are placed in air to form two-dimensional lattices with a lattice spacing of ***a***. Figures 1(a) and 1(b) are the Brillouin regions of the triangular and the hexagonal lattice, respectively. The irreducible part of the Brillouin zone of a triangular lattice is shown in Fig. 1(a) which is a triangle with vertices of Γ, X, M. The irreducible part of the Brillouin zone of a honeycomb lattice is shown in Fig. 1(b), which is also a triangle with vertices of Γ, X, M. The utilized materials properties are given in Table 1.

Table 1. Material parameters used in the analysis

	Composite materials	Air	Tungsten	Water
Speed of sound (m/s)	1420	343	5090	1490
Density (kg/m^3)	690	1.25	19300	1000

For numerical analysis, we used 2D PnC which is most commonly used in the literature [1]. Plane wave signal source with a pressure level of p0=1 [Pa] is applied to PnC with triangular and honeycomb lattices. For normal component of the air particles equal to zero, the absorber boundary condition is applied to the edge of the periodic structure and sound hard boundary conditions is applied to the edge of cylindrical inclusions. To observe the propagation of the acoustic wave in the PnC, we first set the lattice parameter is defined with a value of a=30 mm and the radius of the scatterers is increased from r=7 mm to r=12 mm by a step of 1 mm for both triangular and honeycomb lattice.
We obtained experimentally the acoustic pressure level (dB) of the PnC composed of composite cylinders with radius of 10 mm (r), length of 300 mm (L) and lattice parameter of 30 mm (a) in air. We placed the speaker 1 m away from PnC. The microphone is placed just behind of the PnC for the experimental measurement.
A tone generator software has been used by setting the output frequency100 Hz - 20000 Hz. The sound recorded on the computer by using the M-Audio external sound card and audio-technica pro 44

microphones. The recorded sound data is converted to frequency domain by using standard Fast Fourier Transform (FFT) algorithm as shown in Fig. 3 and Fig5. Due to the circular shape of the equifrequency contour, the group velocity and the wave vector inside the crystal are antiparallel for all directions of propagation inside the crystal [24].

Negative refraction is also investigated for 2D PnC consist of tungsten rods immersed in water with FEM. The group velocity is the gradient of the angular frequency as a function of k point ($v_g = \frac{\partial\omega}{\partial k}$). The group velocity and the incident waves are parallel to wavevector during the incident waves propagating in liquid base.

3. Results and Discussions

3.1. Triangular lattice

We first investigate the case with composite rods placed in the air in triangular lattice with circular cross session. 2D Triangular and honeycomb lattices are considered here and the numerical results are given as follows. For the triangular lattice, the largest absolute PBG is produced for the filling fraction of $f = 0.58$. The maximum band gap (between the 2nd and 3th bands) has the width (gap–mid gap ratio) of $\Delta\omega/\omega g = 0.1355$ as can be shown in fig. 2.

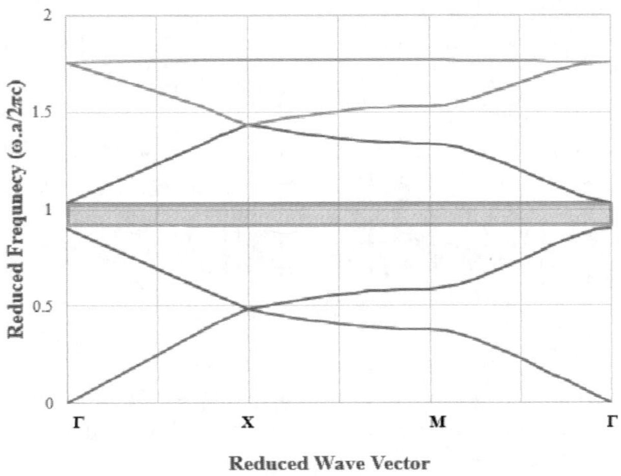

Figure 2. FEM computed band structure for the triangular lattice phononic crystal. The blue shading indicates the presence of resonance band gap.

The upper and lower edges are 0.384 and 0.236 in terms of $\omega a/2\pi c$, where c is the wave velocity in the air ($\Delta\omega = 0.148$).

Figure 3. a) and b shows band gap and experimental measurement results of PnC with the lattice parameter of a=30 mm and the radius of the rods of r=10 mm. It can be seen that that the PnC with triangular lattice have band gap along X-M direction.

IOP Conf. Series: Journal of Physics: Conf. Series **1045** (2018) 012034 doi:10.1088/1742-6596/1045/1/012034

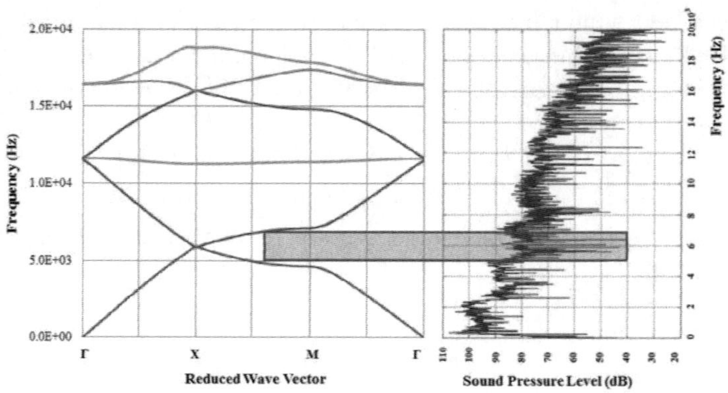

a) b)

Figure 3. a) Band structure b) experimental measurement

3.2. Honeycomb lattice

For the honeycomb lattice, the largest absolute PBG is observed for the filling fraction of $f = 0.38$. The gap between the 1st and 2nd bands has the largest width (gap–mid gap ratio) of $\Delta\omega/\omega g = 0.47$ as can be seen in Fig. 4.

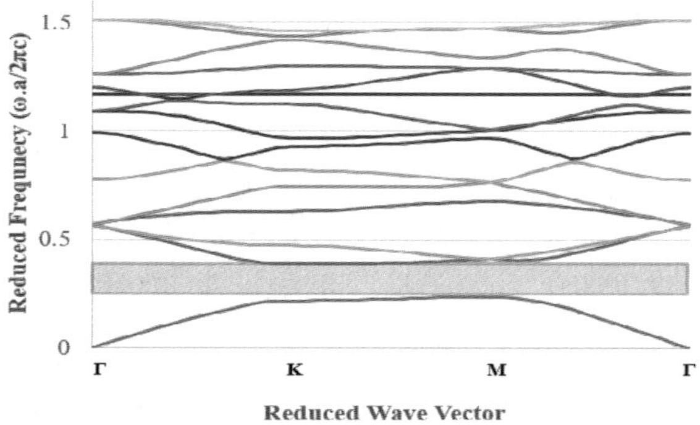

Figure 4. FEM computed band structure for the honeycomb lattice phononic crystal. The blue shading indicates the presence of resonance band gap.

The upper and lower edges are 0.384 and 0.236 ($\Delta\omega = 0.147$) in terms of a $\omega a/2\pi c$, where c is the wave velocity in the air. Band gap and experimental measurement results for the lattice parameter a=30 mm and radius of the rods r=10 mm are shown in Figure 5. It is seen that the PnC with honeycomb lattice have complete band gap.

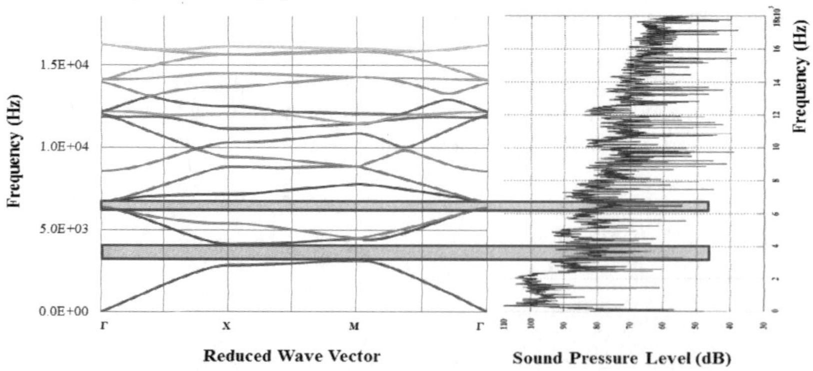

Figure 5. a) Band structure b) experimental measurement

3.3. Acoustic lens effect and negative refraction properties of phononic crystal

The negative refraction can be observed when the waves penetrate to the crystal surface with negative refractive index. If the band structure of the PnC has convex peak at the X point in Fig 2, it could result in negative refraction [2]. The negative refraction properties of PnC may lead to focus acoustic waves by flat lenses.

We investigate the acoustic lens effect and negative refraction properties of the PnC by FEM. PnC consists of tungsten cylindrical inclusion in water. As seen in Fig.6, the PnC with negative refraction properties has acoustic lens effect.

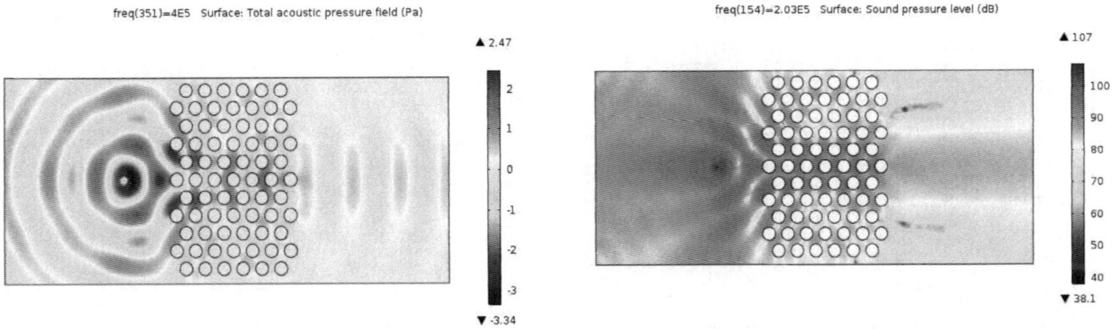

a) b)

Figure 6. Acoustic lens effect of PnC a) Normalised Total Pressure Field (Pa) b) Sound Pressure Level (dB)

The PnC crystal slab is placed in water and a point source is placed at the left side of the slab. Acoustic waves are emitted from the point source and propagate into the PnC slab with negative refractive index. As seen in Fig.6, the transmitted waves are then refocused at the right side of the slab.

4. Conclusion

It's reported that except at very high filling fraction, the triangular lattice is not suitable to realize a complete band gap. When we compare the triangular and the honeycomb PnC, we found that the complete band gaps in the honeycomb case appear at lower frequencies with respect to the triangular case. We have examined the band gap properties of triangular and honeycomb two-dimensional PnC made up of composite cylinders placed in air for sonic frequency. We have also numerically simulated the band gap properties of triangular and honeycomb two-dimensional PnC made up of tungsten cylinders immersed in water for ultrasonic frequency. The same geometric parameters have been used for an accurate comparison of the two lattices.

The honeycomb lattice demonstrates complete band gap at the lower frequency between the first and the second bands. Since, the increasing filling factor of honeycomb lattice is lower than triangular lattice and symmetry. The calculated transmission spectra and band gap are in a good agreement with the experimental results. As a practical engineering design, the proposed structure could be used to fabricate filter devices or noise isolation materials at the low-frequency sonic region. The negative refraction properties of the PnC could lead to design new types of acoustic metamaterials for example acoustic and ultrasonic cloaking [21].

References

[1] Elford D P 2010 Band gap formation in acoustically resonant phononic crystals, *Doctoral Thesis Loughborough University*

[2] Maetani N Kurose T and Tsuruta K 2010 *Memoirs of Faculty of Engineering, -Okayama University* **44** 7

[3] Glusker J P 1985 Crystal Structure Analysis2nd Edition *Oxford University Press*
[4] Brillouin L 1931 Wave Propagation in Periodic Structures *Springer-Verlag* Berlin
[5] Su M F, Olsson R H, Leseman ZC and El-Kady 2010 *Applied Physics Letters* **96** 053111
[6] Wang G Yu D, Wen J, Liu Y, Wen X 2004 *Physics Letters A* **327** 512
[7] Sheng P, Zhang X X, Liu Z, Chan C T 2003 *Physica B* **338** 201
[8] Chen L A, Wang Y S 2006 *Physica B* **392** 369
[9] Feng Z X, Sheng-Chun L, Tao Xu T, Tie-Hai W and Jian-Chun C 2010 *Chin. Phys. B* **19** (4) 044301
[10] Kafesaki M, Sigalas M M and Garcia N 2000 *Phys. Rev. Lett.* **85** 4044
[11] Khelif A B, Djafari-Rouhani J, Vasseur O, Deymier P A, Lambin Pand Dobrzynski L 2002 *Phys. Rev.B* **65** 174308
[12] Khelif A B, Djafari-Rouhani J, Vasseur Oand Deymier P A 2003 *Phys. Rev. B* **68** 024302
[13] Sun Jand, Wu T 2005 *Phys. Rev. B* **71** 174303
[14] Olsson R H, El-Kady F, Su M F, Tuck M R and Fleming J G 2008 *Sens. Actuators* **145–146** 87
[15] Bringuier S A 2011 Acoustic logic gates implemented using a phase-controlling phononic crystal *Masters Of Science The University Of Arizona*
[16] Kurose T, Tsuruta K, Totsuji Cand, Totsuji H 2009 *Materials Transactions* **50** (5) 1004
[17] Alaie S, Su M F, Goettler D F, El-Kady Leseman Z 2013 *J. Appl. Phys.* **113** 103513 doi: 10.1063/1.4790485
[18] Luo C, Johnson S, Gand Joannopoulos J D 2002 *Phys. Rev. B* **65** 201104
[19] Lu Met al. 2007 *Nat. Mater.* **6** 744
[20] Brûlé S, Javelaudv E H, Enoch S, Guenneau S 2013 Seismic metamaterial: how to shake friends and influence waves? http://arxiv.org/abs/1301.7642 consulted 31.12.2015.
[21] Chung GPhan D T 2010 *Journal of the Korean Physical Society* **57** (3) 446
[22] Torres M, Montero de Espinosa F R, Garcia-Pablos N, Dand Garcia 1999 *Phys. Rev. Lett.* **82** 3054
[23] Pendry J B 2000 *Phys. Rev. Lett* **85** 3966
[24] Sukhovich A, Jing L and Page J H 2008 *Physical Review B* **77** 014301

IOP Conf. Series: Journal of Physics: Conf. Series **1045** (2018) 012035 doi:10.1088/1742-6596/1045/1/012035

Influence of silicon oxide (SiO_2) and sulfosuccinic acid (SSA) loading on properties of poly(vinyl alcohol) (PVA) derived composite membranes.

Tomáš Remiš, Jaroslav Kadlec

New Technologies - Research Center, University of West Bohemia, Plzeň 306 14, Czech Republic

Email: tremis@ntc.zcu.cz

Abstract. New anhydrous proton conducting membranes based on poly(vinyl acohol) (PVA), tetraethyl orthosilicate (TEOS) and sulfosuccinic acid (SSA) were prepared in a single step using the solution casting method, with the aim to improve the mechanical properties of PVA membranes. SSA has been used as a sulfonating agent for crosslinking the membrane structure and also as a source of protons. TEOS has been used as crosslinking agent to increase the thermal and mechanical properties of the membranes and membranes stability in aqueous environment. In order to verify that all the substances were immobilized into the matrix, the membranes were analyzed by means of Fourier transform infrared spectroscopy (FTIR).The thermal, mechanical and rheological properties of the membranes were investigated by means of thermogravimetry (TGA), dynamic mechanical analysis (DMA) and Ares G2 rheometer. Water uptake (Wu) of composite membranes was determined. The properties were investigated for various PVA solutions and for each dried membrane. The analysis of mixed PVA solutions exhibited unique behaviour of viscosity with increasing the crosslink density. TGA and DMA measurements showed increased thermal and mechanical resistance of membranes depending on the extent of their crosslinking. Due to incorporation TEOS, the resistance of PVA membranes to the aquatic environment has increased. These properties can increase the resistance of the membranes during the processes occurring in the fuel cells.

1. Introduction

PVA membranes are used in many applications, especially in organic/inorganic separation, pervaporation, biomedical applications, catalytic processes, controlled release of drugs, etc. These membranes are very hydrophilic, which is considered an advantage for some applications, but mostly becomes as problem due to the high swelling of the membranes in the water. The properties of new proton conductive membranes based on PVA, tetraethylortosilicate (TEOS) and sulfosuccinic acid (SSA) have been determined for different stoichiometric ratios. SSA has been used as a sulfonating agent, for crosslinking the membrane structure and also as a source of protons. SiO_2 has been used to increase the thermal and mechanical properties of membranes and to reduce membrane solubility in water. Thermal and rheological properties of the membranes were investigated using thermogravimetry (TGA) and Ares G2 rheometer. The absorption of the membrane and its resistance to the water has been investigated.

IOP Conf. Series: Journal of Physics: Conf. Series **1045** (2018) 012035 doi:10.1088/1742-6596/1045/1/012035

2. Experimental

2.1. Material

PVA (hydrolyzed 99+%, M_w = 86000–98000) and SSA (70 % in water) were obtained from Sigma-Aldrich (USA). TEOS and hydrochloric acid (HCl, 37 wt %) were purchased from Merck Chemicals (Germany). All the chemicals were of analytical grade and used as received. Deionized water was utilized for all the membrane preparation experiments.

2.2. Membrane preparation

Composite PVA membranes were prepared by the solution casting method. At first, a 5 wt% solution of PVA in water was prepared by continuous stirring at 90 °C until the complete dissolution was achieved. The PVA solution was mixed with SSA for 24 h at room temperature. Further, TEOS mixture was prepared by mixing H_2O, HCL and TEOS in the molar ratio of 4:0.1:1, which was stirred at room temperature for 2 h [1] and added to the PVA-SSA solutin. The composition of each solution is specified in Table 1. The prepared solution was poured into the automatic film applicator (TQC, Germany) equipped with a plexiglass plate and was fabricated into membranes with uniform thickness and dried at 40 °C.

2.3. Characterization

FTIR reflectance spectroscopy studies were performed on the samples using NICOLET 380-FTIR spectrometer, in a spectral range of 400 – 4,000 cm^{-1}.

The rheological studies of the PVA solutions were measured on a rheometer ARES G2 from TA Instruments. The polymer solutions were inserted between two paralel plate geometries (d =25 mm). For each measurement, the shear rate from 0.1 to 100 s^{-1}, for a duration 30 s was carried out at room temperature.

TGA was used to evaluate the relative thermal stability of modified PVA membranes. The TGA curves were obtained with a TGA Q500 thermogravimetric analyser (TA Instruments, USA). Samples of about 5 mg were heated at a rate of 10 °C/min in N_2 atmosphere. The weight of the samples was measured as a function of temperature.

The temperature dependence of the storage modulus was determined by DMA using a DMA Q800 instrument (TA Instruments, USA). Membrane samples with a size of 20 x 6 mm and a thickness of about 0.1 mm were used to measure the mechanical properties. Experiments were conducted in N_2 atmosphere at frequency 1 Hz, starting from ambient temperature up to 100 °C with a heating rate of 5 °C/min.

The water uptake studies were carried outat 24, 48 and 120 h at room temperature. The membranes were placed in deionized water for specified time. The wet membranes were then wiped with a filter paper and weighted.The water uptake was calculated using the following equation:

$$W_u = \frac{W_s - W_d}{W_d}$$

where W_s and W_d are the weights of swollen and dry membranes, respectively.

IOP Conf. Series: Journal of Physics: Conf. Series **1045** (2018) 012035 doi:10.1088/1742-6596/1045/1/012035

Table 1. Composition of polymer solutions, PVA, SSA, TEOS.

Sample	**Solution mixture**	**Composition**
5PVA	5% PVA	25ml H2O, 1.25g PVA
5PVA_0.5SSA	5% PVA, 0.5%SSA	25ml H_2O, 1.25g PVA, 0.125ml SSA
5PVA_1SSA	5% PVA, 1%SSA	25ml H_2O, 1.25g PVA, 0.25ml SSA
5PVA_2SSA	5% PVA, 2%SSA	25ml H_2O, 1.25g PVA, 0.5ml SSA
5PVA_5SSA	5% PVA, 5%SSA	25ml H_2O, 1.25g PVA, 1.25ml SSA
5PVA_1TEOS	5% PVA, 1%TEOS	25ml H_2O, 1.25g PVA, 0.25ml TEOS (0.00625ml HCl, 0,0625ml H_2O)
5PVA_2TEOS	5% PVA, 2%TEOS	25ml H_2O, 1.25g PVA, 0.5ml TEOS (0.0125ml HCl, 0,125ml H_2O)
5PVA_5TEOS	5% PVA, 5%TEOS	25ml H_2O, 1.25g PVA, 1.25ml TEOS (0.032ml HCl, 0,32ml H_2O)
5PVA_10TEOS	5% PVA, 10%TEOS	25ml H_2O, 1.25g PVA, 2.5ml TEOS (0.064ml HCl, 0,64ml H_2O)
5PVA_0.5SSA_1TEOS	5% PVA, 0.5%SSA, 1%TEOS	25ml H_2O, 1.25g PVA, 0.125ml SSA, 0.25ml TEOS (0.00625ml HCl, 0,0625ml H_2O)
5PVA_0.5SSA_2TEOS	5% PVA, 0.5%SSA, 2%TEOS	25ml H_2O, 1.25g PVA, 0.125ml SSA, 0.5ml TEOS (0.0125ml HCl, 0,125ml H_2O)

3. Results and discussion

3.1. FT-IR

Figure 1 shows the FT-IR spectra of the M1 and M7 membranes. It was reported that the absorption band of ester (–COO–) appeared at 1730–1735 cm^{-1}. This results indicate the spectral changes are an evidence of a crosslinking by the esterification between –OH in PVA and –COOH in SSA. The absorption band at 2369 cm^{-1} indicate the presence of sulfonic acid group, which is formed by the introduction of SSA [2].The band at approximately 1040 cm^{-1} is assigned to stretching vibrations of Si–O–Si. The absorption band, at 900 cm^{-1} can be attributed to the stretching vibration of Si–OH or Si–O– groups. A broad absorption band, situated between 3,000 and 3,600 cm^{-1}, are assigned to O–H stretching and O–H bending vibrations, respectively [3].

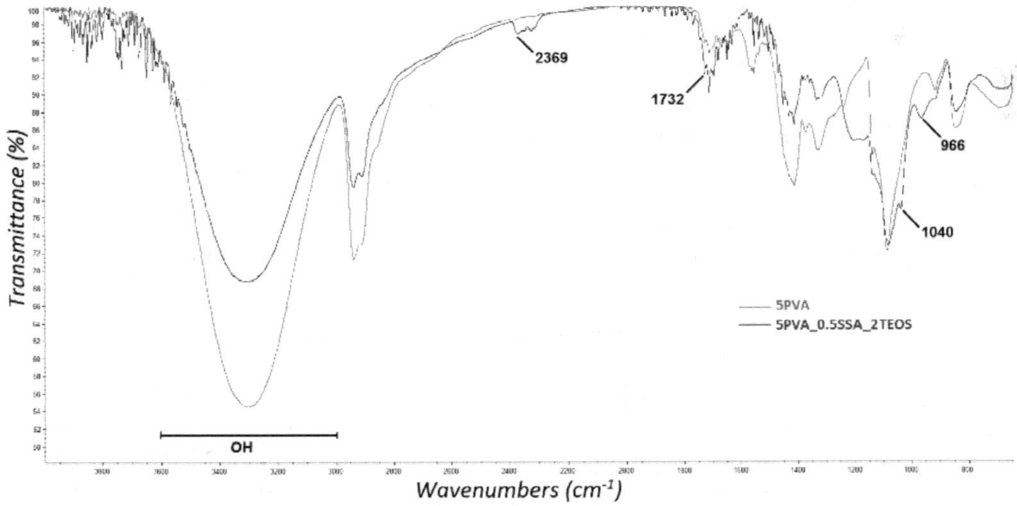

Figure 1. FT-IR spectra of composite PVA membranes

3.2. Rheology

The viscosity of the pure PVA solution changed after the incorporation of SSA (Figure 2). At low SSA content, the viscosity was initially increased due to the interaction between the acid groups of SSA and PVA. However, the viscosity was reduced slightly when the SSA loading percentage exceeded 0.5 wt%. The slight reduction in our observed viscosity can be explained by the limited dispersion behaviour of SSA in the PVA matrix. Agglomeration of particles resulted in the reduced surface area between the SSA and PVA, and caused the decrease in viscosity.

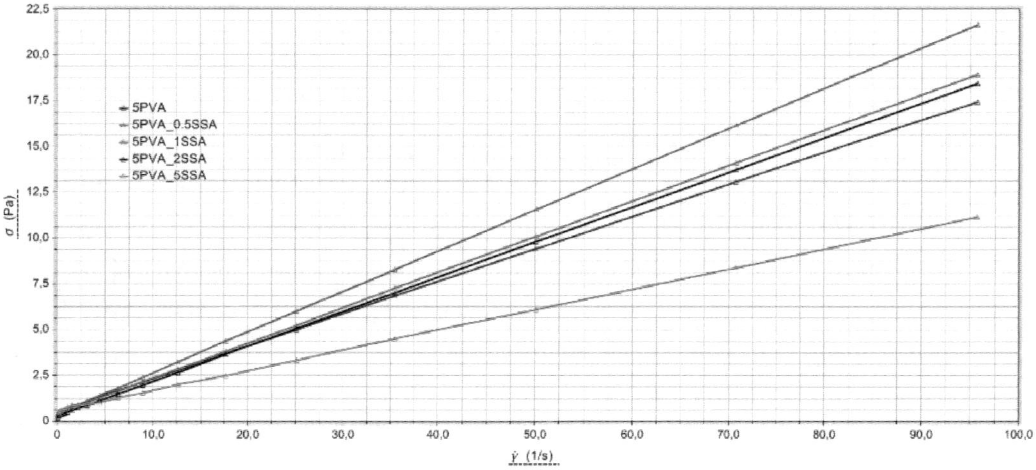

Figure 2. Rheological spectra of pure PVA solution and different SSA loadings of PVA/SSA solutions

3.3. TGA

Figure 3 shows thermogarmof composite PVA membranes. Thermal stability increase with each additional added component,which means higher crosslinking extentsfor the composite membrane with SiO_2, with SSA and with combination of SiO_2-SSA particles, respectively. This results correspondwith rheological properties of modified membranes and show better thermo-mechanical properties of membranes compared to pristine PVA membrane.

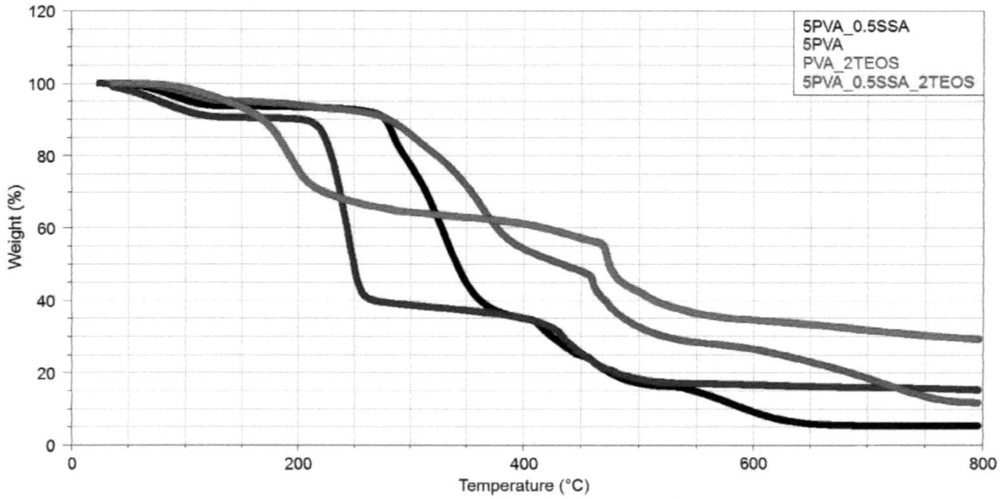

Figure 3. TG thermograms of composite PVA membranes

IOP Conf. Series: Journal of Physics: Conf. Series **1045** (2018) 012035 doi:10.1088/1742-6596/1045/1/012035

3.4. DMA

DMA analysis was performed to investigate the thermomechanical properties of the composite PVA membranes. The temperature dependence of the storage modulus was measured in the range from 30 to 100 °C. In the case of composite membranes, storage modulus increased with the increasing SiO_2 and SSA loading, respective (Figure 4). Thanks to SSA content in composite 5PVA_0.5SSA_2TEOS membrane, storage modulus decrease compared 5PVA_2TEOS membrane. This indicate stronger crosslinking extents for the composite membrane with silica particles.

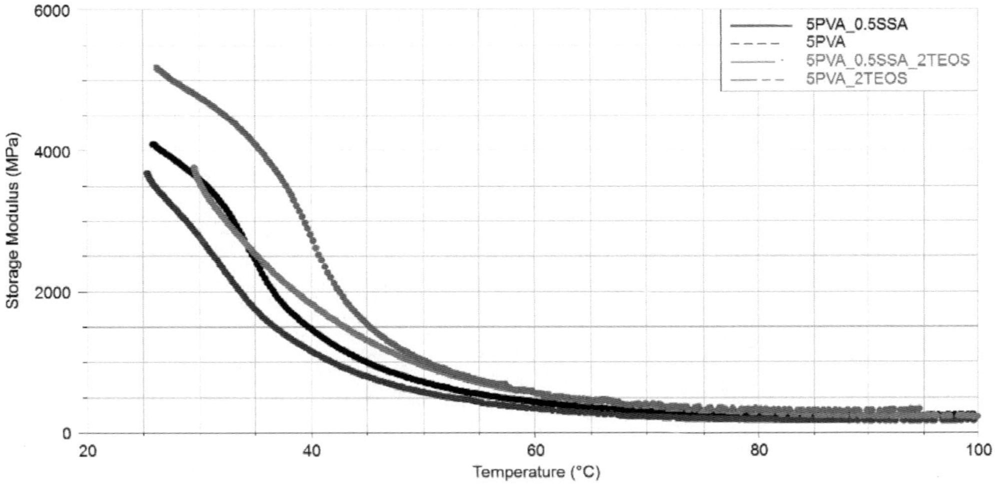

Figure 4. DMA of composite PVA membranes

3.5. Water Uptake

Figure 5 shows the decreasing ability of PVA composite membranes to absorb water depending on the increasing SiO_2 content, while increasing the water resistance of PVA-SiO_2 membranes. SSA increases W_u of composite membranes.

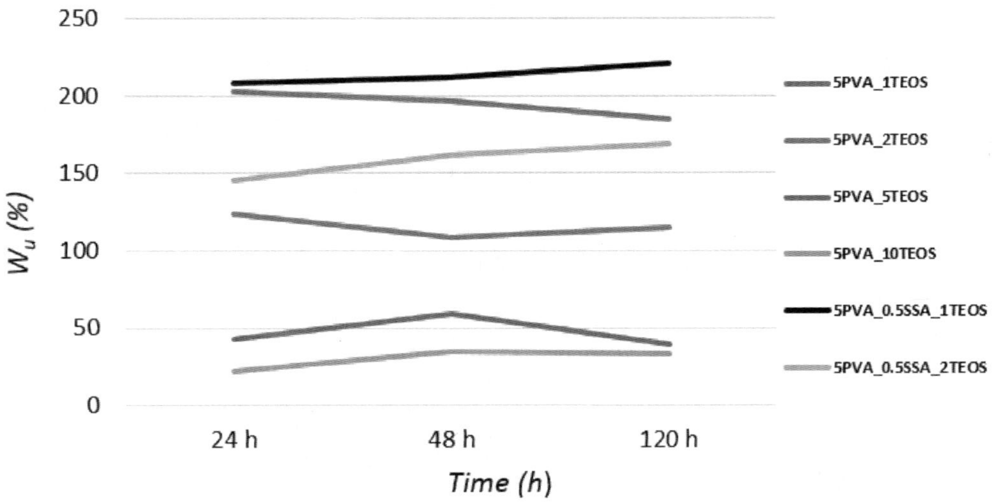

Figure 5. Water uptake of PVA-SiO_2 and PVA-SiO_2-SSA membranes

4. Conclusion

The results of this study can by useful for next research of composite PVA membranes. Rheological studies show that ideal motion of used SSA is 0.5 wt%. Incorporation of SiO_2 in the PVA matrix increases the mechanical and heat resistance of membranes and water resistance of PVA membranes. SSA increases $W_u$of PVA membranes. Advantageous is a combination of SiO_2 and SSA in the PVA matrix.

Acknowledgements

The result was developed within the CENTEM project, reg. no. CZ.1.05/2.1.00/03.0088, cofounded by the ERDF as part of the Ministry of Education, Youth and Sports OP RDI program and, in the follow-up sustainability stage, supported through CENTEM PLUS (LO1402) by financial means from the Ministry of Education, Youth and Sports under the "National Sustainability Program I.

References

[1] Kim, Dae, et al 2007 *Macrom. Res.* **15**
[2] Rhim, Ji-Won, et al 2004 *J. Memb. Sci.* **238** 143
[3] Katumba G, Mwakikunga B W, and Mothibinyane T R 2008 *Nanosc. Res. Lett.* **3** 421

IOP Conf. Series: Journal of Physics: Conf. Series **1045** (2018) 012036 doi:10.1088/1742-6596/1045/1/012036

The effect of milled basalt fibers addition on rheological and mechanical properties of alkali activated binder

David Rieger, Tomáš Kovářík, Michal Pola, Petr Franče and Jaroslav Kadlec

New Technologies - Research Centre, University of West Bohemia, Univerzitni 8, 306 14 Pilsen, Czech Republic

Email: DavidPJ@ntc.zcu.cz

Abstract. In this study, the effects of milled basalt fibers addition on rheological behaviour of alkali activated geopolymer binder were determined. This binder was synthesized by alkaline activation of calcined clay stone powder and milled blast furnace slag by potassium silicate solution. Milled basalt fibers as filler were used in range of 0-20 wt% addition to powder part of binder and the effects on rheological and mechanical properties were evaluated. The rheological properties were determined in accordance to flow properties by measurements on Ford viscosity cup and by strain controlled small amplitude oscillatory rheometry measurements of hardening process. The final properties were documented by measurements of flexural strength after 28 days and by scanning electron microscopy. The results indicate that milled basalt fibers addition has effect on flow properties of binder. The flow time increased three times and the complex viscosity is also coherently increased. The reinforcement led to two times higher values of flexural strength and the best figure of merit lied in the range of 1-5 wt% addition of milled basalt fibers.

1. Introduction

Inorganic polymer cements can be synthesized by alkali-activation of a variety of materials including thermally activated clays and coal fly ashes to produce material with mechanical and thermal properties suitable for wide range of industrial applications [1, 2]. Davidovits entitled this type of materials as "geopolymers" because of polymer like chemical structure and introduced pioneering work on alkali-activated binders based on calcined clays [3]. Geopolymers are products of chemical reaction between aluminosilicate material and liquid alkaline environment where chemical cleavage of the Si-O and Al-O bonds in parent material leads to saturation of liquid solution and subsequent polycondensation of amorphous aluminosilicate matrix [4]. In some cases slag is used as an additive for improvement of setting time and mechanical strength [5, 6].

However, by ceramic-like nature have brittle fracture character [7]. Using of fibers can change the character of fracture into quasi-brittle and prevent the crack propagation in material. Wide range of fibers is used for this purpose from organic natural and synthetic materials as PVC, PP, PVA and cellulose [7-9], inorganic carbon-based materials and mineral materials as basalt, glass, zirconia and others [10-12]. Fiber reinforcement is mainly used in form of short fibers for casting processes or as woven and non-woven composites prepared by impregnation.

This experimental research focuses on rheological and mechanical properties of geopolymeric matrix reinforced by milled basalt fibers. The binder comprises of calcined clay stone powder and milled blast furnace slag. Milled basalt fibers were applied in range of additional 0-20 wt% related to powder partand

IOP Conf. Series: Journal of Physics: Conf. Series **1045** (2018) 012036 doi:10.1088/1742-6596/1045/1/012036

the effect on flow properties and final mechanical strength was evaluated. Furthermore, the hardening process of individual compositions was evaluated by the small amplitude oscillatory rheometry measurements [13, 14].

2. Materials

The mixture of calcined clay stone powder and milled blast furnace slag (CCS) was used as binder. This binder was activated by potassium silicate solution with silicate module 1.61.These materials were supplied by České lupkové závody a. s, Czech Republic and their chemical composition is presented in Table 1. The basalt fibers were provided by Kamenny Vek, Russian Federation. The diameter of fibers was 14 μm and fibers were milled for 30 seconds in laboratory vibrational mill to final length. Figure 1 and Figure 2 represents morphology of binder and milled fibers respectively taken by scanning electron microscope (SEM). The overall granularity curves measured by laser light scattering are presented on insets in individual figures.

Table 1. Chemical composition of raw materials [mass %]

Composition	SiO_2	Al_2O_3	CaO	MgO	TiO_2	Fe_2O_3	K_2O	LoI	Others
CCS	46.37	28.72	15.46	3.40	2.01	0.93	0.56	1.92	0.63
Activator	17.60	-	-	-	-	-	17.13	65.27	-

Figure 1. SEM image and distribution of binder

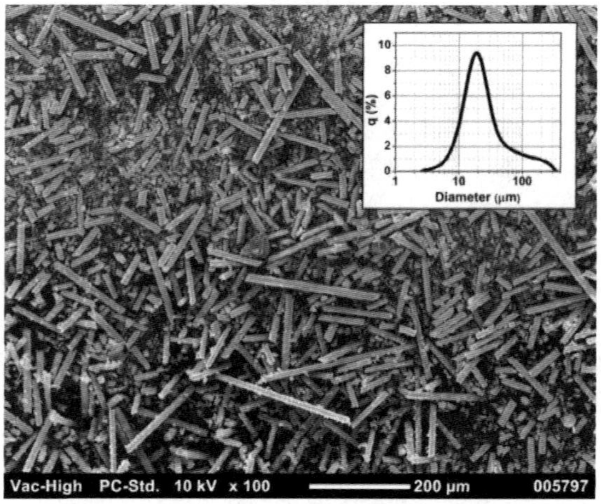

Figure 2. SEM image and distribution of milled basalt fibers

3. Procedures and results

3.1 Preparation of pastes

Binder pastes were prepared by mixing of 100 weight parts of binder powder together with 80 weight parts of activator solution. Subsequently, different amount of fibers in the range of 0 – 20 wt% was added into the paste and mixed together for 5 minutes in laboratory vacuum mixer. Pastes with different content of fibers were casted into moulds with dimensions of 120x20x20 mm. After 2 days were specimens unmoulded and aged for 28 days in polypropylene bags.

3.2 Rheological measurements

In this study the complex viscosity of prepared pastes in the beginning of hardening process was determined by time-resolved oscillatory rheometric measurements with small amplitude on a strain controlled rheometer TA Instruments Ares G2 in plane-plate geometry of 40 mm in diameterwith frequency of 1 rad/s and strain controlled amplitude of 0.01 %. Measurements were performed after

15 minutes from beginning of mixing. The effect of fibers addition can be clearly seen in Figure 3 where individual curves represent increase of complex viscosity in early age of pastes hardening.

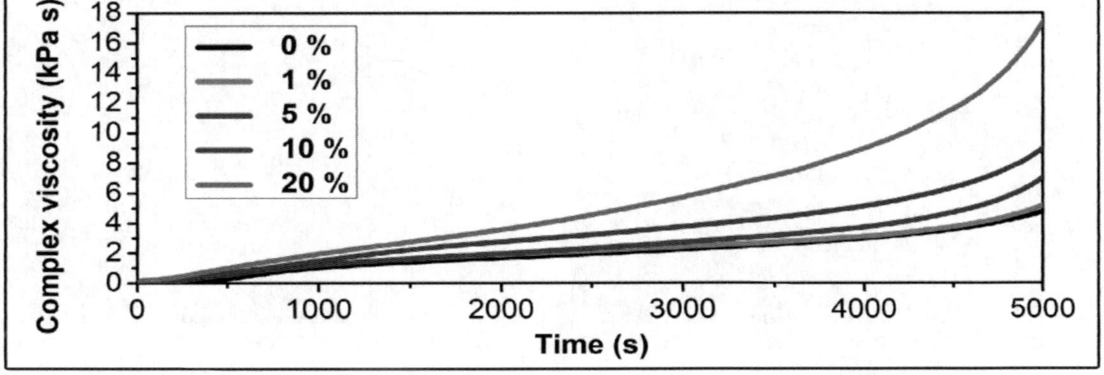

Figure 3. Complex viscosity at beginning of hardening process for individual fiber contents

As a quantification of workability the flow properties were determined on fresh pastes immediately after mixing by measurements on Ford viscosity cup with diameter of 6 mm. The results can be seen in Figure 4 described as time required to flow. It can be compared with results of complex viscosity measurements and in both cases the increasing content of fibers increases viscosity. The flow time is increased approximately three times over the investigated range. This trend is the same in measurements of complex viscosity, but this effect is not so noticeable because of different conditions of measurement when the small amplitude and frequency are applied. However, it can be concluded that even at highest investigated filling is retained good workability of paste.

3.3 Mechanical properties and microstructure

The flexural strengths of individual compositions were measured after 28 days on 120x20x20 mm samples by 3 point bending test with support span of 50 mm. Results can be seen in Figure 5 and at 20 wt% the flexural strength reaches twice as high value of unfilled binder.However, the flexural strength increasing with lower slope than time to flow through viscosity cup and thus the best figure of merit can be found in between of 1 and 5 wt% composition. The microstructure of fracture surfaces was evaluated by SEM and pure hardened paste without addition of fibers can be seen in Figure 6.

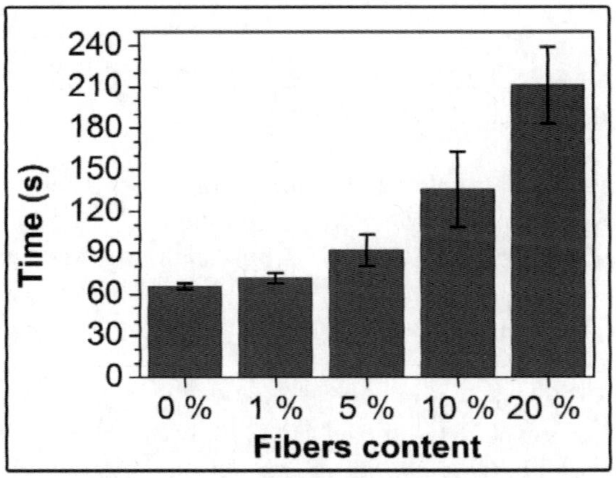

Figure 4. Time toflow through 6 mm Ford viscosity cup

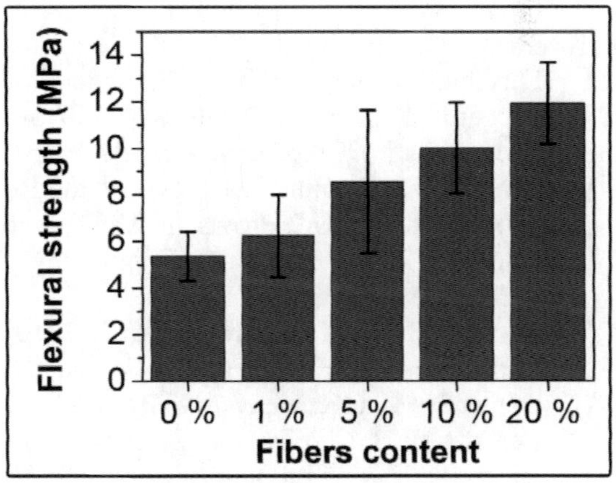

Figure 5. Flexural strength of composites after 28 days

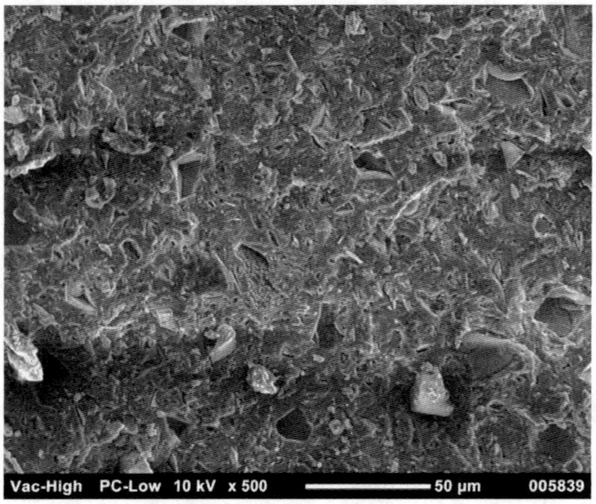

Figure 6. SEM image of hardened paste without fibers addition.

Figure 7. SEM image of hardened paste with 20 wt% of basalt fibers.

The brittle fracture surface can be seen with irregular holes after unreacted slag particles. The composition with 20 wt% of basalt fibers is presented in Figure 7 with marked uncovered fibers and imprints on fracture surface. Thus it can be concluded that reinforcing effect is predominantly produced by the pull out mechanism.

4. Conclusions

It can be summarized that with increasing content of basalt fibers in geopolymer paste are rising both the flexural strength and viscosity. At 20 wt% addition of basalt fibers rise the flexural strength 2 times and the flow time 3 times in contrast to values of unfilled paste and the complex viscosity follows the same trend. However, at highest investigated filling is still retained good workability of binder. From fracture progress can be concluded that the reinforcing effect of milled basalt fibers in binder is predominantly produced by the pull out mechanism.

Acknowledgments

The result was developed within the CENTEM project, reg. no. CZ.1.05/2.1.00/03.0088, cofounded by the ERDF as part of the Ministry of Education, Youth and Sports OP RDI programme and, in the follow-up sustainability stage, supported through CENTEM PLUS (LO1402) under the NSP I.

References

[1] Duxson P, Provis J L, Lukey G C and Van Deventer J S J 2007 *Cem Concr Res* **37** 1590-97
[2] Rieger D, Kullová L, Čekalová M, Novotný P, Pola M 2017 *J Phys Conf Ser* **790**012030
[3] Davidovits J 1991 *J Therm Anal* **37** 1633–56
[4] Kouassi S S, Tognonvi M T, Soro J and Rossignol S 2011 *J Non-Cryst Solids* **357** 3013-21
[5] YipChK, Lukey GC, Provis JL, Van Deventer JSJ 2008 *Cem Concr Res* **38**554-564
[6] Kadlec J, Rieger D, Kovářik T, Novotný P, Franče P, Pola M 2017 *IOP Conf Ser Mater Sci Eng* **175** 012052
[7] Toutanji H, Xu B, Gilbert J, Lavin T 2010 *Const Build Mater* **24** 1
[8] Yan L, Kasal B, Huang L 2016 *Composites Part B* **92** 94
[9] Alzeer M, MacKenzie K 2013 *Appl Clay Sci* **7 5–76** 148
[10] Peigang He SY, Jia D, Yang Z, Duan X, Wang S, Zhou Y 2016 *Ceram Int* **42** 7837
[11] Rieger D, Kadlec J, Pola M, Kovářik T, Franče P 2017 *IOP Conf Ser Mater Sci Eng* **175** 012054
[12] Pernica D, Reis P N B, Ferreira J A M et al. 2010 *J Mater Sci* **45** 744
[13] Steins P, Poulesquen A, Diat O and Frizon F 2012 *Langmuir* **28** 8502
[14] Rieger D, Kullová L, Čekalová M, Kovářík T 2017 *J Phys Conf Ser* **790** 012029

IOP Conf. Series: Journal of Physics: Conf. Series **1045** (2018) 012037 doi:10.1088/1742-6596/1045/1/012037

Comparison of hardening kinetics by small amplitude oscillatory rheometry in the case of alkali activated kaolinitic clays

David Rieger, Michal Pola, Petr Frančе, Jaroslav Kadlec and Tomáš Kovářík

New Technologies - Research Centre, University of West Bohemia, Univerzitni 8, 306 14 Pilsen, Czech Republic

Email: DavidPJ@ntc.zcu.cz

Abstract. In this study, the rheological behaviour of geopolymeric inorganic binders were determined. These binders were synthesized by alkaline activation of aluminosilicate material. Five kaolinite clays were chosen for investigation and as an alkaline activator of hardening process, the potassium silicate solution with silicate module of 1.61 was used. For increase of reactivity, the kaolinite clays were calcined by heat treatment at 600 °C for 4 hours and converted to metakaolin state. The particle size of individual materials was determined by measurement of laser light scattering. The rheological properties were determined in accordance to flow properties by measurements on Ford viscosity cup and by strain controlled small amplitude oscillatory rheometry measurements at 30 °C isotherm and the hardening process evolution was documented by measurement of complex viscosity. The effects of granularity and accessory minerals on flow and hardening properties are discussed.

1. Introduction

The inorganic polymeric binders, Davidovits introduced as a pioneering work on alkali-activated binders based on calcined clays [1,2] and entitled them as "geopolymers" because of theirs polymer like structure. Geopolymers are products of chemical reaction between aluminosilicate material and liquid alkaline environment where chemical cleavage of the Si-O and Al-O bonds in parent material leads to saturation of liquid solution and subsequent polycondensation of amorphous aluminosilicate matrix [3-5]. Geopolymers can be synthesized from thermally activated clays and coal fly ashes and can be modified by slags with consideration of mechanical and rheological properties [6,7].

In this experimental research we examined rheological properties of different kaolinitic clays. Five clays were selected from local resources and calcined to metakaoline state at 600 °C [8-10]. The flow properties and early age of consolidation process were documented after initiation of geopolymerizaton process by measuring through flow time on Ford viscosity cup and by the small amplitude oscillatory rheometry measurements respectively [10-12].

2. Materials

Five selected kaolinite clays were examined as a binder. These materials were supplied by Sedlecky kaolin a. s, Czech Republic and chemical compositions of individual materials are presented in Table 1 according to data provided by supplier. Obtained materials were thermally activated by transformation of kaolinite content into the metakaoline state. The heat treatment comprised of heating to the 600 °C

IOP Conf. Series: Journal of Physics: Conf. Series **1045** (2018) 012037 doi:10.1088/1742-6596/1045/1/012037

by the rate of 3 °C/min and of dwell time of 4 hours. After the dwell time the material was cooled to room temperature naturally.

For alkaline activation of geopolymerization reaction an aqueous solution of potassium silicate with silicate module 1.61 was used. The morphology of individual clays after heat treatment taken by scanning electron microscope (SEM) are presented in Figures 1-5. The overall granularity curves were measured by laser light scattering and are presented in Figure 6. The main accessory mineral is mica in the case of T-79, MK and MK3 clays and in the case of Sedlec Ia and OT 82 it is combination of mica and quartz minerals.

Table 1. Chemical composition of raw materials [mass %]

Composition	SiO_2	Al_2O_3	CaO	MgO	TiO_2	Fe_2O_3	K_2O	LoI
SedlecIa	47	37	0.3	0.2	0.2	0.8	0.9	13
T 79	47	37	0.3	0.3	0.2	1.2	1.2	13
OT 82	46	37	0.3	0.2	0.7	1.3	0.8	13
MK	50	34	0.1	0.2	0.5	1.5	1.6	11
MK 3	48	36	0.2	0.2	1.5	1.6	1.0	12
Activator	17.6	-	-	-	-	-	17.1	65.3

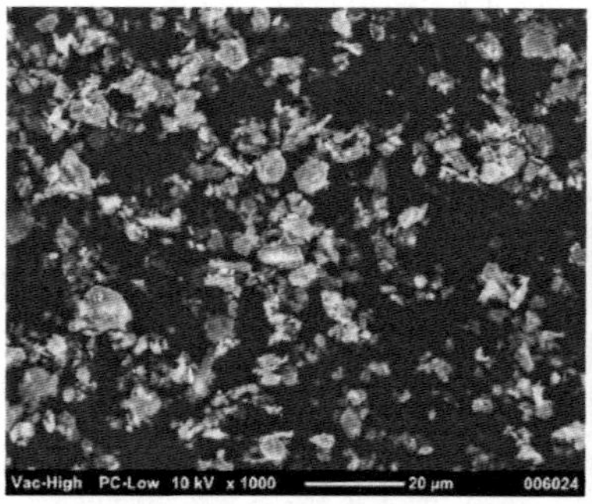

Figure 1. SEM image of calcined clay Sedlec Ia.

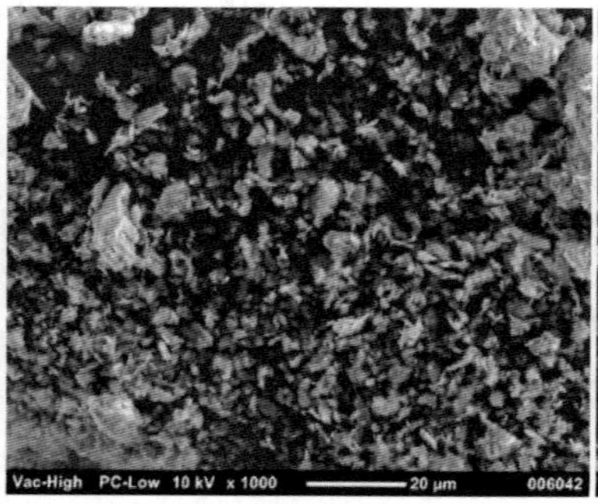

Figure 2. SEM image of calcined clay T 79.

Figure 3. SEM image of calcined clay OT 82.

Figure 4. SEM image of calcined clay MK.

Figure 5. SEM image of calcined clay MK 3.

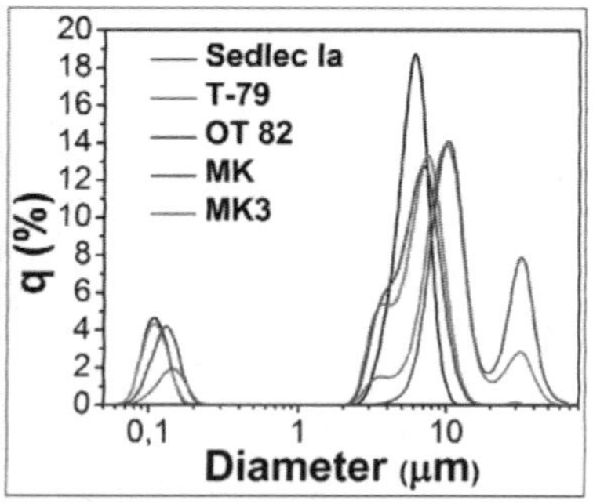

Figure 6. Particle distributions of individual calcined clays.

IOP Conf. Series: Journal of Physics: Conf. Series **1045** (2018) 012037 doi:10.1088/1742-6596/1045/1/012037

3. Procedures and results

3.1. Preparation of individual binders
In the case of individual materials after thermal treatment the geopolymerization process was activated by mixing of 100 weight parts of metakaolin powder with 100 or 120 weight parts of potassium silicate activator in laboratory vacuum mixer for period of 5 minutes and then subjected to analysis.

3.2. Flow properties
The flow properties were determined immediately after mixing by measurements on Ford viscosity cup with diameter of 6 mm and the results can be seen in Figure 7 and 9 described as time required to flow through the viscosity cup. In the case of 100:100 weight percent compositions only the materials MK and MK 3 exhibited measurable flow. It is because of considerably higher amount of coarse particles and lower specific surface. However, MK material exhibited slightly higher viscosity which can be attributed to higher content of mica particles around 30 μm. In relation, all 100:120 compositions exhibited measurable flow where MK and MK 3 materials exhibited lowest viscosity. It can be concluded that amount of potassium silicate solution was sufficient for liquefaction of all studied materials. The effect of different mineral composition can be deduced from different viscosities of T-79 and OT 82 materials. The main accessory mineral of T-79 is mica and the viscosity is almost twice as high as of OT 82 which contains also quartz. Although granularities are almost identical the different shape of accessory minerals strongly influences the flow properties [13].

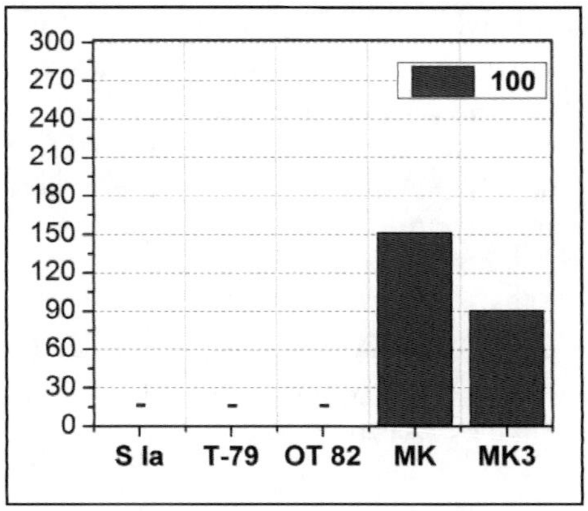

Figure 7. Time to flow of mixture 100:100 through 6 mm Ford cup.

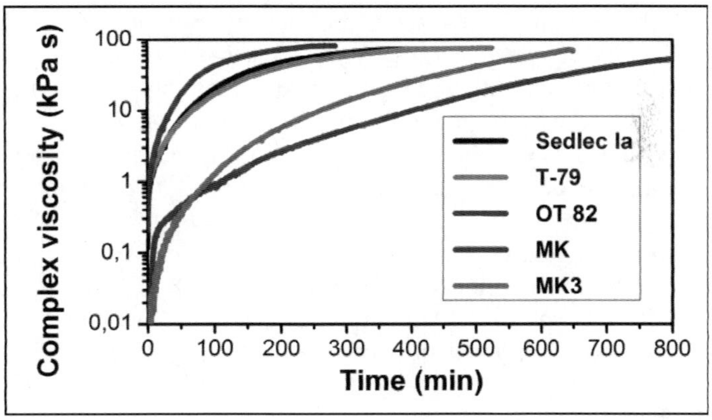

Figure 8. Complex viscosity at beginning of hardening process for mixtures 100:100.

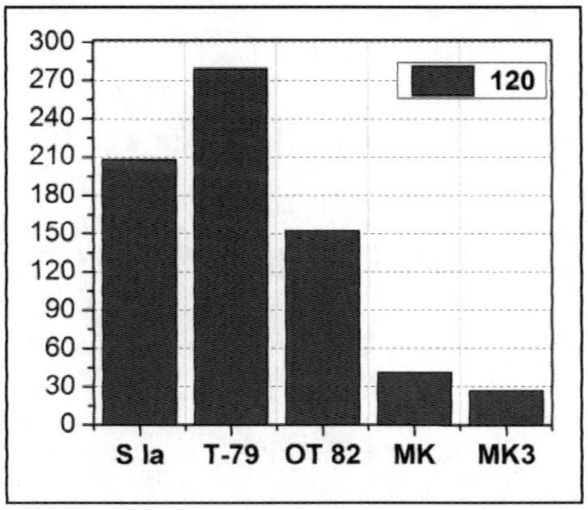

Figure 9. Time toflow of mixture 100:120 through 6 mm Ford cup.

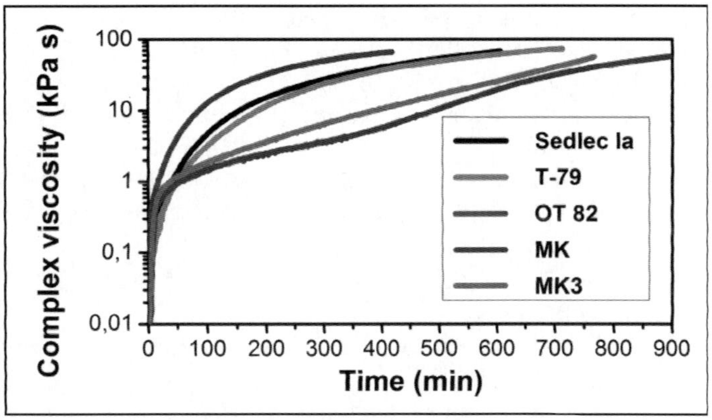

Figure 10. Complex viscosity at beginning of hardening process for mixtures 100:120.

3.3. Complex viscosity measurements

Changes of complex viscosity in the beginning of hardening process was determined by time-resolved oscillatory rheometric measurements with small amplitude on a strain controlled rheometer TA Instruments Ares G2 in plane-plate geometry of 40 mm in diameter. Measurements were performed after 15 minutes from beginning of mixing. The radial velocity of 10 rad/sec and strain 0.01% were applied to ensure a minimal influence on the hardening process. The results of 100:100 and 100:120 compositions expressed as complex viscosity can be seen in Figure 8 and 10 respectively. The changes of complex viscosity measured by low strain oscillations describe two processes. In the early time of measurement the particles in suspension come to equilibrium position after high shear mixing procedure and it contributes to fast increase of initial viscosity. In the later time, the subsequent rising of viscosity can be attributed to dissolution and polycondensation process which leads to hardening of mixture. Comparing Figure 8 and 9, it can be clearly seen that increasing content of activator solution leads to delay in saturation and polycondensation process, because of increasing the liquid to solid ratio. In both concentrations the MK and MK 3 materials exhibited lower reactive rate due to lower content of reactive metakaolin where it can be expected that mica under these conditions do not contribute to the geopolymerization process [14, 15]. From results the OT 82 can be evaluated as the most reactive material in this study and the reactivity of T-79 and Sedlec Ia is almost the same.

4. Conclusions

In this study the effect of mineral composition and granularity on geopolymerization process and flow properties of prepared suspensions were investigated in the case of five calcined kaolinite clays. It can be concluded that content of coarse particles in clays have beneficial effect on flow properties and reduction of viscosity and it can be attributed mainly to the content of accessory minerals which are in this study mica and quartz. On the contrary, these minerals do not contribute to the geopolymerization reaction and to the hardening process. In the case of this study the clay marked OT 82 represents best compromise between hardening kinetics and flow properties in the range of investigated conditions.

Acknowledgments

The result was developed within the CENTEM project, reg. no. CZ.1.05/2.1.00/03.0088, cofounded by the ERDF as part of the Ministry of Education, Youth and Sports OP RDI programme and, in the follow-up sustainability stage, supported through CENTEM PLUS (LO1402) under the NSP I.

References

[1] Davidovits J 1991 *J Therm Anal* **37** 1633
[2] Davidovits J 1982 *U.S. Patent* 4, 349, 386
[3] Kouassi S S, Tognonvi M T, Soro J and Rossignol S 2011 *J Non-Cryst Solids* **357** 3013
[4] Shi C, Fernández-Jiménez A, Palomo A 2011 *Cem Concr Res* **41** 750
[5] Šesták J and Foller B 2012 *J Term Anal Calorim* **109** 1
[6] Rieger D, Kullová L, Čekalová M, Novotný P, Pola M 2017 *J Phys Conf Ser* **790** 012030
[7] Kadlec J, Rieger D, Kovářik T, Novotný P, Franče P, Pola M 2017 *IOP Conf Ser Mater Sci Eng* **175** 012052
[8] Ilić B R, Mitrović A A, Miličić L R 2010 *Hem Ind* **64** 351
[9] Kovářík T, Bělský P, Novotný P, Říha J, Savková J, Medlín R, Rieger D and Holba P 2015 *Const Build Mater* **80** 98
[10] Rieger D, Kovářík T, Říha J, Medlín R, Novotný P, Bělský P, Kadlec J and Holba P 2015 *Const Build Mater* **83** 26
[11] Steins P, Poulesquen A, Diat O and Frizon F 2012 *Langmuir* **28** 8502-10
[12] Rieger D, Kullová L, Čekalová M, Kovářík T 2017 *J Phys Conf Ser* **790** 012029
[13] Nosrati A, Addai-Mensah J, Skinner W 2011 *Chem Eng Sci* **66** (2) 119
[14] Vigil de la Villa R, García R, Martínez-Ramírez S, Frías M 2017 *App Clay Sci* **150** 1
[15] Lammers K, Smith M M, Carroll S A 2017 *Chem Geol* **466** 149

Computer Modeling of Intervertebral Disk Endoprosthesis

Yu A Rikun[2],T V Kolmakova[1, 2], S P Buyakova[1, 2] and S N Kulkov[1, 2]

[1]Institute of Strength Physics and Materials Science, Tomsk, Russia
[2]Tomsk State University, Tomsk, Russia

E-mail: yulirik.93@mail.ru

Abstract. The article describes the model of the intervertebral disc endoprosthesis. The algorithm for constructing the geometric model of the intervertebral disc prosthesis is developed in the ANSYS programming language APDL, which allows to automatically rebuilding the shape and size of the prosthesis depending on the size of the vertebrae and the required intervertebral spaces of different parts of the spine of particular individual. The geometric model of the intervertebral disc prosthesis is based on the experimental sample of a ceramic endoprosthesis developed at the Institute of Strength Physics and Materials Science SB RAS.

1. Introduction

One of the most important constructions of the human body is the spine. Its structure allows you to perform the functions of support and movement. The spine is divided into cervical, thoracic, lumbar and sacral regions [1, 2]. An intervertebral disc (IVD) separates each vertebra, except in the upper cervical spine (C1 & C2), and in the sacrum and coccyx, where the vertebrae are fused together [3]. The discs resists spinal compression, permits limited bending, twisting, and sliding between vertebral bodies [2].

Degenerative changes of the intervertebral discs are the major cause of pain in the spine and neck in those of the middle and elder age [4]. A radical method of treatment in such cases is the replacement of the intervertebral disc with an implant. Adequate development and individual selection of implants play a decisive role in the treatment of a person. Improper selection of materials and design of implants can lead to deterioration of the bone tissue and the functioning of the spine as a whole. Ceramics are identical to the inorganic bone matrix in the type of chemical bond. Traditionally, brittleness, low mechanical fracture toughness and low resistance to impact have limited the applications of the ceramic materials. Nevertheless, a strong interest in the use of ceramics for biomedical engineering applications has been developed from the end of the 20th century. New ceramics [5-9], with very improved properties, contributed to increasing the possibilities of using ceramics as osteoimplants.

The problems of mechanics arising in the creation and selection of implants of biological tissues are solved on the basis of studying the structure, mechanical behavior and properties of the biological tissues themselves and their interaction with implants. The use of computer modeling methods makes it possible to more deeply understand the patterns of functioning of the spine in norm, at pathologies, at interaction with the implant, which facilitates the selection of a suitable prosthesis for a particular individual.

The paper considers one of the steps on the way to solving the important scientific problem of choosing the individual intervertebral disc prosthesis witch is mechanically compatible with organism,

IOP Conf. Series: Journal of Physics: Conf. Series **1045** (2018) 012038 doi:10.1088/1742-6596/1045/1/012038

namely the construction of a geometric model of the intervertebral disk endoprosthesis with the purpose of the subsequent aided study of the prosthesis behavior in the spine segment under physiological loads and the study of its influence on the stress-strain state of segment.

2. Materials and methods

The algorithm for constructing the geometric model of the intervertebral disc prosthesis is developed in the ANSYS programming language APDL. The geometric model of the intervertebral disc prosthesis is based on the experimental sample of a ceramic endoprosthesis developed at the Institute of Strength Physics and Materials Science SB RAS.

Figure 1 shows the geometrical model of the intervertebral disc prosthesis built in the ANSYS. The Z axis is directed along the axis of the prosthesis (spine axis), the Y axis in the anteroposterior direction, the X axis in the lateral direction. Geometrical model includes the two opposing plates: upper plate and lower plate. The upper and lower plates contain elements forming a pair of conjugation (Figure 2). It is assumed that the upper plate of the prosthesis will be attached to the lower surface of the upper vertebra, and the lower one to the upper surface of the lower vertebra respectively.

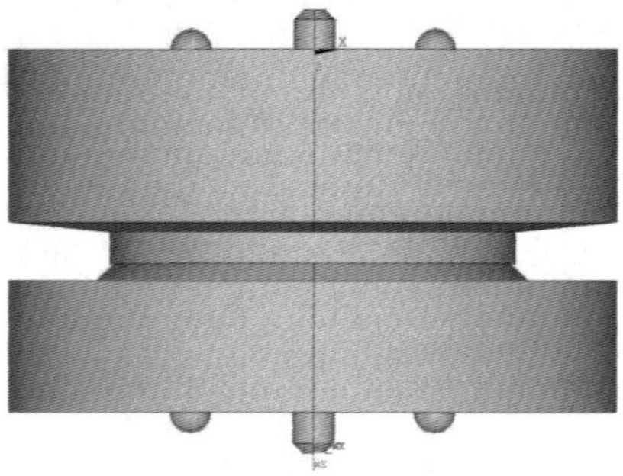

Figure 1. Geometrical model of the intervertebral disc prosthesis (front view)

(a) (b)

Figure 2. Geometric models of the structural components of the model prosthesis of the intervertebral disc: (a) the upper plate of the prosthesis, (b) the lower plate of the prosthesis

The developed algorithm allows to automatically rebuilding the shape and size of the prosthesis depending on the size of the vertebrae of different parts of the spine of a particular individual and the size of the required intervertebral spaces. The shape of the upper and lower surfaces of the opposing prosthesis plates accordingly can vary from round to ellipsoidal. Figure 3 shows the geometrical model of the intervertebral disc prosthesis with ellipsoidal shape of the upper and lower planes of the prosthesis plates.

The normal spine has three natural curves, the cervical (neck) curve, the thoracic (middle back) curve, and the lumbar (lower back) curve. The cervical and lumbar sections curve forward (lordosis), while the thoracic section curves backward (kyphosis). This curvature allows even distribution of weight and the withstanding of the applied loads [2]. The presence of lordosis and kyphosis of the spine provides a different height of the intervertebral space in front and behind, occupied by the intervertebral disc, which must be replaced with an implant.

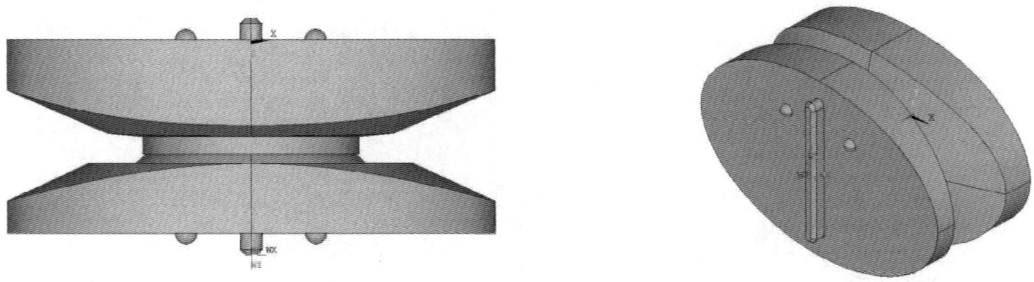

Figure 3. Geometrical model of the intervertebral disc prosthesis with ellipsoidal shape of the upper and lower planes of the prosthesis plates

For example, Table 1 and Table 2 present the anterior and posterior heights of the intervertebral disc of the cervical and thoracic spine measured by different authors [3, 10-13] and the ratio of these lengths.

Table 1. Intervertebral disc heights of the cervical spine

IVD heights	C2-3	C3-4	C4-5	C5-6	C6-7	C7-T1	References
Anterior l_1 (mm)	4.8	5.3	5.5	5.4	5.2	4.7	
Posterior l_2 (mm)	3.4	3.3	3.0	3.0	3.3	3.5	[3, 10]
l_1/l_2	1.4	1.6	1.83	1.8	1.58	1.34	
Anterior l_1 (mm)	5.2	5.3	5.2	4.6	4.9	-	
Posterior l_2 (mm)	3.4	3.4	3.7	3.9	4.3	-	[3, 11]
l_1/l_2	1.53	1.56	1.41	1.18	1.14	-	
Anterior l_1 (mm)	4.67	4.67	3.71	3.38	2.84	-	
Posterior l_2 (mm)	2.91	2.47	2.34	2.13	1.78	-	[12]
l_1/l_2	1.6	1.89	1.59	1.59	1.6	-	

Table 2. Intervertebral disc heights of the thoracic spine [13]

Disk location	Anatomical measurements		l_1/l_2	Radiographic measurements		l_1/l_2
	Anterior l_1 (mm)	Posterior l_2 (mm)		Anterior l_1 (mm)	Posterior l_2 (mm)	
C7-T1	4.5	4.5	1.0	3.96	3.20	1.24
T1-2	4.5	4.3	1.05	3.69	3.04	1.21
T2-3	3.4	3.5	0.97	3.23	1.95	1.66
T3-4	3.3	3.2	1.03	3.07	2.09	1.47
T4-5	3.0	3.3	0.9	2.85	2.06	1.38
T5-6	3.5	3.6	0.97	3.36	2.10	1.6
T6-7	4.1	4.1	1.0	3.7	2.18	1.69
T7-8	4.2	3.6	1.16	4.06	2.47	1.64
T8-9	5.6	5.0	1.12	4.52	2.49	1.82
T9-10	5.4	4.2	1.29	4.88	2.97	1.64
T10-11	7.2	5.8	1.24	5.91	3.66	1.61
T11-12	6.0	4.8	1.25	5.9	3.23	1.83

The developed algorithm for constructing the geometric model of the intervertebral disc prosthesis allows to take into account the ratio of the anterior and posterior heights of the intervertebral space. Figure 4 shows the geometrical model of the intervertebral disc prosthesis with l_1/l_2=1.5.

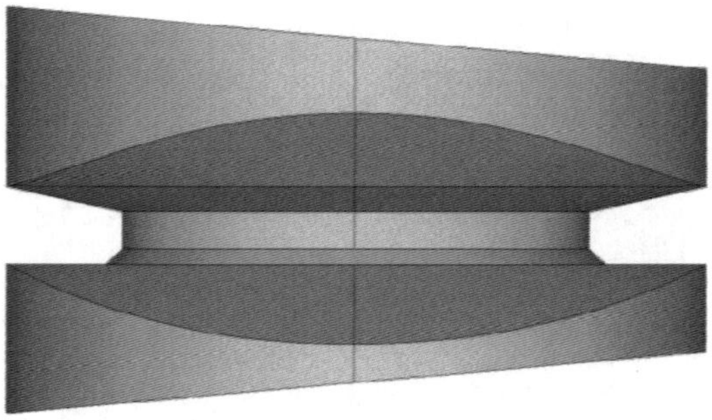

Figure 4. Geometrical model of the intervertebral disc prosthesis with various anterior height l_1 and posterior height l_2 (side view, l_1/l_2=1.5)

3. Conclusion

An algorithm for constructing the geometric model of intervertebral disk prosthesis in the ANSYS programming language APDL is developed. The algorithm allows to automatically rebuild the shape and size of the prosthesis depending on the sizes of the vertebrae of different parts of the spine of a particular individual and the size of the required intervertebral spaces. The geometric model of the intervertebral disc prosthesis is based on the experimental sample of a ceramic endoprosthesis developed at the Institute of Strength Physics and Materials Science SB RAS.

The constructed computer model is designed to study the behavior of the prosthesis in the segment of the spine under physiological loads and to study its effect on the stress-strain state of the spinal segment.

References

[1] Bendoukha M, Mosbah M 2017 *Épitőanyag-JSBCM* **69** (4) 122 http://dx.doi.org/10.14382/epitoanyag-jsbcm.2017.22
[2] Latif M J A 2011 Characterisation and Modelling of Spinal Facet Joints *Ph.D. thesis* The University of Leeds 336
[3] Panzer M B 2006 Numerical modeling of the human cervical spine in frontal impact *Master thesis* Waterloo, Ontario, Canada 248
[4] Wang Y, Chen H, Zhang L, Liu J, and Wang Z 2012 *Chin. J. Traumatol.* **15** 67
[5] Gömze A L 2016 Applied materials science I. Compilation of Selected Scientific Papers *IGREX Ltd*, Hungary 189
[6] Gömze A L and Gömze N L 2017 *IOP Conf. Ser.: Mater. Sci. Eng* **175** 012001 https://doi.org/10.1088/1757-899X/175/1/012001
[7] Kurovics E and Gömze A L 2017 *MultiScience - XXXI. microCAD Int. Multidiscip.Sci.Conf.* University of Miskolc, Hungary
[8] Buyakov A S, Kulkov S N 2017 *IOP Conf. Ser.: Mater. Sci. Eng.* **175** 012025 https://doi.org/10.1088/1757-899X/175/1/012025
[9] Kulkov S N, Buyakova S and Gömze A L 2017 *J.Phis.: Conf. Ser.* **790** 012015 https://doi.org/10.1088/1742-6596/790/1/012015
[10] Gilad I and Nissan M 1986 *Spine* **11** 154
[11] Przybylski G J, Patel P R, Carlin G J and Woo S L-Y 1998 *Spine* **23** 893
[12] Sohn H M, You J W and Lee J Y 2004 *J. Korean Med. Sci.* **19** (1) 101
[13] Kunkel M E, Herkommer A, Reinehr M, Bockers T M and Wilke H-J 2011 *J. Anat.* **219** 375

Structure and Properties of Al – ZrW_2O_8 Pseudo Alloys

V S Shadrin[1, 2], S N Kulkov[1, 2]

[1]Strength and Design Department, Physical-Technical Faculty, Tomsk State University, Lenin ave. 36, Tomsk, 634050, Russia
[2]Laboratory of Physics of Nanostructured Functional Materials, Institute of Strength Physics and Materials Science SB RAS, Akademicheskii ave. 2/4, 634055, Tomsk, Russia

E-mail: kulkov@ms.tsc.ru

Abstract. Al – ZrW_2O_8 powder mixture was sintered to obtain alloys were Al – ZrW_2O_8 pseudo alloys. The structure formation process of the pseudo alloys obtained was investigated. During sintering process zirconium tungstate was found to decompose into constituent oxides with consequent formation of WAl_{12} and $ZrAl_3$ intermetallic compounds. Increasing the sintering time up to 5 hours leads to re-synthesis of the zirconium tungstate represented with elongated particles - microfibers. The formation of internal stresses in the material was shown using XRD and Williamson-Hall analyses. The value of internal compression stresses was estimated to be 90 MPa. Vickers hardness of the pseudo alloys obtained increases by two times with increasing sintering time from 1h to 5h.

1. Introduction

The development of modern technology, science and industry is impossible without the creation of novel materials with improved physical-mechanical properties. Zirconium tungstate ZrW_2O_8 possessing isotropic negative thermal expansion coefficient (NTE) in a wide temperature interval [1] is a perspective material for development of new class of metallic materials with controlled thermal expansion.

Low-temperature decomposition process of zirconium tungstate into consistuent oxides WO_3 and ZrO_2 upon heating was shown in recent works [2]. Therefore taking this effect into account one can expect some specific behavior of ZrW_2O_8 during sintering near the melting point of metal. Aluminum based materials are lightweight, exhibit excellent mechanical properties, thermal and electrical conductivity and therefore are promising candidates for development of multifunctional Al – ZrW_2O_8 pseudo alloy with controlled thermal expansion. Low melting temperature of aluminum will make it possible to observe this effect most simply.

The work presented is aimed at investigating the structure and mechanical properties of Al – ZrW_2O_8 pseudo alloys.

2. Experiment

Commercially pure aluminum powder ASD-6 and zirconium tungstate powder obtained by hydrothermal synthesis [3] were used as initial components. The amount of ZrW_2O_8 addition was 10 wt%. High energy ball milling for 1 minute using AGO-2 planetary ball mill with acceleration 60g was performed to achieve homogeneous distribution of ZrW_2O_8 in aluminum powder and for mechanical activation of the components. Mechanically activated (MA) Al – ZrW_2O_8 powder mixtures were uniaxially cold pressed into ~ ⌀20mm×4mm samples under 100 MPa and sintered at 600 °C in argon atmosphere for 1h and 5h. Surface of the specimens was etched in HF acid. The microstructure of the specimens obtained was observed using TESCAN Vega3 scanning electron microscope. Structure and phase composition were investigated using XRD analysis with Cu Kα irradiation. After sintering of the specimens the Vickers hardness test was performed, using PMT-3equipment.

3. Results and Discussion

3.1. High energy mechanical activation of Al – ZrW_2O_8 powder mixtures

Aluminum powder was represented with spherical particles with average size of 3 μm. ZrW_2O_8 particles were rod-shaped with average length of 5 μm and are agglomerated, Figure 1. Morphology and particle sizes of ZrW_2O_8 are in a good agreement with [3, 4].

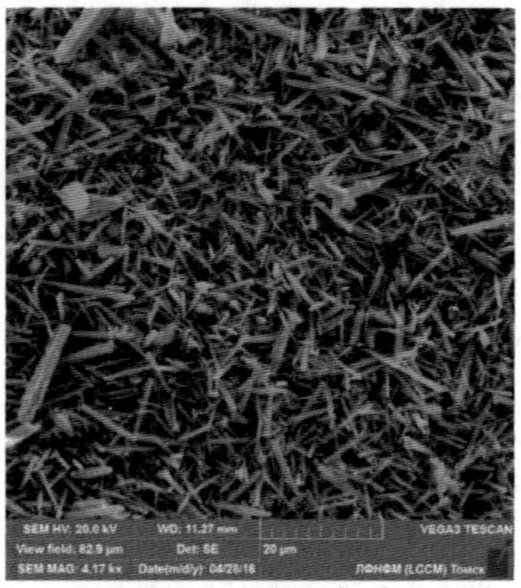

Figure 1. Morphology of ZrW_2O_8 powder

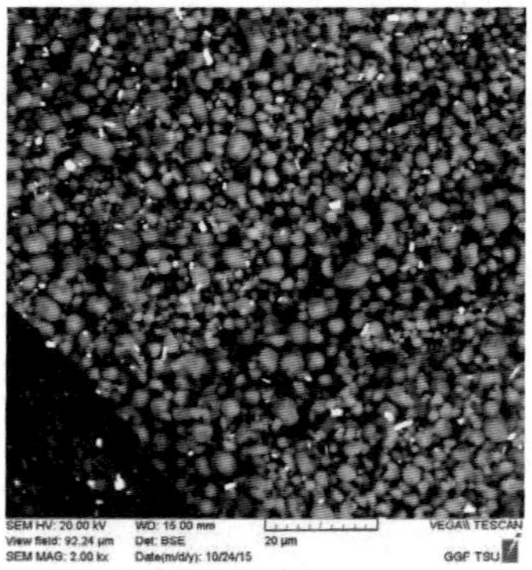

Figure 2. Morphology of Al - ZrW_2O_8 powder mixture after 1 min MA.

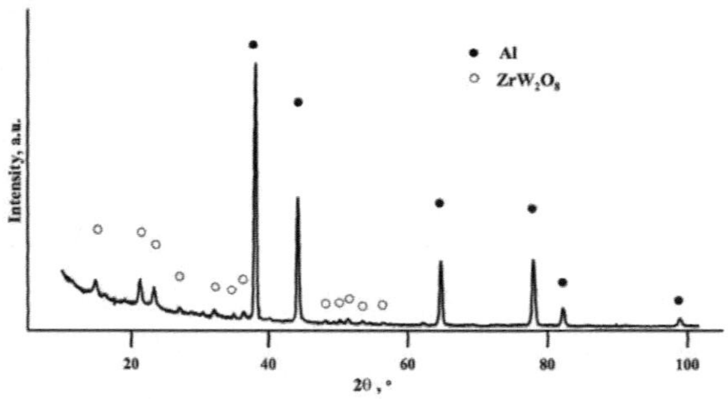

Figure 3. XRD pattern of Al – ZrW_2O_8 powder mixture after 1 minute of MA.

SEM images of the powder mixtures after high energy mechanical activation is shown in Figure 2. One can observe homogeneous distribution of ZrW_2O_8 particles in aluminum powder. Also the length of rod-shaped particles reduced during high energy mechanical activation to 1.5 μm, and their transverse size did not change.

XRD studies of the powder mixture after mechanical activation showed that there are no reactions between the mixture components during MA. In the Figure 3 one can observe only peaks corresponding to aluminum and cubic modification of ZrW_2O_8.

IOP Conf. Series: Journal of Physics: Conf. Series **1045** (2018) 012039 doi:10.1088/1742-6596/1045/1/012039

3.2. Al – ZrW_2O_8 specimens after sintering

In Figure 4 SEM image of the etched surface of material after 1h of sintering is presented. The structure is not homogeneous, one can observe elongated dense areas (2), porous matrix (1) and white particles distributed both in the matrix and elongated dense areas. The average size of white particles is 0.5 μm. Obviously during the sintering process ZrW_2O_8 is redistributed in the matrix. And since, after mechanical activation no irregularities in the initial powder mixture are found, one can assume that the formation of inhomogeneous structure takes place during the sintering process.

EDAX analysis showed that white particles consist of Al, W and Zr atoms. On the XRD pattern one can observe peaks corresponding to Al and $ZrAl_3$, WAl_{12} intermetallic compounds, Figure 6. Calculated lattice parameters of intermetallics are in a good agreement with ASTM diffraction data cards: a=7. 5803 Å for WAl_{12} (8-331) and a=4.0130 Å for $ZrAl_3$ (2-1093). The lattice parameter of aluminum a= 4.0517 Å is slightly increased in comparison with 4-787 ASTM data card. The coherent diffraction domain (CDD) value and lattice micro distortion for aluminum were calculated by Williamson-Hall analysis D=50 nm, $\varepsilon = 2.5 \cdot 10^{-3}$.

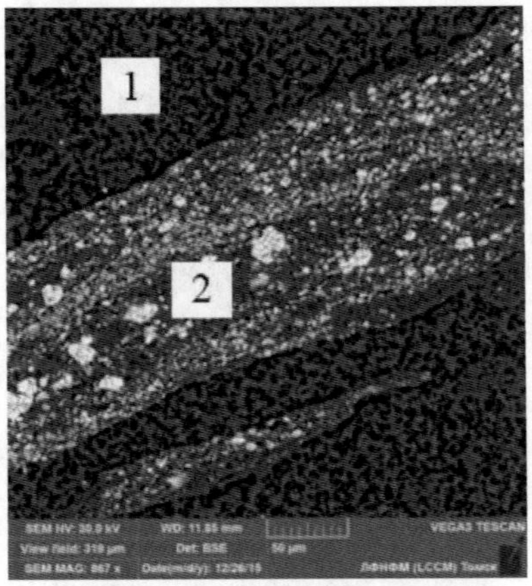

Figure 4. Etched surface of Al – ZrW_2O_8 sintered for 1h.

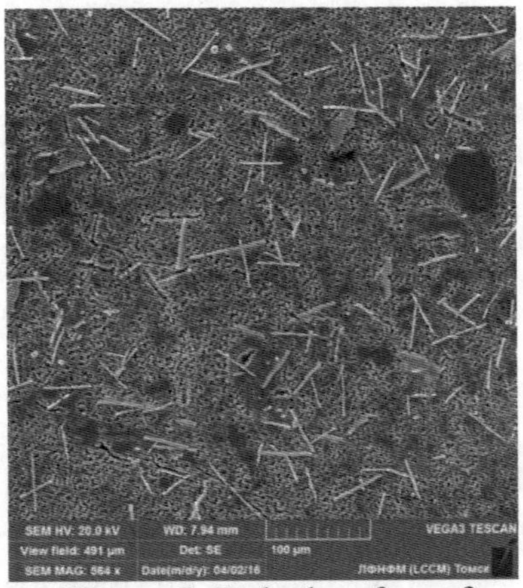

Figure 5. Etched surface of Al – ZrW_2O_8 sintered for 5h.

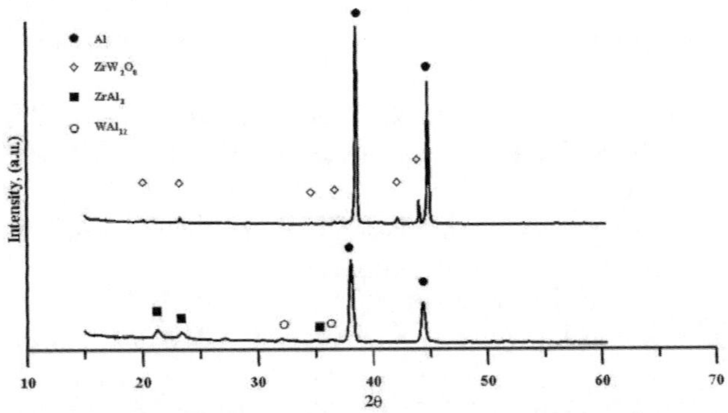

Figure 6. XRD patterns of Al – ZrW_2O_8 specimen

After 5h of sintering the structure of the material is represented with microfibers with average length of 30 μm homogeneously distributed in aluminum matrix, Figure 5. On the XRD pattern, Figure 6, one can observe peaks corresponding to Al and cubic ZrW_2O_8. Lattice parameter of aluminum increased in comparison with 1h isothermal hold, a= 4.0544 Å. ZrW_2O_8 lattice parameter a=9.2187 Å is increased in comparison with literature data [5]. CDD value and lattice micro-distortions for Al D=500 nm and $\varepsilon = 7 \cdot 10^{-3}$ correspondingly are increased in comparison with ASTM data and results for specimens sintered for 1 hour.

Vickers hardness tests were performed to evaluate mechanical properties of the pseudo alloys obtained after 1h and 5h sintering. A sample obtained with an isothermal hold for 1 h has the microhardness of the dense areas (2) containing Zr and W atoms of 455 MPa, which is higher than the microhardness of the aluminum matrix (1) equal to 195 MPa (Figure 4). Microhardness of the material increases with increasing sintering time. After 5h isothermal sintering microhardness value of the pseudo alloy obtained increased up to 530 MPa [6].

The data obtained using scanning electron microscopy and XRD analysis allow us to suggest that formation of the material proceeds through decomposition of zirconium tungstate into consistuent oxides WO_3 and ZrO_2 and synthesis of WAl_{12} and $ZrAl_3$ intermetallic compounds after 1 hour of isothermal hold [7, 8]. The further increase of isothermal holding time up to 5h leads to subsequent formation of ZrW_2O_8 microfibers. Lattice parameter and microdistortions of aluminum increase with increasing isothermal holding time. After 1h of sintering when the decomposition of zirconium tungstate occurred the lattice parameter increased slightly but after 5 hours of isothermal holding a significant increase of the lattice parameter occurred. This effect may indicate the formation of internal compressive stresses in the material after synthesizing of zirconium tungstate. Their value is estimated to be 50 MPa after 1 hour of isothermal hold and 90 MPa, and after 5 hours of isothermal hold.

4. Conclusions

It was shown that during sintering of Al - ZrW_2O_8 powder mixture zirconium tungstate decomposes into its constituent oxides WO_3 and ZrO_2 with subsequent formation of WAl_{12} and $ZrAl_3$ intermetallic compounds. Increasing the sintering time up to 5 hours leads to re-synthesis of the zirconium tungstate represented with elongated particles - microfibers. The formation of internal stresses in the material was shown using XRD and Williamson-Hall analyses. The value of internal compression stresses was estimated to be 90 MPa. Vickers hardness of the pseudo alloys obtained increases by two times with increasing sintering time from 1h to 5h.

5. Acknowledgments

The work presented was carried out in accordance with plan of Siberian Branch of Russian Academy of Science, Project III.23.2.3, Tomsk State University's competitiveness program and with partial support of RFBR, #16-08-00037.

References

[1] Evans J S O 1999 Negative thermal expansion materials, The Royal Society of Chemistry, Dalton Trans. 3317

[2] Wu G, Zhou C, Zhang Q, Pei R 2015 *Scripta Materialia* **96** 29

[3] Dedova E S, Shadrin V S, Gubanov A I, Kulkov S N 2014 *Inorganic Materials: Applied Research* **5** (5) 471

[4] Kulkov S N, Dedova E S, Pedraza F, Erdelyi J 2014 *Épitőanyag-JSBCM* **66** (2) 35 http://dx.doi.org/10.14382/epitoanyag-jsbcm.2014.7

[5] Mary T A, Evans J S O, Vogt T, Sleight A W 1996 *Science* **272** 90

[6] Fan G H, Wang Q W, Geng L, Wang G S, Feng Y C 2012 *Journal of Alloys and Compounds* **545** 130

[7] Kulkov S N, Buyakova S P, Gömze A L 2014 *Épitőanyag-JSBCM* **66** (1) 2 http://dx.doi.org/10.14382/epitoanyag-jsbcm.2014.1

[8] Gömze A L, Gömze N L 2010 *Épitőanyag-JSBCM* **62** (4) 98 http://dx.doi.org/10.14382/epitoanyag-jsbcm.2010.18

IOP Conf. Series: Journal of Physics: Conf. Series **1045** (2018) 012040 doi:10.1088/1742-6596/1045/1/012040

Determination of polymer melts flow-activation energy a function of wide range shear rate

G Toth[1], D Nagy[1], A Bata[1], and K Belina[1]

[1]Department of Materials Technology, GAMF Faculty of Mechanical Engineering and IT, Kecskemet, Izsaki str. 10, Hungary, John von Neumann University

E-mail: toth.gergely@gamf.uni-neumann.hu

Abstract. The majority of the processing of thermoplastic polymers is in melt condition. For example, extrusion, or injection moulding, which are the two biggest plastic processing technologies. Nowadays, there is a need to simulate these processes. The simulations are helpful for the plastic product and mold designers. In these simulations one of the most important internal parameters is the shear viscosity, which depends on the temperature, pressure, molecular weight or structure. So easy to see how important is the top priority precise knowledge of the shear viscosity of plastic melts. The importance of polymer melt flow properties during processing has required continuous research of polymer melt rheology. In a previous article, we described how to measure shear viscosity of polycarbonate from 100 to 105 shear-rate regions with standard and non-standard equipments. In this paper a method is shown to calculate the activation energy of flow of high density polyethylene (HDPE), polystyrene (PS) and polycarbonate (PC). It was shown that shear viscosity and activation energy of flow of the polymers depend on the chemical structure. It was found that more rigid molecular structure results higher values.

1. Introduction

The influence of temperature on the shear viscosity of polymer melts is usually expressed in terms of flow activation energy. In our previous article [1], a method was described to measure and calculate viscosity of polycarbonate melt. Using our raw results, we can calculate activation energy of flow based on known models. An approximation model for kinetic activity in chemistry was developed in the general form by Svante A. Arrhenius who has introduced "activation constant" [2]. The Arrhenius relation in the form of a η(T) fitting function describes the change of viscosity for both increasing and decreasing temperatures:

$$\eta\,(T) = c_1 \cdot \exp\,(-c_2/T) = c_1 \cdot \exp\,[(E_A/R_G)/T] \qquad (1)$$

with the temperature (T) in [K], and the material constants C1 [Pas] and C2 [K] of the sample where C2=EA/RG is the activation energy of flow EA is in kJ/mol and the gas constant RG=$8.314 \cdot 10^{-3}$ in kJ/(mol K).

At a certain temperature, the activation energy of flow EA characterizes the energy needed by the molecules to be set in motion against the frictional forces of the neighboring molecules. This requires exceeding the internal flow resistance, with other words, a material-specific energy barrier, the so-called potential barrier [3].

IOP Conf. Series: Journal of Physics: Conf. Series **1045** (2018) 012040 doi:10.1088/1742-6596/1045/1/012040

Our target was compare polyethylene, polystyrene and polycarbonate polymer types flow activation energy considering the macromolecular segment size. In theory this types of polymers looks like linear freely-jointed chains. The freely-jointed chain is the simplest model of a polymer. The Figure 1 shows differences between the polymer types monomer construction. Size of the segment is in non linear contact with the size of the monomer.

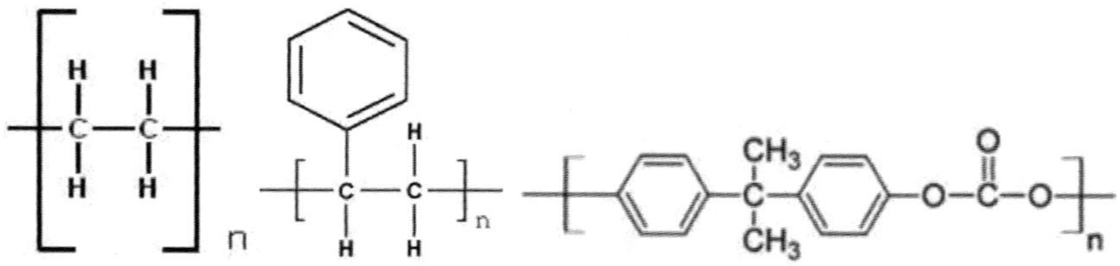

Figure 1. Monomer types of PE, PS, PC

The basic assumption that the flow activation energy based on viscosities at constant shear rate decreases as the segment size increases. Furthermore, based on viscosity decreases the shear rate the flow activation energy will increase.

The samples were PC- sample was Dow Calibre 3030 EP, PS- Eni Edistir N 3380, PE- MolTipplen BS-501.

2. Method

Our basic results we got from three shear viscosity investigation methods, which covers a wide range of shear rate and temperature steps near the processing temperature [1]. These were in the low shear rate range of the rotational viscometry (TA Ares G2 range: 0.05 – 30 s-1), in the middle-high range of the capillary viscometry (Göttfert RG25 range: 100 – 10,000 s-1) and in the high range we used a special technique (range: 10,000 – 240,000 s-1) which was based on the capillary viscometry method.

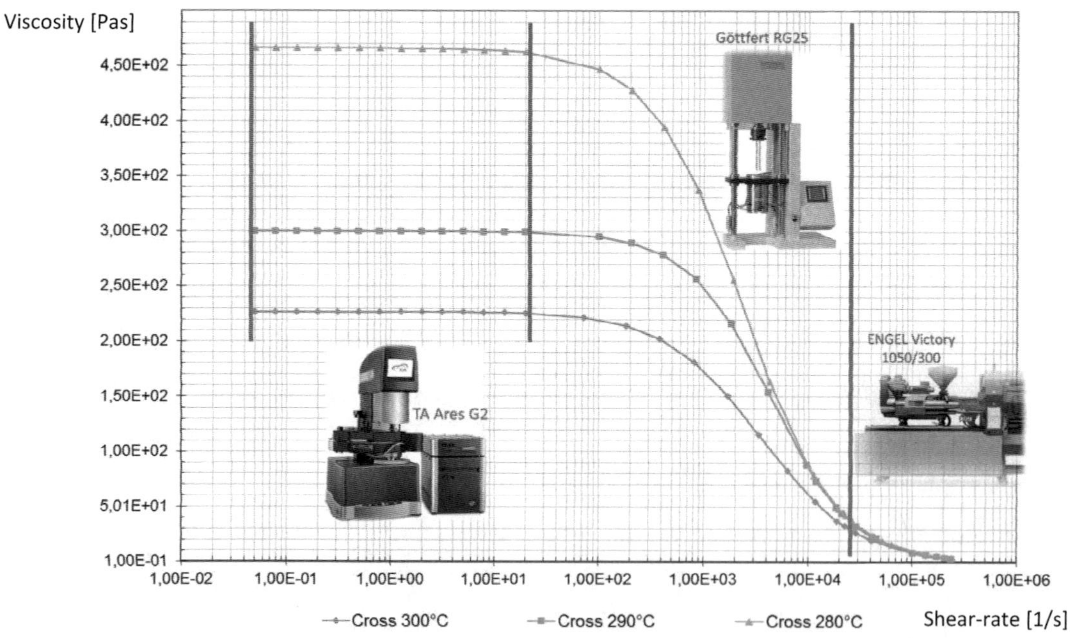

Figure 2. Polycarbonate shear viscosity raw results separated by equipments [1]

From the raw results (Figure 2) all of the viscosity curves describe by Cross equation [3-10].

$$\eta = \frac{\eta_0}{1+\left(\frac{\eta_0 \dot{\gamma}}{\tau^*}\right)^{[1-n]}} \qquad (2)$$

Where η is the shear viscosity (Pa*s); $\dot{\gamma}$ deformation rate (s-1); η_0 zero shear viscosity (Pa*s); τ^* stress limit between the Newtonian and non Newtonian region (Pa); $[1-n]$ slope of the transient region (-).

It was necessary for calculation shear viscosity on constant shear-rate. In this case the constant shear rate mean 100 s-1 value. The measured shear viscosity curves compared with the Cross equation curves it shows good correlation (Figure 3).

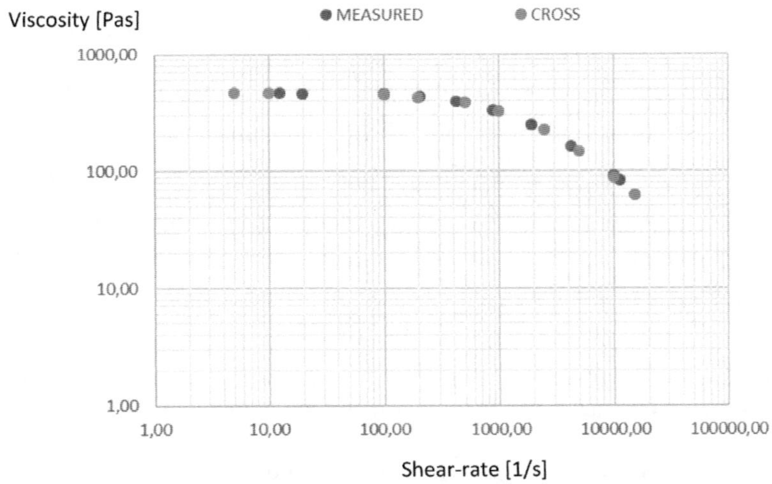

Figure 3. Measured and Cross viscosity curves compare

From the viscosity curves (described with Cross equation) we need just one point per a temperature. To calculate flow activation energy we need to calculate the shear viscosity on constant shear-rate on three temperature in 40-50°K range T >Tg + 100K (with the glass-transition temperature Tg).

3. Results

The shear viscosity of polymer melts is temperature dependent. The relationship between the shear viscosity and the temperature of the sample melts can be obtained by plotting ln(η)*RG against 1/T (T is in the absolute temperature scale), as illustrated in Figure 4.

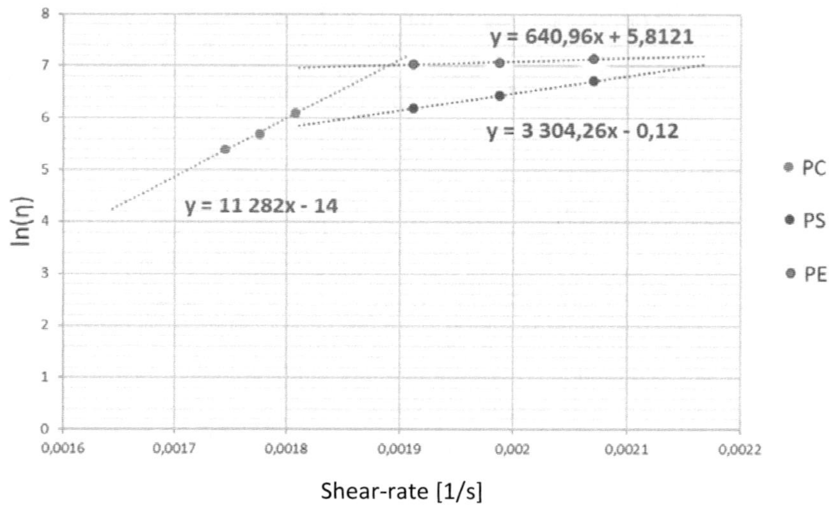

Figure 4. Samples ln(η) and 1/T equation

Under same shear rate, the curves of ln(η) versus 1/T are closely linear, the linearity suggesting that the dependence of the samples melt shear viscosity on temperature can be described by the Arrhenius equation general format, i.e.:

$$\eta\,(\mathrm{T}) = \mathrm{Aexp}\left(\frac{E_A}{R_G}\right) \qquad (3)$$

where A is the constant related to the melt viscosity, EA is the activation energy for viscous flow, and RG is the universal gas constant. The calculated activation energy values is in Table 1.

Table 1. Samples activation energy

Polymer type	Flow activation energy [J/mol]
PE	5,33E+03
PS	2,75E+04
PC	9,38E+04

The flow of polymer is caused by the movement of the whole molecule, while the molecular movement is caused by the transition of the chain segment. From Figure 5 the melt flow of PC can be seen at wide shear rate range. It is necessary for the thermal motion energy of the molecular chain to be sufficiently high to overcome the internal revolve potential energy since the molecular chain of PC is rigid. As a result, the viscous flow temperature of PC is high. From these results, we can conclude that the activation energy increases with segment size.

As illustrated in Figure 5 the shear rate dependence of samples activation energy shows different behaviors. While the PE and PS samples show near linear dependence PC show nonlinear behavior.

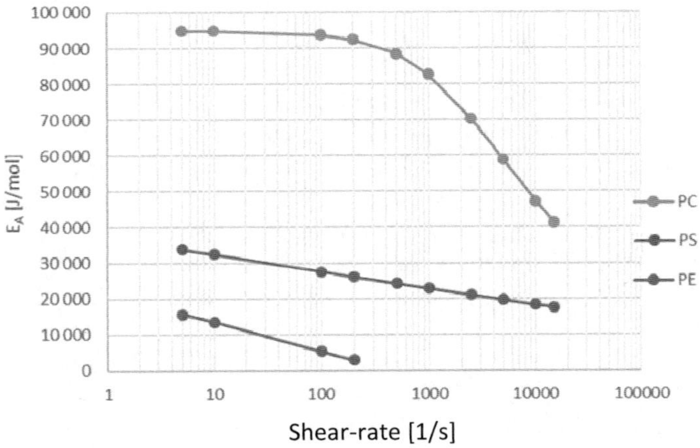

Figure 5. Activation energy shear rate dependence

As illustrated in Figure 5 the shear rate dependence of samples' activation energy shows different behaviors. While the PE and PS samples show near linear (in log scale, but it is exponential) dependence, PC shows nonlinear behavior. It is important that the shear rate dependence is not the same in low and high shear rate range (Figure 5). Until 102 s-1 shear rate, the activation energy is almost constant. Below this value, it decreases like the other two samples. This phenomenon needs further investigation.

It is very important that in Figure 5 the vertical axis shows already the activation energy in linear scale. The horizontal axis shows the shear rate in logarithmic scale. For this plotting mode the experiments should be made at the same temperature steps, but at different shear rate values. So that the determined activation energy for the same temperature ranges (for instance, shown in the Figure 4), is measured and calculated in different shear rate steps. This means that the curve in Figure 4 is shown as a point in Figure 5.

4. Conclusion

Under the present experimental conditions, the shear flow of the sample melts obeys the structural shear viscosity theory. The melt shear viscosity depends heavily on the temperature and the relationship between them can be described by the Arrhenius equation. The shear rate dependence of samples' activation energy shows different behaviors. While the PE and PS samples show near linear dependence, PC shows nonlinear behavior. It is important that the shear rate dependence is not the same in low and high shear rate range. Comparing the samples' segment size and the calculated activation energy the linear relationship has been confirmed.

Investigating the samples in constant temperature ranges a linear equation is shown, while the shear rate dependence of the activation energy is an exponential equation. To determine it, we used Arrhenius equation in constant temperature ranges, but at different shear rate values. With this method the shear rate dependence of the activation energy can be calculated using Arrhenius equation.

Acknowledgment

This research is supported by EFOP-3.6.1-16-2016-00006 "The development and enhancement of the research potential at John von Neumann University" project. The Project is supported by the Hungarian Government and co-financed by the European Social Fund.

IOP Conf. Series: Journal of Physics: Conf. Series **1045** (2018) 012040 doi:10.1088/1742-6596/1045/1/012040

References

[1] Toth G, Bata A, Belina K 2017 *IOP Conf. Ser.: Mater. Sci. Eng.* **175** 012056 https://doi.org/10.1088/1757-899X/175/1/012056
[2] Arrhenius S A 1884 Recherches sur la conductibilitée galvanique des electrolytes; 1887 Über die innere Reibung verdünnter wässriger Lösungen, *Z. Physik. Chem.* **1**
[3] Meichsner G, Mezger T, Schröder J 2003 Lackeigen schaften messen und steuern *Vincentz* Hannover
[4] Malcolm M Cross 1965 *Journal of Colloid Science* **20** (5) 417
[5] Liang J Z, Tang C Y 1997 *Journal of Materials Processing Technology* **63** 501
[6] Cogswell F N 1981 Polymer melt rheology, *George Godwin Ltd* London
[7] Kovács J G 2010 *Anyagvizsgálók lapja* **1-2**
[8] Tóth G, Bata A, Dugár Zs, Dugár T, Szabó T 2015 *GRADUS* **2** (2) 112
[9] Mózes Gy, Vámos E 1968 Reológia és Reometria, *Műszaki könyvkiadó*, Budapest
[10] Haddout and Villoutreix G 2000 *International Polymer Processing* **15** (3) 291

Institute of Physics
Dirac House, Temple Back
Bristol BS1 6BE UK

ISSN: 1742-6588
ISBN 978-1-5108-6787-1

9 781510 867871

CURRAN ASSOCIATES INC.
proceedings.com